6G Security

IEEE Press

445 Hoes Lane
Piscataway, NJ 08854

6G Security

New Frontiers in Mobile Network
Technologies and Verticals

Edited by

Luiz A. DaSilva

Commonwealth Cyber Initiative
Virginia Tech
Arlington, Virginia
USA

IEEE Press
WILEY

Published by John Wiley & Sons, Inc., Hoboken, New Jersey.

Library of Congress Cataloging-in-Publication Data has been applied for:

Hardback ISBN: 9781394292462
ePDF ISBN: 9781394292486
ePUB ISBN: 9781394292479
oBook ISBN: 9781394292493

Cover Design: Wiley
Cover Image: © Yuichiro Chino/Getty Images

We dedicate this book to the Commonwealth of Virginia, which plays a vital role in identifying, mitigating, and protecting its residents from cybersecurity threats that may significantly impact our individual and collective safety and privacy. The Commonwealth established and invested in the Commonwealth Cyber Initiative (CCI), an unprecedented coalition of more than 40 Virginia universities and colleges, to address cybersecurity needs in workforce development, innovation, and research.

Contents

List of Contributors

Luiz A. DaSilva
Commonwealth Cyber Initiative
Virginia Tech

Laura Gazda
Commonwealth Cyber Initiative
Virginia Tech

Eric Burger
Commonwealth Cyber Initiative
Virginia Tech

Mayukh Roy Chowdhury
Nokia Bell Labs

Md Habibur Rahman
Commonwealth Cyber Initiative
Virginia Tech

Arham Siddique
Cybersecurity Engineering
Department
George Mason University

Naveen Sapavath
Electrical and Computer Engineering
Department
Northeastern University

Aloizio P. DaSilva
Commonwealth Cyber Initiative
Virginia Tech

Vijay K. Shah
Electrical and Computer Engineering
North Carolina State University

Duminda Wijesekera
Cybersecurity Engineering
Department
George Mason University

Yalin Sagduyu
National Security Institute
Virginia Tech

Tugba Erpek
National Security Institute
Virginia Tech

Safdar Hussain Bouk
School of Cybersecurity
Old Dominion University

Sachin Shetty
Virginia Modeling
Analysis, and Simulation Center
Old Dominion University

Rui Ning
Computer Science
Old Dominion University

Hongzhi Guo
School of Computing
University of Nebraska-Lincoln

Chunsheng Xin
Department of Computer Science
Iowa State University

Rahul Varma Chintalapati
Electrical and Computer Engineering
Virginia Tech

Paul Masur
Electrical and Computer Engineering
Virginia Tech

Nishith D. Tripathi
Electrical and Computer Engineering
Virginia Tech

Jeffrey H. Reed
Electrical and Computer Engineering
Virginia Tech

Jacek Kibilda
Commonwealth Cyber Initiative
Virginia Tech

Parth Pathak
Department of Computer Science
George Mason University

Vijay K. Shah
Electrical and Computer Engineering
North Carolina State University

Kai Zeng
Department of Electrical and
Computer Engineering
George Mason University

André Gomes
Department of Computer Science
Rowan University

Atul Khatri
College of Engineering and Computing
George Mason University

Ruizhi Cheng
College of Engineering and Computing
George Mason University

Songqing Chen
College of Engineering and Computing
George Mason University

Bo Han
Department of Computer Science
George Mason University

Khaled N. Khasawneh
Department of Electrical and
Computer Engineering
George Mason University

Sai Dinakarrao
Department of Electrical and
Computer Engineering
George Mason University

Sunanda Roy
Department of Electrical and
Computer Engineering
George Mason University

Anhton Nguyen
Department of Electrical and
Computer Engineering
George Mason University

Pavlos Zouridakis
Department of Electrical and
Computer Engineering
George Mason University

Saikat Das
Department of Electrical and
Computer Engineering
Virginia Commonwealth University

Salih Sarp
Department of Electrical and
Computer Engineering
Virginia Commonwealth University

Zhifang Wang
Department of Electrical and
Computer Engineering
Virginia Commonwealth University

Yanxiao Zhao
Department of Electrical and
Computer Engineering
Virginia Commonwealth University

Shiyi Liu
Department of Computer Science
University of Virginia

Tanmoy Sen
Department of Computer Science
University of Virginia

Haiying Shen
Department of Computer Science
University of Virginia

About the Editor

Luiz A. DaSilva is the inaugural executive director of the Commonwealth Cyber Initiative (CCI), a statewide initiative in Virginia that brings together 47 institutions of higher education with a mission of research, workforce development, and innovation in cybersecurity. He is also the Bradley Professor of Cybersecurity at Virginia Tech.

Dr. DaSilva previously held the chair of Telecommunications at Trinity College Dublin, where he served as the Director of CONNECT, a national networking research center comprising 10 universities in Ireland.

Dr. DaSilva's research focuses on wireless networks, including being a frequent keynote speaker on security in wireless networks and the societal impacts of network technologies. He has authored seminal papers and a book on the application of game theory to model resource decisions in these networks. He is a Fellow of the Institute of Electrical and Electronics Engineers (IEEE) for his contributions to cognitive networks and resource management in wireless networks.

Dr. DaSilva has published two books and more than 300 peer-reviewed articles and is a frequent keynote speaker around the world. He has served as an IEEE Communications Society Distinguished Lecturer (2015–2018) and a Fellow of Trinity College Dublin. In the last two years, he has received the Golden Goose Mentoring Award, recognizing his contributions to mentoring and excellence in graduate education, and the Dean's Excellence in Service award.

Foreword

Walking down the street or in any office, building, or restaurant, wireless connections to mobile networks through cellular phones, "cash" registers, laptop computers, headphones, and more exist in such abundance that the sheer number of connections feels impossible to grasp. Beyond these visual reminders of our wireless mobile connections, the vast hidden network of devices, systems, and infrastructure ensures quick, clear, and effortless access to all that the network and mobile communications have to offer. Who doesn't have a cell phone in hand and earbuds in their ears as they navigate their day? Much like Haraway's modern cyborg, 5th-generation wireless and the promises of the 6th generation make connectivity not just ubiquitous but also an extension of ourselves [1]. With the high demand for easy, inexpensive, safe, and quick access anywhere in the world, invisible wireless infrastructure security becomes paramount, especially with so much personal data being produced that is easily identifiable and close to each user. Securing 6G will prove critical to safeguard not only this vital infrastructure and our devices but also each individual user and this new way of connected living.

6G Security: New Frontiers in Mobile Network Technologies and Verticals highlights, showcases, and explains forward-thinking research, releases, and technologies for securing 6th-generation wireless. Bowker and Star note in their Science, Technology, and Society (STS) infrastructure research the importance of making the complex, invisible infrastructure and classification systems (like the 3rd Generation Partnership Project) visible to more than the technical experts as the easier these networks are to use and the larger they become, the more difficult the infrastructure becomes to see [2]. This book makes network infrastructure visible through readable, actionable, format-specific ways to secure the hardware, software, devices, and data in 6G wireless network technologies. Its user-friendly language and definitions, detailed examples, algorithms, experiments, and citations enable the reader of *6G Security: New Frontiers in Mobile Network Technologies and Verticals* to use the book as a researcher, professor, or student easily. Whether preventing spoofing or jamming, developing secure open radio access networks (Open RAN), or using AI for cybersecurity or cybersecurity for AI, these 10 chapters each introduce, define, and explain key takeaways for 6th-generation wireless.

One can easily read this book from cover to cover as a course textbook or to develop an understanding of ways to secure 6G, but the book is also divided into chapters and two identifiable parts. Chapter 1 provides a historical overview of wireless technologies and securing them through the generations, and then looks ahead to 6G security issues. The book's first half explains new technologies leading wireless cybersecurity researchers expect to be incorporated into 6G and by the 3GPP's (3rd Generation Partnership Project) standard setting or releases that align with what the wireless technologies community calls 6G.

The second half of the book explores how 6G features will impact a new generation of applications and verticals. This organization of the book allows the reader to use each chapter or section as a standalone or a collection of articles. Unlike most computer science/engineering books, *6G Security: New Frontiers in Mobile Network Technologies and Verticals* may be used by a wider audience, from a cybersecurity policymaker looking to understand and develop policy and funding around securing 6G to a social science researcher or student exploring sociotechnical imaginaries or network ideologies [3, 4].

Understanding and using the technical artifacts and the research contained within this book leads to a critical examination of power dynamics, geopolitics, possible inequities, social dynamics, and the social construction of wireless technologies. Together with computer scientists and engineers researching and developing the artifacts themselves, social and political scientists can collaborate to ensure safe access and use of 6G for all. Dirk van Laak details in *Lifelines of Society: A Global History of Infrastructure* that the need for securing 6G infrastructure includes a promising future for the wireless technologies and users to develop more opportunities, better living conditions, participate actively in more social and political decision-making, and increase one's prosperity [5]. On the flip side, 6G can be very expensive, especially if insecure. 5G's ubiquity has already caused user dependence on wireless infrastructure. This dependency of wireless technologies will further shape user and geopolitical control and conduct [5]. The borderless existence of 6G wireless networks demands security prioritization since 6G will further globalization and interconnectedness.

Though many authors in *6G Security: New Frontiers in Mobile Network Technologies and Verticals* describe their research and chapters as a paradigm shift in securing wireless technologies, I caution using this term and ask the readers to look at its use through an informed and critical lens. Thomas Kuhn, in *The Structure of Scientific Revolutions*, defined a paradigm as a widely accepted scientific model that gains its status and acceptance in the scientific field because the paradigm is successful in solving critical problems the scientific community deems problematic and acute [6]. Kuhn further contends that the majority of normal science (traditional basic scientific research) consists of researching to articulate and show the theories and phenomena the paradigm already provides [6]. Therefore, a scientific revolution, known as a paradigm shift, is when the former paradigm no longer resolves critically acute problems.

Thus, a new one must be developed to resolve the crisis. Normal science is problem-solving within the current paradigm, while the paradigm shift is when the scientific community finds the crisis intolerable, rejects the previous paradigm, then substitutes it and accepts the new paradigm [6]. I caution the acceptance of a 6G security paradigm shift for a few specific reasons, including:

1. Kuhn argues that textbooks are produced after a scientific revolution to spread and develop this new paradigm tradition. Textbooks become produced and used as a resolution of revolutions [6].
2. These chapters exemplify the puzzle-solving of normal research within a paradigm, as they do not reject the previous paradigm and present a new one.
3. The issue of securing 6G in these chapters, though problematic and a worrisome concern that needs to be the priority of 6G, is not so problematic that it is currently unresolved and not a true crisis.

With these reasons in mind, this book could represent the beginning of a paradigm shift, and the reader should wonder and examine whether the chapters blur the paradigm, which then loosens the paradigm rules for normal research [6]. Since the book is at the forefront of 6G, outlining these possible new frontiers could test the paradigm boundaries, as described in President Franklin D. Roosevelt's quotation used in the National Science Foundation's 75th Anniversary Edition (2020) of Vannevar Bush's *The Endless Frontier*, that new frontiers of the mind lie ahead and could help us build a better, more fulfilling life (Bush).

Additionally, regardless of interest or use of the book, readers should read the second half of it with the social construction of technology (SCOT) in mind. These wireless artifacts have politics, meaning that technical things, including wireless networks and technologies, embody authority and power due to their construction [7]. Many actors help in the development of science and technology through their active participation or silence. Decisions about what will be funded, researched, and developed by these actors shape the basic and applied sciences. Though it may be common and accepted practice to look at normal and applied science through a positivist lens, meaning that all science and technology are discovered through an objective view, empirical data, and experimentation, SCOT argues that scientific knowledge is not produced so linearly. Instead, it is multidirectional, with not just the normal science of research and experimentation but also the relevant social group's usage and the technology itself influencing the development [8]. Through these interactions, there is interpretive flexibility. Flexibility occurs in how one interprets or thinks about how the technology works and develops how it is designed [8]. Once the interpretation and use of the technology stabilize or succeed as an accepted paradigm, there is closure as the controversy/problems with the construction of the technology have ended. Like Kuhn's paradigms, this closure can be reopened when a relevant social group has an unresolved problem or crisis. Questions to ask throughout the book include:

1. Who are the relevant social groups, and how may they construct 6G?
2. What are their interpretations of the technologies and verticals? How have or could these contributed to the success or failure of previous constructions?
3. What stakes do relevant social groups have in this next generation of wireless technology, and how may this influence the stabilization and closure of 6G security?

I challenge readers to go beyond their role as engineers, students, industry partners, or professors and consider the layers embedded within each chapter's argument and the applications for securing 6G. Situate yourself as a reader, not as some objective view from above, but from below and within all the roles, places, and ideas you inhabit [9]. These partial perspectives can be more powerful in developing an understanding of scientific knowledge than one may realize, especially in understanding and developing critical, secure infrastructure technologies.

Wireless technologies have become a global need tied to everyday life. When the infrastructure breaks and security is breached, life-and-death consequences shift from imagined to real. Additionally, when access, whether due to location, affordability, or availability, is insecure, prosperity, safety, and active social and political participation can be at risk. We have all experienced a break in our connectivity, a breach of data, or a breach of wireless security, whether by traveling beyond the last mile of networks, being outside our cellular network, or experiencing constant spoofed calls to scam us out of money. Therefore, securing 6G mobile technologies and verticals proves to be a critical public problem. This book does more than outline possible solutions: it calls to action and enhances visibility for securing 6G wireless networks. As a forward-looking book, I hope that the reader explores the solutions outlined in the chapters and expands their vision to look at the social construction of this technology, the potentially hidden layers of power, control, and authority embedded within, and ways to engage the relevant social groups to develop a secure 6G for all.

Laura Gazda
Commonwealth Cyber Initiative
Virginia Tech, Arlington, VA, USA

References

1 Haraway, D. (2013a). Simians, cyborgs, and women: The reinvention of nature. Routledge.

2 Bowker, G. C. and Star, S. L. (2000). Sorting things out: Classification and its consequences. MIT Press. [Bush] Bush, V. United States Government Printing Office.

3 Jasanoff, S. and Kim, S.-H. (2019). Dreamscapes of modernity: Sociotechnical imaginaries and the fabrication of power. University of Chicago Press.

4 Ten Oever, N. (2023). 5G and the notion of network ideology, or: The limitations of sociotechnical imaginaries. Telecommunications Policy, 47(5):102442.

5 Van Laak, D. (2023). Lifelines of our society: A global history of infrastructure. MIT Press.

6 Kuhn, T. S. (1997). The structure of scientific revolutions, volume 962. University of Chicago press Chicago.

7 Winner, L. (2017). Do artifacts have politics? In Computer Ethics, pages 177–192. Routledge.

8 Bijker, W. E., Parke Hughes, T., and Pinch, T. (2012). The social construction of technological systems, anniversary edition: New directions in the sociology and history of technology. MIT Press.

9 Haraway, D. (2013b). Situated knowledges: The science question in feminism and the privilege of partial perspective 1. In Women, Science, and Technology, pages 455–472. Routledge.

Acknowledgments

This work was supported in part by the Commonwealth Cyber Initiative, an investment in the advancement of cyber R&D, innovation, and workforce development.

For more information about CCI, visit www.cyberinitiative.org.

Thank you to Dr. Joao Santos for his ongoing technical and collaborative LaTeX support.

Chapter 3: We would like to thank Dr. Yi Shi and Dr. Daniel Jakubisin for valuable comments and discussions.

Chapter 5: This research was supported in part by NSF CNS Award No. 2235139.

Chapter 6: This material is based upon work supported by the National Science Foundation (NSF) Collaborative Grant SECURIS No. 2318796 and 2318798. Additionally, Drs. Kibilda and Gomes were supported by NSF under Grants No. 2326599 and 2412872. Dr. Zeng was supported by NSF under Grant No. 2131507, and by the Army Research Office under Grant No. W911NF-21-1-0187. The authors also thank InterDigital for providing them with their mmW equipment.

Chapter 10: This research was supported in part by U.S. NSF grants NSF-2319988, NSF-2206522, NSF-1822965, FHWA grant 693JJ31950016, and Microsoft Research Faculty Fellowship 8300751.

Acronyms

3GPP	3rd Generation Partnership Project
AF	Application Function
AI	Artificial Intelligence
AKA	Authentication and Key Agreement
AMF	Access and Mobility Management Function
AML	Adversarial Machine Learning
AMPS	Advanced Mobile Phone System
API	Application Programming Interface
AR	Augmented Reality
ARQ	Automatic Repeat Request
ASR	Attack Success Rate
AUSF	Authentication Server Function
BNF	Backus Normal Form
BPMN	Business Process Meta Language
BPSK	Binary Phase Shift Keying
BS	Base Station
CA	Certificate Authority
CDMA	Code Division Multiple Access
CFMF	Communication Function Management Function
CPS	Cyber Physical System
CQI	Channel Quality Indicator
CRAN	Cloud Radio Access Network
CRC	Cyclic Redundancy Check
CSI	Channel State Information

CSI-RS	Channel State Information-Reference Signal
CSMF	Communication Slice Management Function
CU	Centralized Unit
D-AMPS	Digital Advanced Mobile Phone Systems
DCI	Downlink Control Information
DDoS	Distributed Denial of Service
DDQN	Double Deep Q-Network
DL	Deep Learning
DMRS	Demodulation Reference Signal
DN	Data Network
DNN	Deep Neural Network
DoS	Denial of Service
DQL	Deep Q Learning
DQN	Deep Q-Network
DRL	Deep Reinforcement Learning
DU	Distributed Unit
EM	Element Manager
eMBB	Enhanced Mobile Broadband
EPC	Evolved Packet Core
ETSI	European Telecommunications Standards Institute
FDD	Frequency Division Duplexing
FEC	Forward Error Correction
FGSM	Fast Gradient Sign Method
FL	Federated Learning
GAT	Graph Attention Network
GEO	Geo-Synchronous Earth Orbit
GSM	Global System for Mobile
GNN	Graph Neural Network
GNSS	Global Navigation Satellite System
HAPS	High-Altitude Platform Station
HARQ	Hybrid Automatic Repeat Request
HEO	Highly Elliptical Earth Orbit
iDEN	Integrated Digital Enhanced Network
IDS	Intrusion Detection System
ILP	Integer Linear Programming
IMEI	International Mobile Equipment Identity
IMSI	International Mobile Subscriber Identity
IoT	Internet of Things
IP	Internet Protocol

IPsec	Internet Protocol Security
ISAC	Integrated Sensing and Communication
ISL	Inter Satellite Link
ITU-T	International Telecommunication Unit – Standardization sector
JSAC	Joint Sensing and Communication
KTRA	Knowledge Transfer-based Resource Allocation
LEO	Low Earth Orbit
LTE	Long Term Evolution
MAC	Medium Access Control
MANO	Management, Automation, and Network Orchestration
MARL	Multi Agent Reinforcement Learning
MDP	Markov Decision Process
MEC	Mobile Edge Computing
MEO	Medium Earth Orbit
MIA	Membership Inference Attack
MIB	Master Information Block
MIMO	Multiple-Input Multiple-Output
MIQCP	Mixed Integer Quadratically Constrained Program
ML	Machine Learning
MME	Mobility Management Entity
mMTC	Massive Machine Type Communications
MNO	Mobile Network Operator
MOCN	Multi-Operator Core Network
MOGA	Multi Objective Genetic Programming Algorithm
MORAN	Multi-Operator RAN
MVNO	Mobile Virtual Network Operator
NAS	Non-Access Stratum
Near-RT RIC	Near-Real-Time RAN Intelligent Controller
NEF	Network Exposure Function
NF	Network Function
NFV	Network Function Virtualization
NFVI	Network Function Virtualization Infrastructure
NFVO	NFV Orchestrator
NGAP	Next Generation Application Protocol
NGSO	Non-Geosynchronous Orbit
NR	New Radio
NRF	NF Repository Function
NSMF	Network Slice Management Function

NSSF	Network Slice Selection Function
NSSMF	Network Slice Subnet Management Function
NTN	Non-Terrestrial Networks
ONAP	Open Network Automation Platform
ORAN	Open Radio Access Network
PBCH	Physical Broadcast Channel
PCF	Policy Control Function
PDCP	Packet Data Convergence Protocol
PDU	Protocol Data Unit
PER	Packet Error Rate
PKI	Public Key Infrastructure
PLMN	Public Land Mobile Network
PNR	Perturbation-to-Noise Ratio
PoS	Proof-of-Stake
PoW	Proof-of-Work
PRACH	Physical Random Access Channel
PSS	Primary Synchronization Signal
PUSCH	Physical Uplink Shared Channel
QLRA	Q-Learning Resource Allocation
QoE	Quality of Experience
QoS	Quality of Service
QPSK	Quadrature Phase Shift Keying
RA	Registration Authority
RACH	Random Access Channel
RAN	Radio Access Network
RAR	Random Access Response
RAT	Radio Access Technology
RB	Resource Block
RF	Radio Frequency
RIC	RAN Intelligent Controller
RIS	Reconfigurable Intelligent Surface
RL	Reinforcement Learning
RLC	Radio Link Control
RRC	Radio Resource Control
RRM	Radio Resource Management
RSRP	Reference Signal Receive Power
RSRQ	Reference Signal Receive Quality
RSSI	Received Signal Strength Indicator
RTT	Round Trip Time

RU	Radio Unit
SA	Service and System Aspects
SAMA-RL	Single-Agent Multi-Agent RL
SBA	Service-Based-Architecture
SDN	Software-Defined Networking
SDU	Service Data Unit
SGD	Stochastic Gradient Descent
SIB	System Information Block
SIM	Subscriber Identity Module
SINR	Signal-to-Noise-and-Interference Ratio
SLA	Service Level Agreement
SMC	Secure Multi-Party Computation
SMF	Session Management Function
SMO	Service Management and Orchestration
SMTC	SS/PBCH Measurement Timing Configuration
SON	Self Organizing Network
SNR	Signal-to-Noise Ratio
SRI	Satellite Radio Interface
SSB	Synchronization Signal Block
SSL	Secure Sockets Layer
SSS	Secondary Synchronization Signal
SVD	Singular Value Decomposition
SVM	Support Vector Machine
TA	Tracking Area
TDMA	Time Division Multiple Access
TLS	Transport Layer Security
TPM	Trusted Platform Module
TRL	Transfer Reinforcement Learning
TTI	Transmission Time Interval
UAS	Unmanned Aircraft System
UDM	Unified Data Management
UE	User Equipment
UPF	User Plane Function
URLLC	Ultra Reliable Low Latency Communications
VM	Virtual Machine
VNF	Virtual Network Function
VNFD	VNF Descriptors
VoLTE	Voice over LTE
VR	Virtual Reality

1

Securing 6G: New Frontiers

Luiz A. DaSilva and Eric Burger

Commonwealth Cyber Initiative, Virginia Tech, USA

1.1 Introduction

Mobile network security resembles a game of cat-and-mouse: each new generation introduces new features and technologies, which are exploited by new attacks; the next generation fixes these vulnerabilities, while introducing new features, which are then exploited, and so on. Sometimes a new technology is effective in helping protect the network while at the same time exposing new attack surfaces. This is the case, for example, with artificial intelligence (AI): AI for cybersecurity and cybersecurity for AI are both rich areas for research and innovation.

This book focuses on security mechanisms and challenges for the sixth generation of mobile networks, or 6G, which at the time of writing is in its early stages of standardization.

This introductory chapter reviews the evolution of security mechanisms over the past five decades, since the beginning of commercial mobile service, with analog wireless networks devoted to voice communications, to today's 5G, meant to support a variety of services in public and private networks. We also introduce an overall vision for 6G, the technologies that will gain importance in this upcoming generation, and what opportunities and challenges they will open up in securing the network.

The remainder of the book discusses 6G security issues related to specific technologies and services, as the title indicates. We devote chapters 2 through

6G Security: New Frontiers in Mobile Network Technologies and Verticals, First Edition.
Edited by Luiz A. DaSilva.

6 to security features needed for specific technologies: softwarization, network slicing, blockchain, non-terrestrial networks, and millimeter and Terahertz communications. Chapters 7 through 10 then cover new use cases and verticals that will be supported by 6G and their related security requirements and solutions: cyber-physical systems, power-constrained devices, microgrids, and autonomous vehicles.

1.2 Generational Evolution of Networks and Security Mechanisms

Before getting into the specific technologies and security threats expected in 6G, it is useful to take stock of how mobile networks have evolved in the last 50 years or so, accompanied by the evolution of security mechanisms (and new attacks).

The first generation of cellular networks, which we nowadays would call 1G, was introduced in the 1980s. In the US, this technology was known as Advanced Mobile Phone System (AMPS). These were analog systems meant for voice communications, with no support for encryption. Eavesdropping on those voice calls required simply a scanner – as is often the case, policy lagged technology, and eventually the government made it illegal to sell those scanners. Since mobile identification numbers were sent in the clear, it was also possible to clone a mobile phone by capturing this control information.

The next generation, 2G, emerged in the early 1990s and included several standards using digital communications, based on Time Division Multiple Access (TDMA) or Code Division Multiple Access (CDMA). Most of the world adopted the Global System for Mobile (GSM) standard, although other systems such as Digital Advanced Mobile Phone Systems (D-AMPS) and Integrated Digital Enhanced Network (iDEN) are also considered part of this generation. In addition to voice, 2G supported text messaging.

Communications could now be encrypted, although 2G uses weak encryption that is now possible to break in real time. In GSM, a 128-bit master key is stored in each Subscriber Identity Module (SIM) card and used when the network authenticates the user. It has been shown, however, that a few million interactions with the SIM card were sufficient for an attacker to obtain this master key [1].

Another known security vulnerability is that, while in GSM the cell tower authenticates the mobile, the mobile does not authenticate the cell tower. This opens the door to a man-in-the-middle attack [2]: The attacker broadcasts the identifier of the subscriber's home network. The mobile device will always connect to the base station from which it receives the strongest signal, so it may connect to the fake base station. The attacker can also impersonate the mobile device to the network, by replaying the authentication information received by the mobile.

Further, it can request the mobile to drop encryption, and subsequently all communications will be conducted in cleartext.

In 3G, introduced in the early 2000s, data starts to come of age. 3G provides digital transmission of voice and data, with significant increases in data rate now allowing for web browsing. This generation also brings significant improvements in network security. Network access security maintains confidentiality about the user's identity, supports mutual authentication between the user and the network, provides data confidentiality, and protects the integrity of signaling messages [3].

Nevertheless, 3G still presents security vulnerabilities. There are instances in which the International Mobile Subscriber Identity (IMSI), which is the permanent identifier for a subscriber, is still transmitted in cleartext and is vulnerable to capture. Network-wide encryption is not always possible, and there is no general-purpose mechanism to protect data traffic [3].

The next generation of mobile networks, 4G, starts to emerge in the late 2000s, with most deployments occurring in the 2010s. This is the era when smartphones achieve full market dominance, with Long Term Evolution (LTE) as the primary representative of this generation. 4G supports broadband connectivity in an all-Internet Protocol (IP) network. In these networks, an evolved NodeB, the base station responsible for radio interface functions such as radio resource management, scheduling, and admission control, connects to an Evolved Packet Core (EPC). We should note WiMax as another architecture considered to be part of 4G, albeit with much lower market adoption.

Additional security mechanisms are introduced in 4G. LTE defines both an equipment identifier, the International Mobile Equipment Identity (IMEI), and a subscriber identifier, the IMSI. Temporary identifiers are used to avoid transmitting permanent unique identifiers over the air and to reduce the ability of an adversary to steal and make use of legitimate identities [4]. LTE also supports secure signaling between the user equipment (UE) and the Mobility Management Entity (MME), responsible for managing UE contexts and controlling authentication functions. For the first time, authentication and key agreement protocols are defined to support secure internetworking between 3GPP and non-3GPP technologies.

Notwithstanding these advances in security, types of attacks on 4G reported in the literature include location leak attacks, IMSI capture, vulnerabilities in the LTE access procedure, and vulnerabilities in the handover procedure [5].

The current generation of mobile networks, 5G, arrives in the late 2010s and early 2020s. It is motivated by three broad types of services: Enhanced Mobile Broadband (eMBB) is supported by user experience data rates in the order of 100 Mbits/sec; Ultra Reliable Low Latency Communications (URLLC) requires data plane latency below 1 ms and several "nines" or reliability (e.g., 99.999%); and Massive Machine Type Communications (mMTC) supports connectivity to a million or more devices per square kilometer.

5G introduces security features in various domains, including network access security, user and control plane security mechanisms, and application domain security. These improve the ability of the serving network authenticating the UE and vice-versa, and of the subscriber's serving network working together with the home network to authorize the user. Virtualization and softwarization in 5G have meant an increasingly complex supply chain, where it is difficult to establish a trusted perimeter. In such an environment, zero-trust architectures gain importance, such as the one proposed in [6]. On the positive side, softwarization makes it easier to introduce and update software cybersecurity solutions, such as the latest anomaly and intrusion detection mechanisms leveraging AI.

Finally, we arrive in 6G, which is very much a work in progress, going through ideation and initial standardization phases. We look at what is coming with 6G in the next section.

1.3 What Is Coming in 6G?

Before diving further into 6G, it is worth understanding the drivers for the *n*G release cycle noted in the previous section as well as what has driven the softwarization of the RAN.

By designing purpose-build hardware, network equipment manufacturers could produce systems that had at least ten times the performance or ten times the capacity of software solutions running on commodity hardware. This resulted in five-year life cycles for telecommunications network equipment.

This lifecycle made sense: if commodity hardware doubled in performance every 18 months, five years later commodity hardware would be delivering ten times the performance as it did in the base year. Thus, network equipment manufacturers could continuously add value if they refreshed their hardware at least every five years to be that much ahead of software solutions on commodity hardware.

This was one of the drivers pushing 3G to an all-Internet Protocol (IP) network core from the time-domain multiplexed core of 2G. There was the expectation that by virtue of being on IP, a standard from the information technology (IT) world, service providers would be able to ride the commodity hardware cost curve. However, two things stood in the way of making this a reality. The first was that carriers needed the density that could only be economically provided by bespoke hardware solutions. As an example, the SnowShore Networks N20 Media Server[1] could process 20,000 simultaneous media streams in an 8-rack unit (14-inch) box. At the time of introduction, it would have taken many racks of general-purpose computers to service the same number of customers. The second is that many misunderstood where the economies of scale in the IT market came from. They

1 Burger was a founder of SnowShore Networks

did not come from IP packet transport versus time-domain multiplexing. Rather, the economies of scale came from the ability to process IP packets on hardware directed to the IT market.

The internals of most network equipment were substantially the same as what we see in the IT market: Intel processors, Linux-based operating systems, and so on. The differences were often hardware accelerators, such as digital signal processors (DSPs), dedicated network processors, and improved internal communication busses to provide faster data flow over what we called the Fast Path.

A good year at SnowShore would be selling 1,000 of our bespoke boxes. Thus, we could amortize our engineering costs over 1,000 units. Given the millions of dollars of engineering required to design hardware and integrated software, that meant that to break even, each unit needed thousands of dollars in revenue to pay for design engineering. This figure does not begin to cover the components or manufacturing costs. Moreover, as we would be using components mostly from the time of design, the cost of the box would be stable or possibly increasing as old parts become harder to source. Another factor is competition. In the network equipment space, there are tens of relatively small competitors. Each have the same issue of relatively small volume shipments, so none would be at a competitive disadvantage to manufacture the same hardware for years until the above-mentioned performance between IT software and bespoke hardware crossed.

Compare this to the IT marketplace. If IBM, HP, or Dell sold only 1,000 units a week, it would be considered a very bad week. That means the engineering costs for IT hardware are on the order of dollars per unit. On the competitive side, because of the fungibility of Intel/AMD computers running Linux or Windows, the moment Intel or AMD released a new processor, there would be a strong competitive driver for the IT equipment manufacturers to quickly integrate the new processor to provide higher performance or lower cost. It is the combination of hardware amortization and a vibrant, competitive market for IT equipment that relentlessly drives the availability of more and more capable commodity equipment. The last factor driving the wireless market to commodity IT hardware is the scale required by carriers. At any point in time, it is likely that bespoke telecom hardware could still provide ten times the performance of software running on commodity IT hardware. However, it is no longer the case that carriers need that density. Given the costs being driven down in the IT space, we have arrived at the point where running specialized software on commoditized IT hardware is good enough for telecom.

1.3.1 Standardization Timeline

The 3rd Generation Partnership Project (3GPP) specifies the technical standards for global radiocommunication capability and interoperability. Seven telecommunications standards development organizations (SDO), namely ARIB (Japan),

ATIS (North America), CCSA (China), ETSI (Europe), TSDSI (India), TTA (Korea), and TTC (Japan) are the organizational partners of 3GPP. Industry, government, and academia participate in 3GPP through one or more of the SDOs.

Given the 5- to 10-year lifecycle of network equipment noted above, 3GPP major releases (3G, 4G, 5G, and now 6G) occurred on a decadal schedule, with 3G around 1999 (Rel. 99)[7], 4G around 2009 (Rel. 8)[8], and 5G around 2018 (Rel. 15)[9]. Thus, it would be no surprise that 2030 is the target for 6G. Like 4G and 5G, we expect to see some of the features of the 'next' G to trickle into the current G, as we did with LTE-Advanced (Rel. 10). One driver of this decade-long cycle is to fully leverage the new features, often the entire network needs upgrades, from the radio access network (RAN) to the core of the network. Because of the hundreds of vendors at the time of 3G and even the tens of vendors surviving at the time of 5G, it was easier for the industry to bundle the features together into feature releases. Not surprisingly, these releases also happened to match up with the bespoke network equipment lifetime described above. Thus, most if not all vendors could offer and use the new capabilities together.

3GPP convened the first 6G Workshop in Incheon, South Korea, in March 2025 to discuss what the next generation of wireless networks will look like [10]. 3GPP is targeting Rel. 20 for mid-2027. Rel. 20 will include both the standards for 5G Advanced and early studies for 6G.

Rel. 21 will be the first set of 3GPP 6G technical specifications. At the time of this writing, it is too soon to precisely say when Rel. 21 will be. Planning for the schedule is expected to be completed by June 2026. That said, 3GPP has already determined that the freeze date for a stable set of APIs will not be before March 2029.

The broad work plan is to develop several 6G study items, with system architecture work running from mid-2024 through March 2026. The RAN working groups started work in December 2024 and expect to complete their preliminary studies by June 2026. With the results of the March 2025 workshop, the actual work of studying the systems will begin, running through early 2027. Similarly, work on the RAN systems will start in June 2025 with detailed work starting in September 2025. That work is expected to run through mid-2027.

The International Telecommunications Union Radiocommunication Sector (ITU-R) works on a vision statement for the global, interoperable mobile network, called International Mobile Telecommunications, or IMT. IMT-2000 corresponded to 3G, IMT-2010 for 4G, IMT-2020 for 5G, and now the ITU-R is finalizing IMT-2030 for 6G. At the time of this writing in mid-2025, the ITU-T is collecting input on IMT-2030. The ITU-R work plan roughly follows the 3GPP work plan. Work on IMT-2030 started in 2021 with a report on future technology trends and several feasibility studies. The period from 2024 to 2026 will be collecting technical requirements, evaluation criteria, and key performance requirements for IMT-2030. Finally, the work from 2027 to 2030 will be collecting, evaluating, and building consensus around what will be included in IMT-2030.

So far, we have had a decadal cadence of G releases. Does this mean that once released, everyone goes to the next G? Clearly not. Historically, consumers upgraded their phones on an 18- to 24-month cycle. Recently, without compelling new features, this upgrade cycle is being extended. Thus, carriers have to offer at least the current and last G to service their customers. However, leading-edge techies buying the latest Apple or Samsung phone is not the entire market. Another important market is the Internet of Things, or IoT. IoT manufacturers want two things: reliability and low manufacturing cost. This drives IoT devices to use stable networking protocols. At the cusp of a G release, an IoT manufacturer is more likely to select the prior G because of its proven reliability. With nearly ten years of road time, the prior G is going to be very stable. Conversely, the new G is likely to be going through some flux. The other factor, cost, also often drives IoT manufacturers to the prior G. Like network equipment manufacturers, component manufacturers will have already amortized their engineering costs over the prior decade. Finally, many industrial IoT devices have twenty-year lifetimes. This combination of factors means that carriers are likely to need to have three generations of network protocols (e.g., 4G, 5G, and 6G) operating simultaneously. How to do this economically is an active area of research.

1.3.2 Key Features of 6G

1.3.2.1 Delivering on the Promise of 5G

Just as 4G delivered on the promise of 3G, we expect 6G to deliver on the promise of 5G.

The most important feature for 6G, and the one cited by all the participants in the 3GPP's Workshop on 6G, is Security, followed by AI as a close second [11]. There is a consensus that even though 5G improves security from 4G, for various reasons, some geopolitical, 5G does not provide the robust security features required by an always-on, highly interconnected network that many critical infrastructure sectors depend on. The importance of communication networks during times of crisis also drives the need for strong security and continuity of service.

In 2025, government and industry has recognized what academia had been discussing for a few years: networks must implement quantum-safe security mechanisms. Given the long lead time before 6G and IMT-2030 is complete, combined with quantum computers having the time and opportunity to become practical, yesterday's non-quantum-safe security mechanisms will be broken and breached by the time of the launch of 6G. As well, 6G is accelerating the trend mentioned above migrating more and more functions that once resided in silicon or bespoke hardware into IT-centric data centers and cloud computing. This creates new attack surfaces and unstable APIs as the industry gains road time with the new architectures and deployment models.

5G unquestionably delivered enhanced mobile broadband (eMBB). It also has delivered on massive machine to machine communication (mMMC). However, one promise undelivered in 5G is ultra-reliable, low-latency communications (URLLC). While in some scenarios, 5G could deliver 1 ms transmission times, 5G could neither reliably deliver such low latency nor provide end-to-end latencies approaching 1 ms. 6G promises xURLLC, where true end-to-end communication path can have specified quality of service to meet end-to-end millisecond latencies. Whether that strict target is physically possible is still a matter of research.

One promise of 5G that has mixed success so far but will be extended in 6G is mobile edge computing (MEC). In 5G, there was a generic sense that it would be helpful to have compute physically close to the user equipment. In this case, that compute could be as close as the base station or collocated with the centralized radio processing units (e.g., the Open RAN O-DU). Recall that wireless communications propagate nearly at the speed of light in a vacuum, while fiber and copper communications propagate at about 60% the speed of light. For those xURLLC applications that need millisecond end-to-end latency, the compute platform can be no further than a few tens of kilometers from the base station. This physical reality, combined with the expected heavy load presented by AI-native 6G structures, will drive the deployment of MEC. However, what remains to be seen is whether carriers deploy the MEC as part of their RAN deployments or if hyperscalers, other cloud providers, or enterprises will deploy the MEC, as most applications are already cloud-native and not MEC-aware.

1.3.3 Open RAN

While Open RAN was introduced as a 5G technology, we expect 6G to build upon the experience of 5G and accelerate the software innovation unlocked by a thriving xApp and rApp ecosystem. Technically speaking, there is nothing that can be accomplished with Open RAN that cannot be accomplished with a proprietary RAN. However, standard, open interfaces enable developers, vendors, and academics other than the RAN manufacturer to introduce components and concepts. Monolithic RAN vendors' incentives are to focus on features that have mass market appeal to their mass market carrier customers. A dollar spent in R&D towards the Tier-1's would have a much higher ROI for the Tier-1 equipment manufacturers than implementing an unproven technology or a feature targeted to a niche market.

This competition greatly accelerates the pace of innovation. Some features will flow up-market. Likewise, when novel technologies get proven by an upstart, the market can adopt it, driving the legacy vendors to improve their offerings.

1.3.4 AI-native

Just behind security as a key driver for 6G is the desire for AI support. 5G brings concepts of AI to the table. 6G, more especially in conjunction with Open RAN, provides a standard platform for AI-based intelligent network management (e.g., rApps) and AI-based network performance (e.g., xApps). The goals of making 6G AI-native are to use AI for dynamic spectrum access as well as potentially for machines using the physics of the environment combined with the semantics of the communication to generate optimum communication waveforms on the fly [12].

From a cyber perspective, there are three major areas of focus. First, if 6G networks are dependent on AI, we need to ensure that AI is trustworthy. Second, 6G networks will be providing AI services; we need to ensure the services are secure. And Third, AI applications are likely to be collecting data from thousands or millions of sensors. Thus, we need to ensure the integrity of the data as well as the privacy of the data and pattern of life information exposed by wireless networks. Some of the following chapters in this book will explore these themes further.

Given spectrum scarcity, 6G will be *sharing-native*. Instead of an exclusive use mindset, as we had in all of the prior Gs, 6G can let us use unused bands in rural areas for unlicensed use, share communications with radio location services, and share communications with passive sensing.

The introduction of standardized interfaces for xApps introduces the capability for extremely intelligent radio control. We are moving from considering "cognitive radio" being a radio that is software definable, to being a radio that really 'thinks.' Open RAN makes available considerable sensing information and the opportunity to optimize at the radio and network layer. This opens the possibility for novel spectrum sharing solutions. The DARPA Spectrum Collaboration Challenge[2] proved that AI-assisted radios could transparently coexist with legacy waveforms in the spectrum bands. Open RAN provides a platform to translate that capability into a product.

Spectrum sharing for 6G is not only a capability, but a policy imperative. Many bands are desired for communication because of free-space transparency, high frequency carriers, and wide bandwidth. However, those characteristics also make the spectrum desired for radiolocation. There are hundreds of billions of dollars of legacy radar systems deployed, and coexistence would be considerably more economical than buying entire navies' worth of ships.

As hinted above, another use for AI in the network is to improve joint (or integrated) sensing and communication or (JSAC or ISAC). ISAC enables the coexistence of radio navigation with communications in the same band. JSAC uses the radiolocation properties of communication signals, particularly in the midband. The radio unit's emissions reflect off objects in the path of the beam.

---

2 https://www.darpa.mil/research/programs/spectrum-collaboration-challenge

With high-resolution beamforming for finding user equipment, those beams can also be used for sensing. However, unlike a traditional radar system, the waveform is optimized for communications. It is only with significant processing power and the estimation capabilities provided by machine learning is JSAC possible [13]. We will discuss this further below.

1.4 The Road to Resilient 6G

Every new generation of mobile networks is a combination of evolution and revolution, and 6G is no different in that respect. We expect (and hope!) that security will be more than ever front and center in the standardization and deployment of this new generation.

One shift is that, as mobile networks become part of our critical infrastructure, *resilience* is increasingly recognized as a fundamental property of these networks. Resilience refers to the ability of a system to withstand disruptions that could have deliberate (e.g., cyberattacks, sabotage, acts of war) or accidental (e.g., equipment failure, natural disasters) causes. Unlike its cousin, *reliability*, which in the context of mobile networks is often viewed as a service for specific verticals, resilience is expected in all aspects of the network [14, 15]. And security for 6G is, clearly, a prerequisite for resilience.

Some of the new challenges and opportunities to make 6G more secure, and therefore more resilient, are in technologies and trends that are accelerating with this new generation. We group those trends, discussed in the previous section, into three categories:

1.4.1 Introduction of New Technologies

As discussed above, the integration of sensing and communications, referred to as Joint Sensing and Communication (JSAC) or Integrated Sensing and Communication (ISAC), is a unique new feature promised for 6G [16]. The introduction of new technologies needs to consider potential new vulnerabilities that must be addressed. In the case of JSAC, one such vulnerability is information leakage between the sensing and communication functions, which can introduce security and privacy issues. The design of secure JSAC, for example using physical layer security techniques and waveform design that can minimize the risk of an eavesdropper being successful, is an emerging area of research [17, 18].

Another technology that is steadily gaining steam is Reconfigurable Intelligent Surface (RIS). These surfaces can be programmed to control the reflection of electromagnetic waves in order to aid communications, in particular in non-line-of-sight situations. A growing body of work is looking at how RISs can be used to improve information-theoretic security [19], for example, to combat eavesdroppers in the context of JSAC, as described above. Conversely, the technology could be used by malicious actors to disrupt communications, for example, through beam management manipulation [20].

1.4.2 Multi-vendor Operation

Softwarization, virtualization, and disaggregation have started to open the door to network infrastructure that is less monolithic and comprises software and hardware solutions from multiple vendors. The Open Radio Access Network (ORAN) architecture discussed above defines the interfaces and AI-driven controllers to accomplish that goal [21].

On the cyber side, having open interfaces introduces a new attack surface. On the one hand, the interfaces between radio units, control units, and distributed units exist in most proprietary architectures just as they do in Open RAN. However, in monolithic systems, the equipment manufacturer can hand-tailor the interfaces to optimize for their equipment. Moreover, threat actors have to reverse engineer the proprietary interfaces to attack them. On the other hand, Open RAN *presumes* the interfaces will be open. This means that built into the interfaces are authentication and authorization procedures (as in Zero Trust), which will, in fact, make these interfaces more secure than the proprietary interfaces that hope to be secure because the interface is 'secret.' The following chapters in this book will explore this further.

A multi-vendor network relies on a more complex supply chain than a single-vendor, monolithic network. Approaches such as zero-trust should be adopted to address this potential security vulnerability [22].

1.4.3 Artificial Intelligence Everywhere

The pervasiveness of AI in all aspects of 6G, potentially from waveform selection to network management, from resource allocation to energy savings, means that it deserves to be treated as its own category when we consider securing 6G.

Resilience involves the ability to detect, mitigate, and recover from difficulties, and in the context of 6G AI may play a role in all three aspects. AI has been used for anomaly and intrusion detection much before the advent of 6G [23], and its role is likely to continue to grow.

1.5 Technologies and Verticals

There are (at least) two ways to look at the evolution toward a new mobile generation: the technological advances that drive its design and the applications that it enables. The former is critical to achieve ever-evolving gains in spectral efficiency, latency, scalability, resilience, etc. The latter underpins a business case for the massive investments required for the roll-out of a new generation. The first six chapters of this book focus on security issues associated with technologies that we expect to see in 6G; the following four chapters look into the security requirements associated with verticals that are targeted in the vision of 6G.

We start, in Chapter 2, with network softwarization and disaggregation, two trends that are by no means new to 6G but which will significantly impact the architecture of this new generation. Network softwarization decouples the software that implements network functions, protocols, and services from the hardware on which that software runs. This concept is tightly related to that of virtualization, which creates virtual functions (Network Function Virtualization (NFV)) and abstracted instances of storage, processing, and other network resources. Both concepts are well established in 5G, enabling increased flexibility in the provisioning of services [24]. Programmability facilitates the deployment of bespoke services, which is important if next generations are to support completely new classes of service where hyper-reliability, or ultra-low latency, or massive scalability are required. This chapter treats softwarization together with the more recent trend of network disaggregation, where different network elements can be designed and optimized separately and communicate among themselves through standardization Application Programming Interfaces (APIs). The promise of disaggregation is to lower capital and operational costs, decrease vendor lock-in, and spur innovation in network functionality. In the context of the Radio Access Network (RAN), disaggregation is being championed by the ORAN Alliance. Chapter 2 discusses the security, privacy, and resiliency implications of these important trends in the context of the next generation of mobile networks.

Network slicing, which refers to the creation of multiple virtual networks, or slices, on top of shared network infrastructure, is enabled by softwarization and virtualization. This is the subject of Chapter 3. A slice can be optimized for a certain type of service, with particular Quality of Service (QoS) requirements. Increasingly, this will be done through the use of Machine Learning (ML) trained to adapt to variations in demand and resource availability. Conversely, Adversarial Machine Learning (AML) can be used to exploit vulnerabilities in the creation and management of these slices. This chapter focuses on how to establish secure network slices in a world where AI is embedded in all aspects of network control and management, as is expected in 6G.

At the same time that AI is becoming critical for the management of the network, the network is also critical to support AI. This duality is sometimes referred to as *AI for networks and networks for AI*. Chapter 4 investigates Federated Learning (FL), an ML technique whereby the AI model can be trained across a distributed network of devices and services, none of which contributes its own data directly. In this context, we consider the use of blockchain to overcome some shortcomings of FL, including single-node failures and the need for incentives to participate in the distributed learning process. The chapter focuses on how to mitigate security risks in the use of FL, such as potential privacy breaches that can be caused as a result of each node training on its own private data.

Non-Terrestrial Networks (NTNs) have become an integral component of the latest generations of mobile networks. Generally understood as encompassing satellites, High-Altitude Platform Station (HAPS), and drones, NTNs have become an important complement to terrestrial networks, providing coverage in geographical areas where the deployment of terrestrial network assets is difficult or commercially infeasible, and supporting specific use cases such as military and disaster response operations. Chapter 5 is devoted to these networks, explaining how non-terrestrial components are integrated into the network, describing their security features, and discussing potential vulnerabilities and mitigation approaches.

Higher frequency bands have the potential to support significantly higher data rates: first, mmWave communications started to be exploited both in 3GPP and IEEE standards, and more recently, the Terahertz band has been considered for localized deployments. There are tradeoffs, of course, in terms of reduced range and the need for very dense deployments. Chapter 6 discusses how these high-frequency bands may be incorporated into the next generations of mobile networks and what security risks are specific to the use of those bands. The chapter also considers technologies, such as beamforming and the use of Reconfigurable Intelligent Surfaces (RISs), which are often used in conjunction with those bands.

Starting in Chapter 7, our focus transitions from new technologies to the use cases that 6G may enable. Chapter 7 tackles Cyber Physical Systems (CPSs), systems that incorporate physical elements, such as actuators and mechanical structures, and cyber elements, such as compute, software, and networks. Mobile networks that primarily supported human subscribers are increasingly viewed as having a crucial role in communication with and between robots, autonomous and non-autonomous vehicles, industrial equipment, etc. This potential to expand the use base of networks is a major driver for the next generation. In the context of CPSs, the chapter considers homomorphic encryption, a technique that enables computations to be performed on encrypted data without the need to decrypt it first. The chapter discusses the practicalities of homomorphic encryption and other recent advancements in support of security for CPSs.

Chapter 8 investigates the important problem of how to secure power-constrained devices. Massive machine-type communications has been an area targeted by 5G; its potential has not been fully realized in the current generation, and it is still considered in the visioning towards 6G. Many of these devices are characterized by limited processing power and the need to preserve energy in order to extend the device's life. Severe resource limitations, in turn, pose security risks: the limited memory and compute may make sophisticated cybersecurity mechanisms infeasible and limit the viability of security patches and upgrades. There are also cyberattacks that specifically target this type of device, such as battery depletion attacks. The chapter discusses security issues in power-constrained devices and in particular the use of energy-preserving cryptography.

The security of smart electric microgrids is the topic of Chapter 9. Smart microgrids have been evolving rapidly to take advantage of diverse energy generation resources, and in particular of renewable energy generation, and to improve the resilience and reliability of the electric grid. They have also increasingly made use of communications infrastructure to coordinate among their highly distributed elements. The dependence of critical infrastructure on communication networks will continue to increase and are likely to be supported by Service Level Agreements (SLAs) that reflect the unique QoS requirements of those applications. Chapter 9 addresses the security needs associated with smart microgrids, including the impact of data loss or alteration with the intent of manipulating the energy market or impacting the stability of the energy system. The chapter also considers the potential impact of Adversarial Machine Learning (AML) targeting the microgrids.

The final chapter of this book is devoted to the rapidly growing field of autonomous vehicles. Autonomous vehicles are rapidly becoming a commercial reality, with fleets of driverless taxis operating in several countries. Such vehicles rely on a sophisticated array of sensors that collect data to be processed by Machine Learning (ML) algorithms that operate partly in the cloud. Mobile networks are instrumental for communications between the sensors and the locus of processing and decision making, as well as for communications among vehicles and between the vehicles and the road infrastructure for collective decision making. Chapter 10 considers adversarial attacks against ML models for autonomous vehicles, and how to safeguard against and mitigate those attacks.

References

1 Shin, M., Ma, J., Mishra, A., and Arbaugh, W. (2006). Wireless network security and internetworking. *Proceedings of the IEEE*, 94: 455–466.

2 Meyer, U. and Wetzel, S. (2004). On the impact of GSM encryption and man-in-the-middle attacks on the security of interoperating GSM/UMTS networks. In *International Symposium on Personal, Indoor and Mobile Radio Communications*. IEEE.

3 Xenakis, C. and Merakos, L. (2004). Security in third generation mobile networks. *Computer Communications*, 27: 638–650.

4 Seddigh, N., Nandy, B., Makkar, R., and Beaumont, J. (2010). Security advances and challenges in 4G wireless networks. In *International Conference on Privacy, Security, and Trust*. IEEE.

5 Cao, J., Ma, M., Li, H., Zhang, Y., and Luo, Z. (2014). A survey on security aspects for LTE and LTE-A networks. *IEEE Communication Surveys and Tutorials*, 16: 283–302.

6 Olsson, J., Shorov, A., Abdalrazek, L., and Whitefield, J. (2021). 5G zero trust – a zero-trust architecture for telecom. *Ericsson Technology Review*, 2021: 2–11.

7 3GPP (1999). Release 1999 Specifications. Technical Specification (TS) 21.101, 3rd Generation Partnership Project (3GPP). Version 1.0.0.

8 3GPP (2009). Technical Specifications and Technical Reports for a UTRAN-based 3GPP system. Technical Specification (TS) 21.101, 3rd Generation Partnership Project (3GPP). Version 8.0.0.

9 3GPP (2018). Technical Specifications and Technical Reports for a UTRAN-based 3GPP system. Technical Specification (TS) 21.101, 3rd Generation Partnership Project (3GPP). Version 15.0.0.

10 Jain, P., Borsato, R., and Schmitt, P. (2025). Chair's summary of the 3GPP workshop on 6G. In *3GPP workshop on 6G)*. 3GPP.

11 NGA (2022). Trust, Security, and Resilience for 6G Systems. Alliance report, Next G Alliance.

12 Chaccour, C., Saad, W., Debbah, M., Han, Z., and Vincent Poor, H. (2025). Less data, more knowledge: Building next-generation semantic communication networks. *IEEE Communications Surveys & Tutorials*, 27(1): 37–76.

13 NGA (2023). Distributed Sensing and Communications. Alliance report, Next G Alliance.

14 Khaloopour, L., Su, Y., Raskob, F., Meuser, T., Bless, R., and Jansen, L. (2024). Resilience-by-design in 6G networks: Literature review and novel enabling concepts. *IEEE Access*, 12: 155666–155695.

15 Rodriguez, E., Masip-Bruin, X., Martrat, J., Diaz, R., Jukan, A., and Granelli, F. (2024). A security services management architecture toward resilient 6G wireless and computing ecosystems. *IEEE Access*, 12: 98046–98058.

16 Liu, F., Cui, Y., Masouros, C., Xu, J., Han, T., and Y., E. (2022). Integrated sensing and communications: Toward dual-functional wireless networks for 6g and beyond. *IEEE Journal on Selected Areas in Communications*, 40: 1728–1767.

17 Gunlu, O., Bloch, M., Schaefer, R., and Yener, A. (2023). Secure integrated sensing and communication. *IEEE Journal on Selected Areas in Information Theory*, 4: 40–53.

18 Sun, W., Sun, S., Su, X., and Liu, R. (2024). Security-ensured integrated sensing and communication (ISAC) systems enabled by phase-coupled intelligent omni-surfaces (IOS). *IEEE Transactions on Wireless Communications*, 23: 3480–3492.

19 Khoshafa, M., Maraqa, O., Moualeu, J., Aboagye, S., Ngatched, T., and Ahmed, M. (2024). RIS-assisted physical layer security in emerging RF and optical wireless communication systems: A comprehensive survey. *IEEE Communications Surveys and Tutorials.*

20 Gomes, A., de Sena, A., Mahmood, N., Latva-aho, M., DaSilva, L., and Kibilda, J. (2024). Beam management manipulation with adversarial reconfigurable intelligent surfaces. In *Global Communications Conference (Globecom).* IEEE.

21 Abdalla, A., Upadhyaya, P., Shah, V., and Marojevic, V. (2022). Toward next generation open radio access networks: What O-RAN can and cannot do! *IEEE Network*, 36: 206–213.

22 Moudoud, H., El Houda, Z., and Brik, B. (2024). Zero trust security architecture for 6G open radio access networks (ORAN). *IEEE Networking Letters*, 6: 272–275.

23 Hussein, B., Du, Q., Imran, A., and Imran, M. (2019). Artificial intelligence-powered mobile edge computing-based anomaly detection in cellular networks. *IEEE Transactions on Industrial Informatics*, 16: 4986–4996.

24 Condoluci, M. and Mahmoodi, T. (2018). Softwarization and virtualization in 5G mobile networks: Benefits, trends and challenges. *Computer Networks*, 146: 65–84.

2

Securing Softwarization and Disaggregation

Mayukh Roy Chowdhury[1], Md. Habibur Rahman[2], Arham Siddique[3], Naveen Sapavath[3], Aloizio P. DaSilva[2], Vijay K. Shah[4] and Duminda Wijesekera[3]

[1] *Nokia Bell Labs, India*
[2] *Commonwealth Cyber Initiative, Virginia Tech, USA*
[3] *Cybersecurity Engineering Department, George Mason University, USA*
[4] *Electrical and Computer Engineering, North Carolina State University, USA*

2.1 Introduction

The landscape of mobile communication is rapidly evolving, and the emergence of 6G and beyond networks promises revolutionary advancements in connectivity and data services. With an ambitious vision to enable ultra-high data rates, ultra-low latency, and support for diverse applications with stringent Quality of Service (QoS) requirements, 6G networks aim to reshape the digital ecosystem and transform industries.

To achieve these ambitious goals, 6G and next-generation mobile systems are undergoing a fundamental transformation towards softwarized, virtualized, and disaggregated architectures. This paradigm shift leverages cutting-edge technologies such as Software-Defined Networking (SDN) and Network Function Virtualization (NFV). SDN is a revolutionary approach that enables network operators to centrally control and manage network resources using programmable software. It abstracts the control plane from the underlying hardware, allowing for dynamic and flexible network configuration and optimization. By separating the control logic from the physical infrastructure, SDN simplifies network

6G Security: New Frontiers in Mobile Network Technologies and Verticals, First Edition.
Edited by Luiz A. DaSilva.

management and enhances agility in responding to changing traffic patterns and service demands. NFV complements SDN by virtualizing traditional network functions, such as firewalls, load balancers, and routers. These functions are decoupled from specialized hardware and run as software instances on standard servers. NFV enables rapid deployment, deploying 6G on generic computing hardware, making the services dynamically deployable on any computing node belonging to the provider and enabling the scaling of network services, leading to cost savings and operational efficiency.

Another significant aspect of the disaggregation architectures in 5G and beyond is the separation of the core network and the Radio Access Network (RAN) with clearly defined interfaces, allowing multiple providers to work with each other in providing these services. This disaggregation of services into RAN and Core components that can be provided by different vendors leads to a natural trust boundary. We propose formalizing this concept with entailing architectural elements so that caller-callee relationships between Virtual Network Functions (VNFs) is properly established. Given that some orchestrated VNF calling sequences are common, leading to VNF chains, having a pre-computed trusted VNF chain leads to reduced latency during runtime.

While the softwarized, virtualized, and disaggregated 6G architecture brings many benefits, it also introduces new security and privacy challenges. In a multi-vendor, network environment involving various operators and components and multiple trust anchors and levels, ensuring end-to-end security becomes a complex task. The dynamic nature of these networks and the potential vulnerabilities in open-source software and virtualization adds to the complexity of guaranteeing end-to-end security, resulting in the need for a distributed trust management service for VNFs and VNF chains.

To address these security challenges, this chapter proposes a vision for a more secure NFV infrastructure for 6G and beyond networks. The primary focus is on establishing a robust trust management framework that binds the RAN and Core components together. This disaggregated trust management approach ensures secure communication and cooperation among different network elements, even when operating over non-trusted infrastructure. Furthermore, the chapter explores the concept of creating secure VNFs and VNF chains. VNFs are specified with unique identifying attributes, security attributes, and Service Level Agreements (SLAs) to define availability, penalties, and QoS parameters. Each VNF is assigned cryptographic credentials, including public and private key pairs, to ensure their authenticity and integrity.

In the subsequent sections, we delve into the technical details of the proposed disaggregated trust management framework and elaborate on the process of creating secure VNFs and VNF chains. In Section 2.2, some background in the context of NFV and security is provided, along with a discussion of the shortcomings of the state-of-the-art. Subsequently, in Section 2.3, the proposed disaggregated trust management framework is described. Section 2.5 concludes the chapter with insights on the probable research works in this direction.

2.2 Preliminaries

2.2.1 VNF and VNF Chaining

The terminology *Virtual Network Function* [1, 2] was officially defined in 2012 by the European Telecommunications Standards Institute (ETSI). A VNF runs on a Network Function Virtualization Infrastructure (NFVI) (consisting of storage, compute, and network nodes) and is orchestrated by a NFV Orchestrator (NFVO) and a VNF Manager. It has well-defined standardized interfaces to other VNFs via Vnf-1, the VNF Manager, its Element Manager (EM), and the NFVI and a specific functional behavior. A VNF generally implements a network entity or service with interfaces and well-defined behavior described by VNF software components. Figure 2.1 shows the internal architecture of a VNF.

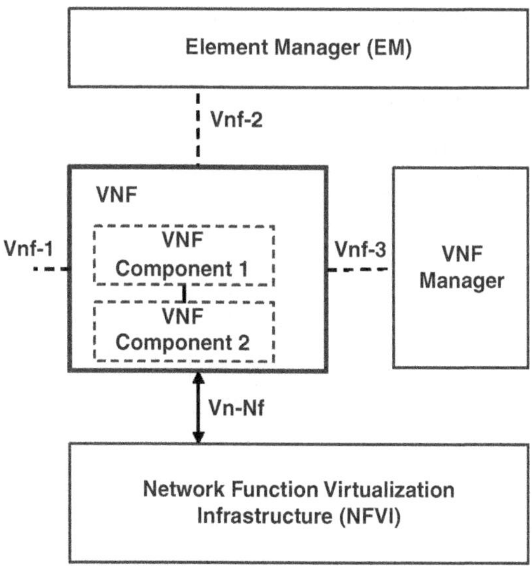

Figure 2.1 VNF internal architecture and interfaces.

In traditional network architectures, Network Functions (NFs) are associated with proprietary hardware components and are completely dependent on them. With the evolution of network softwarization and virtualization, these NFs have been transformed to virtual functions. This transformation has directly contributed to the telecommunications operator digital transformation and virtualization. As a key component of the new virtualized architecture, these VNFs are usually deployed on virtual machines or containers that requires resources allocation, such as, compute, storage and networking. To manage and orchestrate the life cycle of these VNFs, a Management, Automation, and Network Orchestration (MANO) [3] was defined as part of the NFVI. One important distinguishing feature of a VNF is that it can provide a single service or it can communicate with other VNFs to realize specific sub-tasks or provide a specific service. When a VNF needs to communicate or interoperate with other VNFs to realize a service, the MANO identifies this workflow as a VNF chaining [4, 5]. Figure 2.2 illustrates an example of a VNF chain composed of three VNFs deployed across different transport networks in a multi-domain cloud environment.

Launching this chaining of services requires a very careful work to be realized by the MANO, which in coordination with other network components performs the linkage of the functions to form a service or VNF chaining. This is often a very complex procedure that requires synchronization and trust. In the next generation of communication networks, the interoperability and distributed nature of the services, along with the network components and network infrastructures,

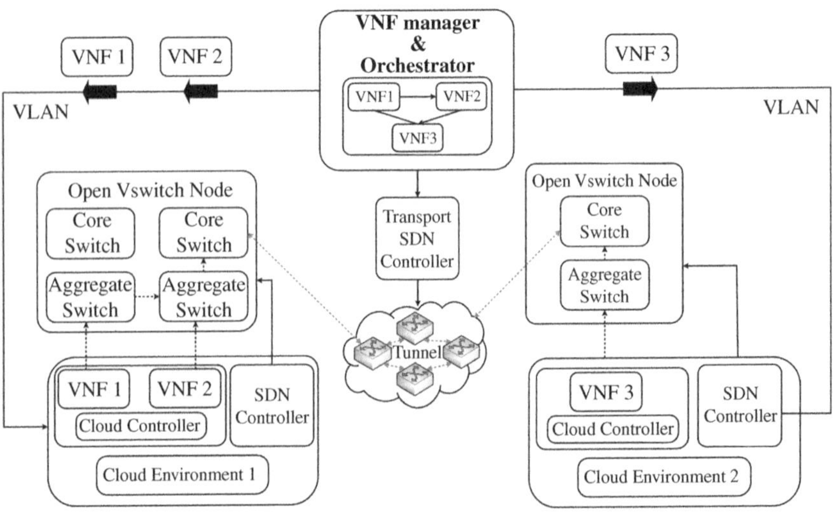

Figure 2.2 Illustration of VNF onboarding and chaining over a multi-site cloud computing environment.

require agility, scalability, lower operational overhead, and trust to realize the end-to-end functionality. The end-to-end service in this context is performed by the VNF chaining that is composed of several instances of different VNFs. However, if one of the VNFs that are part of the chaining is corrupted or becomes malicious, the overall chaining becomes compromised. As a result, the network service providers will deliver unreliable services, or will not be able to respond to customer needs and demands. While VNFs are core pieces of the next generation of networks, they still have limitations (e.g. coexistence in a cloud environment with physical devices, IT management in the abstract level and dynamic scaling) as the operators move toward large-scale 5G or edge deployment that need flexibility, security and interoperability.

An example of VNF and VNF chaining is the 5G Core, which is based on a Service-Based-Architecture (SBA) that implements a cloud-native approach. In this architecture, each VNF offers services to other VNFs via APIs (forming the VNF chaining). Some of the functions included in the 5G Core are User Plane Function (UPF), Data Network (DN), Access and Mobility Management Function (AMF), Authentication Server Function (AUSF), Session Management Function (SMF), Network Slice Selection Function (NSSF), Network Exposure Function (NEF), NF Repository Function (NRF), Policy Control Function (PCF), Unified Data Management (UDM), and Application Function (AF). The next generation of communication networks allows these functions to be disaggregated and consequently deployed in different locations (e.g., edge-cloud) across different transport networks.

Recently there has been similar trend in the RAN after 3GPP proposed the disaggregation of the RAN [6] into Centralized Unit (CU), Distributed Unit (DU), and Radio Unit (RU). This attracted even more interest since the Open Radio Access Network (ORAN) movement, especially by the ORAN Alliance [7], which proposed standardized and open interfaces between disaggregated RAN components. Hence, this distributed context would benefit from a distributed trust mechanism to guarantee service reliability and security.

2.2.2 Security and Trust Management in 5G - State of the Art

5G and beyond cellular networks are built with the the core idea of bringing softwarization, virtualization, disaggregation and intelligence to the network. 5G has led to a paradigm shift from the era of circuit and packet switched networks by bringing in the new SBA. With the SBA also comes the need for additional security in the upper layers like transport and application. NFV and SDN are the primary enabling technologies to achieve completely flexible and autonomous next-generation networks. Softwarization of the network brings the possibility of sharing of resources between multiple stakeholders like network operators, infrastructure and service providers, or even users with varying applications and requirements. With the advent of multi-tenancy in cellular networks also come potential privacy and security threats.

If a VNF in the system is compromised, it can disturb the whole network in multiple ways: (i) by exhausting all the resources and thereby depriving other Virtual Network Function (VNF) from getting resources, (ii) by tampering with the incoming data packets before forwarding them to the intended recipients, (iii) by altering the access levels of users. The level of damage may vary based on the particular VNF that is compromised. For example, a malicious AMF can authenticate unsolicited users, whereas a compromised UPF can expose private data of enterprises. Machine-type devices or Internet of Things (IoT) devices which act as UEs can be exploited to orchestrate attacks like a Distributed Denial of Service (DDoS) in order to cause the exhaustion of resources, leading to unavailability of services. Even intelligent controllers used for autonomous decision making may be victims of security attacks leading to complete disruption of services. Therefore, it is of the utmost importance to build a trust framework between different stakeholders in order to reduce fear and boost confidence among them.

Emerging security protocols which are relevant from the 5G networks perspective include the following:

1. Authentication and Key Agreement (AKA) – user authentication between UE an 5G Core;
2. Internet Protocol Security (IPsec) – to secure the data plane communication between UEand UPF;
3. Transport Layer Security (TLS) – for securing control plane communication.

Security measures to be taken in the 5G SBA, first listed in Release 15 [8], are based on two components :

1. Between NFs: mutual authentication and transport layer security (through TLS v1.2 and v1.3);
2. Between consumer and providers of NF service: Token-based access authorization (through OAuth v2.0).

TLS, an updated version of its predecessor Secure Sockets Layer (SSL), is very much suitable for virtual infrastructures shared between multiple tenants, because it eases the direct termination of security in the NFs, bypassing the need of a gateway [9]. OAuth provides a way to perform token-based authorization. In the 5G SBA, the NRF acts as a central authorization server that grants token for access to consumers (of NF services), to be presented to producers (of NF services) later, while requesting a service.

Being a multi-disciplinary concept, trust has different notions in different areas. But the core idea of trust is defined in [10] as *a relationship in which an entity, often called the trustor, relies on someone or something, called the trustee, based on a given criterion*. The authors in [11] discussed the broad trust surface in 5G and beyond and discussed multiple trust dimensions, such as trust in VNFs, trust in MANO, and trust in NFVI. Some of the emerging technologies and tools which have potential of enabling trust in 5G ecosystem include: blockchain, hardware-based root of trust platforms like Trusted Platform Module (TPM), AI/ML based data analytics, etc.

Public Key Infrastructure (PKI) is central to trust management in 5G and plays a crucial role in usage of both TLS and OAuth in protection of the SBA [12, 13]. A PKI primarily consists of a pair of public-private keys, a certificate for identity verification of the owner of a particular public key, a Certificate Authority (CA), and a Registration Authority (RA). A CA is the authority that can issue X.509 certificates , or delegate to an RA which can forward to another CA for issuing the same [14]. X.509 is an International Telecommunication Unit – Standardization sector (ITU-T) specified standard for defining public key certificate formats.

The ETSI NFV standard [15] suggests multi-tier hierarchy with a root CA in 1st tier and sub-CAs in the following tier, as shown in Figure 2.3. Service providers are free to choose one root trust service along with any combination of sub-CAs shown in Figure 2.3, which is a visualization of exiting trust services.

Authors in [16] described the need for decentralized or federated trust management in 6G, whereas a decentralized certificate management for NFV is described in [17].

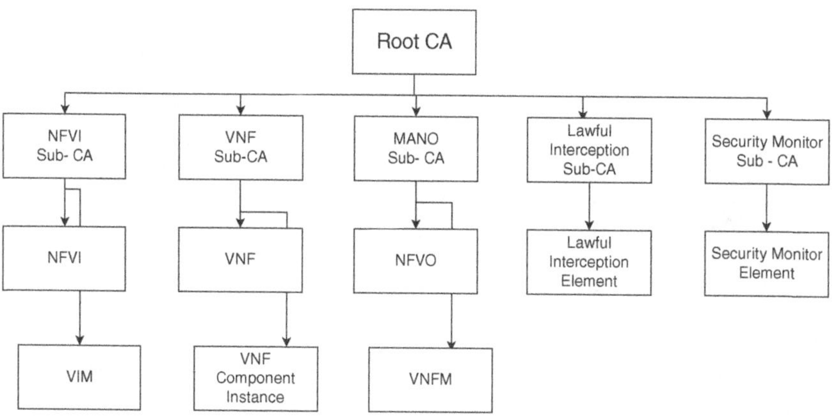

Figure 2.3 ETSI NFV standard for hierarchy in PKI certification.

2.2.3 Limitations of Centralized 5G Trust Management

The current software-based 6G service stack provides fully disaggregated service provisioning, where the services are provided by a stack of software components. These components identify each-other based on pre-established static trust relationships between them. Most mobile service providers own and operate the whole 5G stack, which is run on a virtual environment that is provisioned for them, with their own trust root or certificates and possibly some collection of sub-CAs, as shown in Figure 2.3. The RUs that are used by different providers already have cross certified trust relationships between the tower operators and the telecommunication service providers. Thus, unlike the services, the trust relationships are pre-determined, which limits any RAN services being able to connect to any Core services in the face of failures or service disruptions. If the RAN and Core services are disaggregated and provided as a service, as shown in Figure 2.5, then any operator will be able to dynamically invoke each other's RANs and Cores as orchestrated services. In order to do so, we propose an extension that enables the new RAN/Core as a service, also referred to as Multi-Operator RAN (MORAN) and Multi-Operator Core Network (MOCN) in the telecommunication domain.

MORAN and MOCN signify a paradigm shift in the telecommunications landscape, fostering enhanced efficiency and collaboration among network operators. In a MORAN scenario, multiple wireless carriers pool their RAN resources to

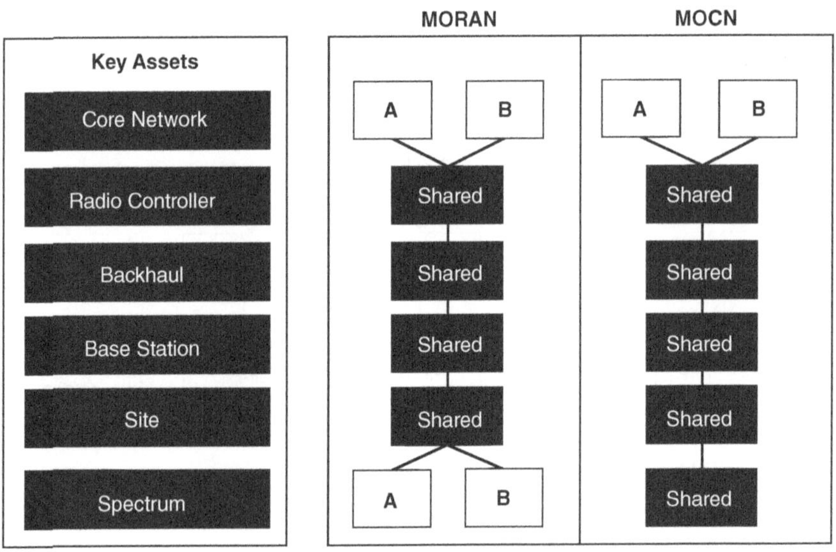

Figure 2.4 Comparison between MORAN and MOCN. Source: Adapted from [18].

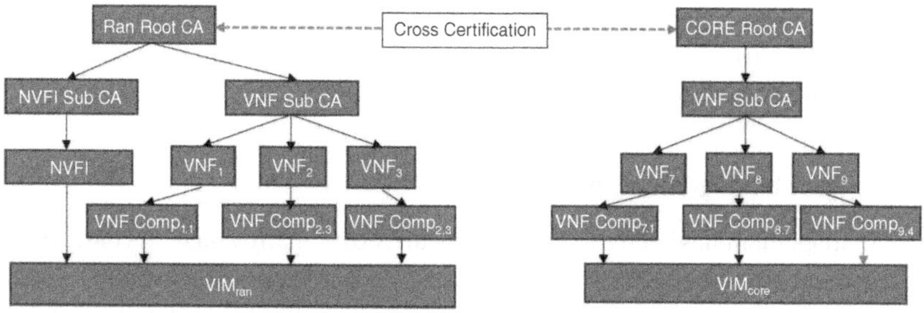

Figure 2.5 Cross certification.

share and optimize the network infrastructure, including base stations and antennas. This approach optimizes spectrum utilization, reduces costs, and diminishes the visual impact of excessive infrastructure. On the other hand, MOCN extends this collaboration to the heart of the network, enabling multiple operators to share core network components such as authentication, policy management, and user data management. This results in streamlined operations, reduced capital expenditures, and quicker service deployment. As the telecommunications industry continues to evolve, MORAN and MOCN represent innovative solutions that facilitate efficient spectrum usage and operational synergy among diverse operators, ultimately leading to improved services for end-users. Figure 2.4 illustrates a comparison of MORAN and MOCN.

2.3 Disaggregated Trust Management Framework

MORAN and MOCN services can use the same service interfaces as currently operating, except that the RAN and Core will have their separate trust roots. One of the consequences of having such trust roots is that the certificates used by the interfaces to identify the end-points will need to be changed to having certificates issued by their respective trust roots, as it happens in web service orchestrations.

Figure 2.5 shows a basic cross-certification schema, applied to a sequence of VNF calls. Recall that Figure 2.3 shows a trust chains anchored in a single trust root of a nextG system. Under the proposed dis-aggregation of trust, RAN and Core VNFs will have separate trust roots as shown in Figure 2.5. Hence, each instance of a VNF from any side will derive its trust certificate from their trust root, through the VNF sub-CA, and will run on Virtual Machines (VMs) that host these VNFs. Now, how would one derive the end-to-end trust for the VNF chain consisting of $\langle VNFComp_{9.4}; VNFComp_{1.1}; VNFComp_{1.1}; VNFComp_{2.3}; VNFComp_{3.2}; VNFComp_{8.7}; VNFComp_{7.1}\rangle$, where the listed components are instances of VNFs from Figure 2.3? Then the trust relationships needed to validate the end-to-end trust of the chain is derived from following the trust derivation arrows shown in Figure 2.3. The trust management system we designed to ensure trust chains like the the ones above is described in Section 2.3.1.

2.3.1 Trust Management for NFV

Trust management has been used in cybersecurity over many years. The basic trust management tenet is that that infrastructure selects a so-called trust root, which issues a certificate (mostly of limited duration) that would be used as an identity certificate or an attribute certificate. The most common certificate schema used by the industry is x.509, which is a signed statement using the trust root's private signature and verifiable by the recipient using the public certificate of the issuer. For example, $\{e_A \mid A \mid T\}_{Priv_B}$ is a certificate issued by B, signed by B's private key $Priv_B$, that attests that the entity A has a public key e_A valid during the period T. Any entity in receipt of $\{e_A \mid A \mid T\}_{Priv_B}$ can use the public key of B to validate the attestation issued by B, and may consider the attestation to be valid if its trust certificates issued by B, implying that the entity A has a public key E_A. Using this basic construct, and the assumption of transitive trust, the main trust root can create secondary certificates servers that are entrusted to issue other certificates, as shown in Figure 2.3. If there are multiple trust roots, they can issue cross certificates, so that an entity using a certificate from one trust root (such as the trust root for .com) can use a certificate from another domain (such as .mil).

In using trust management for VNFs, each VNF can be issued a certificate by its creator after it has been tested and deployed, provided that there is corporate permission to so. This would require that the VNF is considered a service in a service-based architecture, which is how 5G and beyond is viewed. Toward achieving that vision, we formulate the problem of providing the modeling and the execution as follows.

The first sub-task is to develop a VNF security framework with the following Service Management and Orchestration (SMO)/MANO components shown in the left side of Figure 2.6. Our proposal for enhancements to the VNF structures is described in Section 2.3.2, and its deployment sequence is described in Section 2.3.3.

1. **VNF Manager:** coordinates the communication between the different intra- and inter-MANO components, including cross-certification.
2. **VNF Trust Manager:** enrolls and provides certificate management functions for VNF interfaces, including updating and revoking certificates.
3. **VNF Orchestrator:** unravels the VNF chains, maps and invokes the VNF chains using the Communication Function Management Function (CFMF) that is invoked by any user of a VNF.

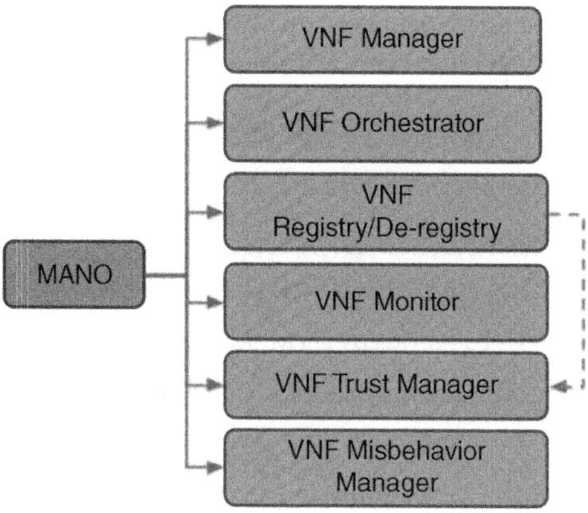

Figure 2.6 VNF component stack.

4. **VNF Registry:** enrolls and exposes VNFs within the trust domain to the other domains, provided VNFs are vetted.
5. **VNF Monitor:** maintains the QoS requirements provided by the VNFs to the user community.
6. **VNF Misbehavior Manager:** receives security incidents and embeds the logic to recommends dis-enrolling VNFs by revoking their certificates.

Communications between 5G components are specified using a Business Process Meta Language (BPMN)-based syntax [19] using VNFs.

2.3.2 Creating Secure VNFs with Disaggregated Trust

Creating security credentials for VNFs and VNF chains consists of multiple steps of specifying, mapping and enforcing the SLAs and cybersecurity of the VNFs and VNF chains. We begin this process by specifying VNFs. We specify identifying attributes, security attributes (including the trust domain and the public keys issued by the trust anchor) and SLAs (including violation penalties). Our SLAs specify their lifetime (including extendability), availability, penalties, and QoS, based on multiple Open Network Automation Platform (ONAP) and ETSI documents [20] and publications [21–23]. VNF Descriptors (VNFDs) define the VNF properties, such as: needed resources (amount and type of virtual compute, storage, networking), software metadata (external and internal connection points, virtual links, etc.), life cycle management behavior (scaling, instantiating, etc.), and life cycle management operations [15, 21]. VNFs are assigned static cryptographic credentials, such as public and private key pairs and certificates, by the VNF trust manager. Following Backus Normal Form (BNF) grammar [24] formalizes VNFs. We formalize the VNF specifications as follows.

$$\langle \text{VNF} \rangle ::= \langle \text{VNF-ID} \rangle \, \langle \text{VNF-SEC} \rangle \, \langle \text{VNF-SLA} \rangle \tag{2.1}$$

$$\langle \text{VNF-ID} \rangle ::= \langle \text{Trust Domain} \rangle \, \langle \text{VNF-POP} \rangle \; \texttt{VNF} - \texttt{Node} \; \langle \text{VNF-DES} \rangle \tag{2.2}$$

$$\langle \text{VNF-SEC} \rangle ::= \langle \text{VNF-pKey} \rangle \, \langle \text{HASH(VNF-ID)} \rangle \tag{2.3}$$

$$\langle \text{VNF-DES} \rangle ::= \texttt{Compute Cycles Memory} \; \langle \text{Meta data} \rangle \tag{2.4}$$

$$\langle \text{VNF-SLA} \rangle ::= \langle \text{Availability Level} \rangle \, \langle \text{Penalty} \rangle \, \langle \text{lifeTime} \rangle \, \langle \text{QoS} \rangle \tag{2.5}$$

$$\langle \text{Availability Level} \rangle ::= \langle \text{AV-High} \rangle \mid \langle \text{Av-Medium} \rangle \mid \langle \text{AV-Low} \rangle \tag{2.6}$$

$$\langle \text{Av-High} \rangle ::= \texttt{100\%} \tag{2.7}$$

$$\langle \text{Av-Medium} \rangle ::= \geq \texttt{99.5\%} \tag{2.8}$$

$$\langle \text{AV-Low} \rangle ::= < \texttt{99\%} \tag{2.9}$$

$$\langle \text{Penalty} \rangle ::= \langle \text{Major-Violation} \rangle \mid \langle \text{Medium-violation} \rangle \mid \langle \text{Minor-violation} \rangle \tag{2.10}$$

$$\langle \text{Major-violation} \rangle ::= \texttt{Terminate} \tag{2.11}$$

$$\langle \text{Medium-violation} \rangle ::= \texttt{Medium Fine} \tag{2.12}$$

$$\langle \text{Minor-violation} \rangle ::= \texttt{Low Fine} \tag{2.13}$$

$$\langle \text{LifeTime} \rangle ::= \langle \text{BeginTime} \rangle \, \langle \text{EndTime} \rangle \, \langle \text{Is extensible} \rangle \tag{2.14}$$

$$\langle \text{BeginTime} \rangle ::= \texttt{Time} \tag{2.15}$$

$$\langle \text{EndTime} \rangle ::= \texttt{Time} \tag{2.16}$$

$$\langle \text{Is Extensible} \rangle ::= \texttt{0} \mid \texttt{1} \tag{2.17}$$

$$\langle \text{QoS} \rangle ::= \langle \text{Latency} \rangle \, \langle \text{Throughput} \rangle \, \langle \text{power} \rangle \, \langle \text{Mobility} \rangle \, \langle \text{Spectrum} \rangle \tag{2.18}$$

$$\langle \text{Latency} \rangle ::= \texttt{time} \; \langle \text{Adjustment} \rangle \tag{2.19}$$

$$\langle \text{Adjustment} \rangle ::= \texttt{time} \tag{2.20}$$

$$\langle \text{Throughput} \rangle ::= \texttt{Decimal per second} \; \langle \text{Elasticity} \rangle \tag{2.21}$$

$$\langle \text{Elasticity} \rangle ::= \texttt{Decimal per second} \tag{2.22}$$

$$\langle \text{Power} \rangle ::= \texttt{kWT time} \tag{2.23}$$

$$\langle \text{Mobility} \rangle ::= \texttt{Distance per hour} \tag{2.24}$$

$$\langle \text{Spectrum} \rangle ::= [\, \langle \text{Low Frequency} \rangle \, \langle \text{High Frequency} \rangle \,] \tag{2.25}$$

VNF chains are created as
$VNF_{chain} = \langle, (VNF - ID_1, \dots , VNF - ID_n), Public_{Chain}, Hash_{Chain} \rangle$.

In order to do so, the creator has to register the VNF chain with the trust root and obtain a public key per the chain, which then becomes a securely created VNF chain (and not a secure execution of a VNF chain). The invocation sequence of a VNF or VNF chain is shown in Figure 2.7. As shown in Figure 2.7, to instantiate a secure VNF chain, the invoker requests communication services from the Communication Slice Management Function (CSMF) as a secure VNF chain. This chain goes to the Network Slice Management Function (NSMF) and the Network Slice Subnet Management Function (NSSMF), if slicing is involved. Because we do not address slicing, we omit the invocation of the NSSMF and assume that the NSMF creates network-specific resources that can execute the VNF proposed by the CSMF. The NSMF has a task manager that creates tasks out of VNF_{chain}, say consisting of $VNF_{TASKS}(VNF_{chain}) = \langle(VNFT - ID_1, \dots, VNFT - ID_m), Public_{NSSMF}, Hash_{NSMF}\rangle$. These are then presented to the mapping function that maps a VNF chain to the underlying network. This creates an optimized VNF chain based on the chosen multi objective optimization function described in Section 2.3.3, which returns an optimized secure network function chain, say $VNF_{OP_{TASKS}}$. The difference between VNF_{chain} and $VNF_{OP-chain}$ is the establishment of the IP routing, firewall, encryption and other attestation functions added to VNF_{Chain}.

2.3.3 Creating a Function That Maps VNF Chains to a 5G Network

Mapping VNF chains to an underlying network architecture has been studied extensively [24–28]. Given that the mapping problem is NP-hard, many heuristic solutions have been proposed in the literature. These include Integer Linear Programming (ILP), genetic programming, a mixture of auctioning and

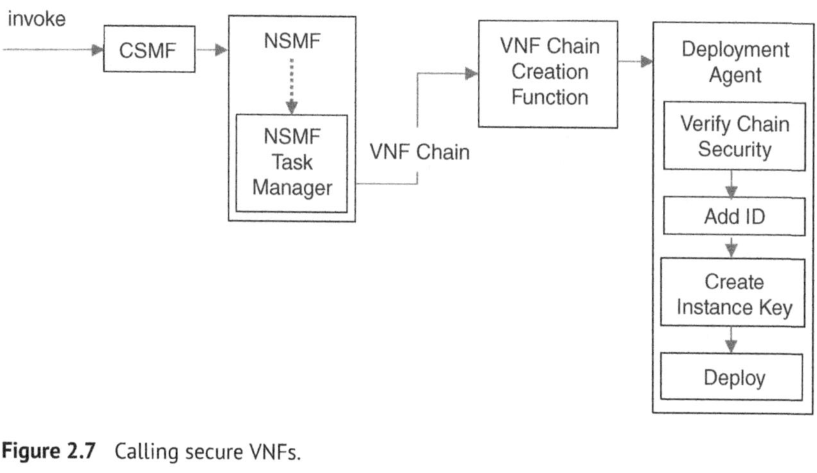

Figure 2.7 Calling secure VNFs.

game theoretic bargaining, and Mixed Integer Quadratically Constrained Program (MIQCP). In our work, we choose to apply MIQCP due to many reasons. First, [25] experimentally shows that genetic programming works better than ILP. Second, MIQCP can be used to formulate the problem of mapping a VNF chain to an underlying network structure with shareable and non-shareable resources (others cannot), and third, because genetic programming can be used within MIQCP formulation in stage by stage search for a VNF to node mapping as a Multi-objective genetic algorithm (MOGA) [29, 30].

2.3.4 Challenges and Open Problems

One of the main issues encountered changing to the disaggregated trust management system is that whenever any two components within 5G establish a secure connection using a secure protocol such as SSL or TLS, the certificates used in the system needs to verify the validity of the certificates. Certificate validation between two entities requires the validating entity to ensure that the trust root of the foreign trust domain has a valid cross-certification. This step takes time for the first validation. If the validating entity retains the certificates for subsequent calls, this time cost can be minimized for subsequent validations. Other issues are the evolving nature of algorithms for challenge/response protocols and hash creations. Both have an impact on the runtime performance of the VNFs and VNF chains.

2.4 Securing O-RAN Systems

While our proposed disaggregated trust management framework in Section 2.3 has primarily focused on securing the RAN and core VNFs, it is equally applicable to securing the disaggregated components of an ORAN architecture. For example, in deployments where an CU from Vendor A interoperates with a DU from Vendor B (or similarly, a DU from Vendor A with an RU from Vendor B), each vendor domain can maintain its own trust root. This enables secure inter-operation across vendor boundaries by enforcing strong isolation and mutual authentication.

The same framework extends naturally to the Near-Real-Time RAN Intelligent Controller (Near-RT RIC) and its associated microservices, such as xApps. In multi-vendor scenarios where xApps originate from different trust domains, each xApp would be anchored to its respective trust root, with secure cross-certification enabling interaction across trust boundaries. On the other hand, if all xApps within the Near-RT RIC belong to a single vendor or trust domain, a unified trust chain anchored in a single trust root would suffice.

2.5 Key Takeaways

ETSI and other work in the literature suggest having a distributed trust management system for 5G and beyond. This distribution can be done vertically, along the service types within the design of a 5G system, or horizontally within a virtualization infrastructure. We have proposed a design of having separate trust roots for Core and RAN components of a 5G system that is complementary to the dimensions proposed earlier. The advantage of have separate trust roots for the RAN and Core components is that our proposal allows any RAN system to co-exist with any Core, and thereby offer RAN as a service with its own encapsulated trust management as a service. Future work will include implementing this trust management services to validate its viability in NextG implementations.

References

1 ETSI (2020). ETSI standards for NFV: network function virtualization solutions. https://etsi.org/technologies/nfv [accessed 22 July 2023].

2 Herrera, G., de Jesus, J., Vega, B., and Felipe, J. (2016). Network functions virtualization: A survey. *IEEE Latin America Transactions*, 14(2): 983–997.

3 ETSI (2021). ETSI Network Functions Virtualization (NFV) Release 2: Management and Orchestration, Architectural Framework Specification.

4 ETSI (2017). ETSI: Network Functions Virtualisation (NFV) Release 2: Acceleration Technologies, vSwitch Benchmarking and Acceleration Specification.

5 Bhamare, D., Samaka, M., Erbad, A., Jain, R., Gupta, L., and Chan, H. A. (2017). Optimal virtual network function placement in multi-cloud service function chaining architecture. *Computer Communications*, 102: 1–16.

6 3GPP (2018). 5G; NG-RAN; Architecture description. Technical Specification (TS) 38.401, 3rd Generation Partnership Project (3GPP). Version 15.2.0.

7 1, O.-R. W. G. (2022). O-RAN Architecture Description. Technical Specification (TS) O-RAN.WG1.O-RAN-Architecture-Description-v06.00, O-RAN Alliance. Version 6.00.

8 3GPP (2019). Security architecture and procedures for 5G System . Technical Specification (TS) 33.501, 3rd Generation Partnership Project (3GPP). Version 15.4.0.

9 Christine Jost, B. S. (2020). Security for 5G Service-Based Architecture: What you need to know. https://www.ericsson.com/en/blog/2020/8/security-for-5g-service-based-architecture. [Online; accessed 19-July-2008].

10 Cho, J.-H., Chan, K., and Adali, S. (2015). A survey on trust modeling. *ACM Computing Surveys.*, 48(2).

11 Benzaïd, C., Taleb, T., and Farooqi, M. Z. (2021). Trust in 5G and beyond networks. *IEEE Network*, 35(3): 212–222.

12 Cao, J., Ma, M., Li, H., Ma, R., Sun, Y., Yu, P., and Xiong, L. (2020). A Survey on Security Aspects for 3GPP 5G Networks. *IEEE Communications Surveys & Tutorials*, 22(1): 170–195.

13 Boubakri, W., Abdallah, W., and Boudriga, N. (2017). Access control in 5G communication networks using simple PKI certificates. In *2017 13th International Wireless Communications and Mobile Computing Conference (IWCMC)*, pages 2092–2097.

14 3GPP (2022). Universal Mobile Telecommunications System (UMTS);LTE;5G; Network Domain Security (NDS); Authentication Framework (AF) . Technical Specification (TS) 33.310, 3rd Generation Partnership Project (3GPP). Version 17.3.0.

15 3GPP (2021). Network functions virtualisation (nfv); trust; report on certificate management. Group Report (GR) 005, 3rd Generation Partnership Project (3GPP). Version 1.2.1.

16 Wang, Y., Kang, X., Li, T., Wang, H., Chu, C.-K., and Lei, Z. (2022). SIX-Trust for 6G: Towards a Secure and Trustworthy 6G Network.

17 Yan, J., Li, N., Yang, B., Li, M., Su, L., He, S., and Liu, X. (2021). Decentralized Certificate Management for Network Function Virtualisation (NFV) Implementation in Telecommunication Networks. *Wireless Communications and Mobile Computing.*, 2021.

18 GSMA (2019). Infrastructure sharing: An overview. www.gsma.com/futurenetworks/wiki/infrastructure-sharing-an-overview/.

19 Group, O. M. (1997-2021). Business process meta-language. bpmn.org.

20 ETSI (2014). Network functions virtualisation (nfv); management and orchestration. Standard, ETSI, Sophia Antipolis Cedex - FRANCE.

21 Atoui, W. S., Grida Ben Yahia, I., and Gaaloul, W. (2019). Virtual network function descriptors mining using word embeddings and deep neural networks. In *2019 IFIP/IEEE Symposium on Integrated Network and Service Management (IM)*, pages 515–520.

22 Durante, L., Seno, L., Valenza, F., and Valenzano, A. (2017). A model for the analysis of security policies in service function chains. In *2017 IEEE Conference on Network Softwarization (NetSoft)*, pages 1–6.

23 Habibi, M. A., Han, B., Nasimi, M., and Schotten, H. D. (2018). The structure of service level agreement of slice-based 5g network. *CoRR*, abs/1806.10426.

24 Mehraghdam, S., Keller, M., and Karl, H. (2014). Specifying and placing chains of virtual network functions. In *2014 IEEE 3rd International Conference on Cloud Networking (CloudNet)*, pages 7–13.

25 Rankothge, W., Le, F., Russo, A., and Lob, J. (2017). Optimizing resource allocation for VNFS in cloud center using gas. *IEEE Transactions on Network and Service Management*, 144(2): 343–356.

26 Lima, D. H. S., Aquino, A. L. L., and Curado, M. (2020). An nfv mano architecture with a resource allocation mechanism based on game theory. In *IEEE INFOCOM 2020 - IEEE Conference on Computer Communications Workshops (INFOCOM WKSHPS)*, pages 1009–1014.

27 Dominicini, C. K., Vassoler, G. L., Meneses, L. F., Villaca, R. S., Ribeiro, M. R. N., and Martinello, M. (2017). Virtphy: Fully programmable NFV orchestration architecture for edge data centers. *IEEE Transactions on Network and Service Management*, 14(4): 817–830.

28 Liu, J., Lu, W., Zhou, F., Lu, P., and Zhu, Z. (2017). On dynamic service function chain deployment and readjustment. *IEEE Transactions on Network and Service Management*, 14(3): 543–553.

29 Fiandaca, G., Fraga, E. S., and Brandani, S. (2009). A multi-objective genetic algorithm for the design of pressure swing adsorption. *Engineering Optimization*, 41(9): 833–854.

30 Zolpakar, N. A., Lodhi, S. S., Pathak, S., and Sharma, M. A. (2020). A multi-objective genetic algorithm for the design of pressure swing adsorption. In Gupta, K. and Gupta, M. K., editors, *Optimization of Manufacturing Processes*, chapter 8, pages 185–200. Springer Series in Advanced Manufacturing, Birmingham, UK.

3

Securing Radio Access Network Slicing

Yalin Sagduyu[1], Tugba Erpek[1], Safdar Hussain Bouk[2] and Sachin Shetty[3]

[1]*National Security Institute, Virginia Tech, USA*
[2]*School of Cybersecurity, Old Dominion University, USA*
[3]*Virginia Modeling, Analysis, and Simulation Center, Old Dominion University, USA*

3.1 Introduction

Network slicing stands as a foundational principle introduced within the realm of 5G and expected to come to full fruition in 6G cellular networks. This concept facilitates the creation of a virtual, distinct, and layered network, complete with well-defined Quality of Service (QoS) criteria, all operating atop a shared pool of physical network resources. Slicing empowers network operators to segment and allot network resources, effectively curtailing energy consumption, minimizing costs, and streamlining network management. This set of services is meticulously tailored to specific applications and necessitates dedicated service-level agreements with providers for each unique network slice. Each slice carries distinct QoS requirements, security protocols, management strategies, and performance benchmarks. Services and procedures within each slice are calibrated to achieve predefined levels of throughput and latency, each guided by unique objectives ranging from dependable data delivery and scalability to efficient content routing.

It is the responsibility of the network provider to offer users reliable services from the provider's data network or Internet to their devices which are referred to as UEs. Looking to the implementation of network slicing in 5G, the network is sectioned as per the network connectivity, technology, and features between UE and the backbone. A 5G network is logically sectioned into *Radio Access Network*

6G Security: New Frontiers in Mobile Network Technologies and Verticals, First Edition.
Edited by Luiz A. DaSilva.

(RAN) and *Mobile Core* that is connected to the Internet. The RAN part of the network consists of the base stations (gNodeBs) that offer and manage the radio spectrum access to the UEs and the backhaul network that connects the distributed gNodeBs. The core network extends UE connectivity to the Internet by offering different functionalities including authentication, IP connectivity, mobility management, subscriber credential and billing management, resource management, and other functionalities.

Open and flexible radio access offered by RAN slicing brings advantages related to resource management and cost reduction. *Machine Learning (ML)* methodologies are progressively finding applications in enhancing the performance of NextG communications systems including RAN slicing. ML plays a significant role in authenticating, optimizing, managing, and dynamically adjusting network slices to meet the diverse requirements of different applications and user groups. ML models can monitor network conditions, user demands, and application requirements in real-time. This information can be used to dynamically adjust and manage slice parameters, such as bandwidth allocation and QoS to optimize performance and user experience. ML can also be used to detect anomalies or authenticate users who will be admitted to the slices so that unauthorized access to the slices is prevented. In this chapter, we consider an ML-based approach for *user authentication* for network slicing as an exemplary use case for investigating security.

As ML algorithms become integral to NextG communications, safeguarding their integrity and preventing malicious exploitation is essential. *Adversarial Machine Learning (AML)* attacks are techniques that exploit vulnerabilities in ML models to manipulate their behavior or output. These attacks aim to deceive the model, causing it to make incorrect predictions or classifications. Adversarial attacks are built upon various motivations, such as causing misclassification, evading detection, or gaining insights into the model's functioning. Some common types of adversarial ML attacks include *evasion (adversarial)*, *exploratory (inference)*, *causative (poisoning)* and *Trojan (backdoor)* attacks. The goal of evasion attacks is to cause the model to make incorrect or unintended predictions by introducing subtle changes to the input data that are imperceptible to humans but effective in misleading the model. Exploratory attacks aim to gather information about a target system, network, or application by leveraging various techniques to explore its vulnerabilities, weaknesses, and potential entry points for further exploitation. Causative (poisoning) attacks involve injecting specially crafted data into the training dataset to intentionally introduce bias, errors, or vulnerabilities in the model's learning process. Trojan (backdoor) attacks involve the deliberate insertion of a hidden and unauthorized behavior into a trained ML model. This hidden behavior, known as a "backdoor", is designed to trigger specific responses or outcomes when the model is exposed to certain inputs or conditions.

AML attacks play a critical role in security of wireless systems, where the adversaries can manipulate signals over the air. Attacks can be launched on different deep learning mechanisms such as *Deep Neural Network (DNN)* or *Reinforcement Learning (RL)* for RAN slicing. In this chapter, we consider adversarial attacks on the DNN-driven authentication process in RAN slicing and describe how RAN slices can be defended against these attacks.

The rest of this chapter is organized as follows. Section 3.2 surveys the RAN slicing capabilities with a focus on the network resource management and security. Section 3.3 provides a general description of AML attacks. Section 3.4 explains how AML attacks are launched against wireless systems. Section 3.5 describes the attacks on RL for RAN slicing. Section 3.6 considers the multi-task learning classifier for user authentication in RAN slicing. Section 3.7 studies adversarial attacks on single-task and multi-task learning for RAN slicing. Section 3.8 reports how defense is applied against adversarial attacks. Section 3.9 provides the key takeaways and concludes the chapter.

3.2 Network Slicing Using Machine Learning

Network slicing has triggered research efforts in different areas including network architecture design, RAN virtualization, network resource management, slice orchestration, and network slicing security. In this section, we focus on the review of efforts focused on RAN slicing resource management and its security.

3.2.1 RAN Slicing Resource Management

Network slicing manages network resources as virtual Resource Blocks (RBs) for NextG RAN. Each communication request comes with Quality of Experience (QoE) requirements such as throughput and latency/deadline, which can be met by assigning RBs, communication power, and processing power to the request. Network slicing aims to multiplex and serve multiple virtualized and independent logical networks on the same physical network infrastructure of NextG. By employing network slicing, NextG RAN can allocate RBs without the need to focus on the physical resources to support RBs.

Reinforcement Learning (RL) is an ML paradigm that involves training agents to make sequential decisions in an environment to maximize a cumulative reward. RL has been successfully applied for resource allocation in network slicing in a dynamic environment in [1]. Once a 5G UE is turned on, it first identifies a 5G cell and initiates a radio connection with a 5G gNodeB. Next, the UE enters the registration process and requests one or more network slices by using the gNodeB signaling connection. The gNodeB selects an Access and Mobility Management Function (AMF) that will support the network slices requested by the UE. There are arrivals of requests over time from the UEs and each request has a dynamic resource requirement along with a priority. The states are the availability of RBs. The gNodeB determines which request to admit or reject and updates the available resources accordingly. The goal is to maximize the number of satisfied requests over time.

A Deep Reinforcement Learning (DRL)-based radio resource management solution for RAN slicing under Service Level Agreement (SLA) guarantees is introduced in [2]. This method is designed with a two-level scheduling structure that works under Open Radio Access Network (ORAN) architecture. At an upper level, a DRL-based inter-slice scheduler is developed to allocate resources to network slices on a coarse time granularity. At a lower level, an existing intra-slice scheduler such as proportional fair working on a fine time granularity to allocate dedicated resources to slice users. This solution is ORAN compliant and can be deployed as an eXtended Applications (xApp) on the Radio Intelligent Controller (RIC).

The multi-agent-based RAN slicing resource allocation using reinforced Q-learning was proposed in [3], named correlated Q-learning-based inter-slice RB allocation. Each slice acts as an intelligent agent and the corresponding agents share Q-values with each other to efficiently utilize the commonly distributed and shared resource, RBs, and achieve inter-slice resource allocation using the model-free correlated Q-learning algorithm. Each learning, intra-slice Q-value sharing, or resource allocation operation is performed within a Transmission Time Interval (TTI) epoch. It was found that the proposed approach has relatively better packet delivery ratio and low latency compared to other schemes in Enhanced Mobile Broadband (eMBB) and Ultra Reliable Low Latency Communications (URLLC) slice scenarios.

In continuation to [3], knowledge transfer-based radio and computational resource allocation scheme by the gNodeB in a cell, called Knowledge Transfer-based Resource Allocation (KTRA) was proposed in [4]. The intelligent gNodeB includes a *Mobile Edge Computing (MEC) server* and *central controller* for intra-slice computation and radio resource allocation tasks. However, each slice, i.e., URLLC and eMBB, has their own controller to perform inter-slice radio resource allocation. The map function has been introduced to realize knowledge transfer in Q-learning without exploring the state-action space to learn the optimal action sequence. The efficiency of Q-learning is improved by using prior expert knowledge. As in [3], every base station (BS) learns and allocates radio and computation resources within a TTI epoch. KTRA's performance is contrasted with the baseline scheme, called Q-Learning Resource Allocation (QLRA) scheme, and simulations show that KTRA outperforms in terms of the throughput and latency in eMBB and URLLC slices.

A DRL-based network computational and radio resource allocation scheme for URLLC services has been proposed in [5]. RAN resource allocation was modeled as a single Markov Decision Process (MDP) agent and DRL was proposed for URLLC RAN slicing. For DRL-based slicing resources allocation, Deep Q-Learning (DQL) with a Q-function approximator using DNN, called Deep Q-Network (DQN) was adapted. Double Deep Q-Network (DDQN) was also adapted to further increase the approximation stability of RB and computational resource slicing. Performance of the proposed scheme has been evaluated in terms of communication delay and training performance (i.e., average loss and cumulative average rewards).

Network slice resource allocation was modeled as a two-level Single-Agent Multi-Agent RL (SAMA-RL) problem in [6]. The first large-time-scale level is modeled as single-agent MDP that uses exponential weight algorithm for exploration and exploitation. On the other hand, the second short-time scale is modeled as a multi-agent MDP that considers DQL algorithm for resource allocation. SDN controller in SAMA-RL scheme allocates RBs to multiple gNodeBs. eMBB and URLLC slicing scenarios were considered to evaluate resource allocation performance (i.e., impact of delay and throughput on number of end-users) of SAMA-RL in a custom simulator.

In [7], it was proposed that the RAN slice resource allocation method adapts the artificial neural network along with the prior expert knowledge to enhance RL. The type of resources considered include radio and RAN cache memory. Transfer learning involves multi-agent based hierarchical Transfer Reinforcement Learning (TRL) method to allocate multiple resources. In this work, Q-value based deep TRL and action selection-based TRL were used for resource allocation. The agents use expert knowledge to minimize training overhead and improve performance of the ongoing tasks. The performance evaluations show that the proposed method has better throughput, lower latency, higher cache hit ratio, and faster convergence speed for eMBB and URLLC slicing use cases compared to the previous work. The training involves a DNN, which requires a longer training and learning time and also faces training stability issues. Additionally, it has low sampling and exploration efficiency and it requires an efficient knowledge transfer function to apply a prior trained model.

A multi-agent DRL-based approach for joint resource allocation in network slicing resource allocation and admission control problems in 5G cloud RANs was proposed in [8]. The DRL approach learns about the network slice dynamics from slice-requests and performs intelligent admission control decisions. The main objective of this joint solution is to improve long-term revenue of the infrastructure provider. Performance of the proposed joint solution is compared against and outperforms the greedy, single-agent, individual network slicing, individual admission control, and node ranking approaches in terms of the infrastructure provider's revenue, average bandwidth, number of in-service slices, computing resource utilization, and number of high-priority slice requests. The main problem with multi-agent DRL-based schemes is to ensure learning robustness under varying network sizes and highly dynamic network conditions.

In [9], a resource allocation solution for ORAN using federated DRL was proposed. Multiple local agents running over independent xApps were considered where model training and resource management were locally executed within the xApp. The study considered two xApps: a) power-control xApp and b) hierarchical radio resource allocation xApp. The local xApp models are trained by each agent and then these trained models are aggregated at the coordination model that maintains the global Q-table for actions. Next, this global Q-table is distributed to local Q-tables maintained by agents for actions. The aim was to increase channel capacity, convergence and reduce delay for eMBB and URLLC slices. The proposed scheme significantly improves the slice performances; however, it requires further investigation to evaluate model aggregation communication overhead, privacy exposure, and model convergence efficiency in the heterogeneous environment and non-independent and identically distributed data.

Past research reveals that it is challenging to formulate a generalized slice resource allocation model that fits all scenarios due to the dynamic nature of RAN environments. To address this problem, a Near-Real-Time RAN Intelligent Controller (Near-RT RIC) was proposed in [10] that considers DRL and federated DRL for radio resource allocation for Mobile Virtual Network Operators (MVNOs) URLLC and eMBB network slices. In this solution, each MVNO trains an RL algorithm for radio resource allocation based on user demands. Then, these trained models are aggregated at Near-RT RIC that updates the global learning model which is further disseminated to MVNOs for execution. The model's effectiveness in terms of reducing the number of unsatisfied service-level agreements was evaluated through simulation that is evident from the results in the research.

Real-time inter-slice resource management scheme can improve spectrum efficiency; however, it is extremely challenging because of the dynamic nature of RAN resources that results from user mobility. In this context, a Graph Attention Network (GAT)-based Multi Agent Reinforcement Learning (MARL) scheme was proposed for efficient resource management in large scale multi-BS networks in [11]. The GAT-based MARL scheme adopts the DRL algorithm (i.e., DQN and DDQN) and advantage agent-critic methods. Every BS in the network acts as a separate agent and GAT is used to aggregate information from the adjacent agents. System performance was measured through a custom simulator that investigated the SLA satisfaction ratio and system utility percentage in Voice over LTE (VoLTE), HD video stream, eMBB, and URLLC services.

A multi-agent DRL-based solution for inter-cell resource partition was proposed in [12]. Here, each DRL agent is associated with an individual slice in the cell and a coordinator is created with each agent that offers inter-slice coordination to indicate resource capacity. The scheme uses twin delayed deep deterministic policy gradient algorithm for DRL. The evaluations of multi-agent coordination DRL scheme were compared with the one without coordination. System performance in terms of resource efficiency was measured over the Nokia Siemens Self Organizing Network (SON).

6G wireless networks are expected to have unique features, including the integration of space, air and ground networks to provide global coverage, resource virtualization using network slicing techniques, and the proliferation of intelligence across the network to provide ubiquitous intelligence. In [13], an AI-native network slicing architecture was proposed for 6G networks to facilitate intelligent network management while supporting emerging AI services. The proposed architecture integrates AI techniques not only in the Software-Defined Networking (SDN) controller but also in network slices as services for end-users. The architecture leverages AI for slicing and slicing for AI, and potential approaches for each phase of the network slicing life cycle and efficient resource management for AI services were introduced.

In the context of 6G, baseband function placement in Cloud Radio Access Network (CRAN) can be performed by the SDN control/management plane or a Radio Intelligent Controller (RIC). Distributed algorithms can be adopted to provide scalable solutions for networks with huge dimensions. Future networks should be able to dynamically adopt the proper resource management strategy and choose the best transmission technology based on network status and QoS requirements. The traditional convex optimization approaches are insufficient for fulfilling the precise QoS requirements of emerging wireless networking applications due to the discrepancies between mathematical tractability and network complexity. To address this issue, a soft actor-critic algorithm learning-based resource allocation framework was proposed in [14] that can provide accurate and stable solutions for scalable networks in dynamic environments.

While DRL has shown significant progress in solving radio resource allocation problems, existing resource allocation methods rely on instantaneous information that is difficult to obtain and have high computing complexity when the number of subnetworks grows. In [14], a novel framework based on multi-agent DRL was proposed for dynamic resource allocation in *6G in-X subnetworks*. The proposed framework shares historical states and actions of all subnetworks during centralized training and uses Received Signal Strength Indicator (RSSI) as input instead of instantaneous channel state information. To extract inter-subnetwork interference relationships from RSSI, a Graph Neural Network (GNN) combined with a two-stage attention network was used. The approach models the subnetwork system as a complete graph and employs improved hard attention to eliminate the impact of unrelated subnetworks.

The challenges associated with the reservation of slice resources for mobile networks were discussed in [15], including different QoS requirements for various services and the need to account for uncertainties. A slice admission control and resource reservation framework was proposed that provided a probabilistic guarantee for SLA satisfaction while minimizing resource operation costs charged to Mobile Network Operators (MNOs). The framework allowed overbooking of resources but bounds the probability of SLA non-satisfaction. A process for anticipating the processing of slice requests depending on their priority level was proposed and the processing of concurrent slice resource reservation requests was formulated as a max-min optimization problem. The solution to the problem provided the maximum earnings for granted slice requests while limiting the impact of reservation on background services. Several reduced-complexity reservation strategies were proposed to solve the problem of slice resource reservation with dynamic resource demands. The proposed approach was consistent with the 3GPP views of the management aspects of network slicing.

3.2.2 RAN Slicing Security

Unauthorized or malicious users may attempt to gain access to RAN slices in a 5G system to exploit isolated resources, disrupt services, or eavesdrop on sensitive data. 5G provides basic authentication mechanisms and uses authentication credentials stored on the SIM card. Zero-trust solutions [16] or their variant solutions have been proposed for application services running on the 5G network. However, these security measures are not suitable for mission-critical and urgent types of network scenarios because these slices must be secured from side-channel and interference attacks. The security solution must be compatible and adapted with minor modifications in the 5G systems, but have the opportunity to be native to 6G networks. Additionally, these solutions must have efficient resource monitoring mechanisms to have autonomous and self-healing characteristics. An architecture referred to as the "Security and Resiliency Techniques for Differentiated 5G Operations" was demonstrated in [17], with implicit zero trust security solution by executing security verification solutions in a real-time manner between UE and the 5G system; attack resilient dynamic slice selection mechanism involving continuous interference and resource monitoring function; architecture to handle hardware-level security solutions from UE to the core network; and the idea of how to identify spoofing attacks by continuously monitoring a UE's physical location and analyzing wireless channel state information to find discrepancies.

A comprehensive literature survey on the security of network slicing in 5G was provided in [18] that discusses the concept of network slicing, its layers and architectural framework, and the use of ML solutions throughout the network slice lifecycle. Prior efforts have shown that ML approaches can be used to improve the security of 5G network slicing. Some vulnerabilities and attacks that could potentially occur in this setting may involve Distributed Denial of Service (DDoS), eavesdropping, and a host of others. In the case of DDoS attacks, for instance, defensive measures can be achieved by detecting traffic that comes from the attacker and distinguishing it from the ordinary traffic. Similarly, Intrusion Detection Systems (IDSs) based on AI/ML techniques can be employed to secure and detect anomalies or system breaches before the attacker can compromise the network. In the case of eavesdropping or sniffing by malicious actors, an ML-based security mechanism can be developed to protect and encrypt the communication channels effectively. Ref. [18] provides information on ML-based solutions to 5G network slicing security and it compares the existing surveys and roadmaps, and identifies significant shortcomings that it aims to improve on.

Security issues greatly impact the effectiveness of network slicing in providing customized services with different requirements. One of the largest challenges is the lack of standardized security solutions, as each entity involved in network slicing defines its own strategies and policies. In addition, specific attacks and security implementation issues can also affect network slicing security. Well-known attacks like Denial of Service (DoS) attacks can severely impact weak slices, and slice information can be exploited by malicious users to identify the slice's status and easily breach it. These security challenges may be addressed through various strategies, including AI and blockchain-based solutions for better isolation and resource management, unique identifiers for sub-slices, and specific security requirements for slice management. Further research and development are needed to create more efficient and intelligent strategies to minimize security vulnerabilities among slices. An extensive insight into network slicing from recent advances to open security issues was provided, and different security solutions were surveyed based on their performance in [19] that discussed challenges and security issues with network slicing in 5G networks, which allow for resource sharing among different network entities for the delivery of smart, critical, and multi-services with distinctive requirements.

Recently, different AI-based models have been applied to ensure end-to-end security in the 5G network. In this domain, in [20] a neural network-based framework, called "Secure5G", was implemented for 5G network offering network slicing services. The framework is designed to proactively detect and eliminate threats based on incoming connections and is resilient enough to quarantine the threats ensuring end-to-end security from UEs to the 5G core network, and to any of the external networks. Secure5G provides an ability to customize each slice based on the service demands (bandwidth, coverage, security, latency, reliability, etc.) and maintains isolation of resources, traffic flow, and network functions between the slices to protect the network infrastructure system from DDoS attacks. Moreover, it is a resilient model that enables the network operators to sell network slicing as-a-service to serve diverse services efficiently over a single infrastructure with high security and reliability. 5G networks security concerns were investigated and the performance of their deep learning neural network model was evaluated over the volume-based flooding and spoofing attack scenarios.

In [21], the DeepSecure framework was presented, which is based on the Long Short Term Memory deep learning technique, for detecting DDoS attacks on network slicing in 5G cellular networks. The framework can detect UE network traffic as either DDoS attack or normal traffic and assign an appropriate slice to a legitimate UE request. Unlike other existing techniques, DeepSecure does not require any predefined threshold, feature engineering constraints or computation overload which can hinder efficient detection of DDoS attacks. It uses a sequence-to-sequence model to capture the temporal dynamics of normal and DDoS traffic, which leads to accurate traffic classification and resource allocation to UE requests. The proposed framework was compared with existing ML and deep learning techniques used in the literature, i.e., Support Vector Machine (SVM). The results showed that DeepSecure outperformed existing techniques such as SVM and Secure5G [20]. The proposed attack detection model and the slice prediction model achieved the highest accuracy rate in detecting DDoS attacks and predicting the appropriate slice requested by a legitimate UE with high accuracy. Therefore, it represented a promising solution to address security threats posed by DDoS attacks on network slicing in 5G and subsequently 6G cellular networks.

3.3 Adversarial Machine Learning

The evolution of wireless communications has reached a pivotal juncture with the advent of 5G and the ongoing development of 6G networks. These next-generation communication systems promise transformative capabilities, including unprecedented data rates, ultra-low latency, massive device connectivity, and enhanced network reliability. However, these remarkable advancements are accompanied by intricate challenges, particularly in ensuring robust security, efficient resource management, and seamless user experiences. In this dynamic landscape, ML emerges as a cornerstone technology, seamlessly integrating with 5G and 6G networks to tackle these challenges and unlock their full potential. By harnessing the capabilities of ML, these wireless ecosystems are poised to revolutionize how we communicate with, connect to, and experience the digital world.

While ML empowers 5G and 6G wireless communication systems with unprecedented capabilities, it also introduces a new dimension of vulnerabilities that adversaries can exploit. ML algorithms, designed to learn patterns and make decisions from data, exhibit certain susceptibilities that malicious actors can leverage to undermine their effectiveness. These vulnerabilities are studied under AML. Understanding how ML becomes vulnerable to attacks is crucial for designing robust defense mechanisms to ensure the integrity, security, and reliability of wireless communications. AML attacks are built upon the following properties of ML.

- *Sensitivity to Perturbations.* ML models, especially deep neural networks, are highly sensitive to small perturbations in input data. Adversarial attacks exploit this sensitivity by introducing carefully crafted perturbations that can lead to misclassification or erroneous decisions. These perturbations are often imperceptible to human observers but can significantly affect model outputs.
- *Lack of Generalization.* ML models may struggle to generalize across different data distributions or scenarios. Adversarial examples that work against one model can sometimes easily transfer to other models with similar architectures or tasks, highlighting the lack of robustness and the potential for widespread impact.
- *Limited Defense.* Defense mechanisms introduced during training can only cover a finite set of potential attacks, leaving the model vulnerable to novel attack strategies not encountered during training.
- *Model Complexity.* Complex ML models often have intricate decision boundaries that can be exploited by adversaries. As the complexity of models increases to handle more intricate tasks, they may inadvertently become more susceptible to adversarial manipulation.
- *Lack of Real-World Context.* ML models make decisions based solely on patterns in their training data, which might not always capture the real-world context. Adversaries can exploit this limitation by crafting inputs that exploit the model's blind spots and biases.
- *Black-Box Attacks.* Even when attackers lack full knowledge of the target model's architecture and parameters, they can still devise strategies using transferable adversarial examples or by querying the model and observing its outputs. This black-box nature of attacks makes them particularly challenging to defend against.
- *Evolving Attack Strategies.* As ML defense techniques advance, attackers adapt by developing new and sophisticated attack strategies. This constant evolution of adversarial tactics requires ongoing research to stay ahead of emerging threats.

In the context of 5G and 6G wireless communication systems, the vulnerabilities due to AML pose significant challenges to ensuring the security, privacy, and reliability of network operations. As these networks become more interconnected and reliant on ML-powered decision-making, it becomes imperative to develop comprehensive defense mechanisms that address the multifaceted nature of AML attacks. Overall, AML attacks can be launched in training and test times and can be categorized as follows.

- *Inference Attacks.* Inference attacks target the information that a model reveals about its training data. These attacks seek to uncover sensitive or private information by exploiting the model's output responses.
  - *Query Attack.* Adversaries query the model with carefully chosen inputs to elicit informative responses that reveal hidden patterns or data characteristics.
  - *Membership Inference Attack.* Adversaries attempt to determine whether a specific data point was part of the model's training dataset. This attack exploits the model's responses to infer the presence or absence of data.
  - *Model Inversion Attack.* Attackers reverse-engineer training data or gain insights into the model's internal workings by analyzing its outputs. Model inversion attacks can be used to reconstruct private information used during training.
- *Adversarial (Evasion) Attacks.* Evasion attacks, also known as adversarial attacks, aim to deceive an ML model during its operational phase. Attackers manipulate input data to cause misclassification or incorrect predictions. Attackers craft perturbed input samples that are minimally different from legitimate examples but lead the model to make incorrect predictions. These adversarial examples exploit the model's sensitivity to small perturbations.
- *Poisoning Attacks (Causative Attacks).* Poisoning attacks involve manipulating the training data to compromise the integrity of the model's learning process. Adversaries inject malicious data points into the training dataset to manipulate the model's behavior.
  - *Data Poisoning Attack.* Attackers inject malicious or misleading data into the training dataset, leading the model to learn incorrect patterns or make biased predictions.
  - *Label Flipping Attack.* Adversaries modify the labels of training examples to skew the model's learning process and cause it to generalize inaccurately.
  - *Data Integrity Attack.* Attackers introduce noise or perturbations to the training data, which can degrade the model's performance and accuracy.
- *Backdoor Attacks (Trojan Attacks).* Backdoor attacks involve embedding a hidden trigger or pattern in the model's parameters during training. When triggered, the model behaves in an unintended manner.

- *Trojan Attack*. Attackers insert a hidden trigger or pattern into the model's training data, which causes the model to produce malicious outputs when exposed to specific input conditions.
- *Model Watermarking*. Adversaries insert a specific trigger or watermark into the model's parameters, which can be activated to produce certain outputs when specific conditions are met.

3.4 Adversarial Machine Learning for Wireless Systems

Wireless communication systems are vulnerable to various malicious activities, including signal jamming, eavesdropping, and unauthorized access. AML attacks demonstrate how subtle manipulations in training and test times can exploit vulnerabilities in ML models, enabling attackers to subvert security measures and gain unauthorized access to networks. Understanding and defending against these attacks is crucial for maintaining the confidentiality, integrity, and authenticity of wireless communications [22–25]. To that end, AML attacks have been studied for emerging wireless systems such as 5G/6G [26] and IoT [27, 28].

Typically, the first step in AML attacks is the inference attack that aims to learn about how the ML-driven decision process works. By leveraging the shared and open nature of wireless medium, inference attack can be effectively launched against ML-based wireless systems [29–31].

In test time, adversarial attacks can manipulate the ML-driven decision process of wireless systems. Examples of these attacks include attacks on wireless signal classification such as modulation recognition [32–35], power control in multiuser communications [36], beam prediction in mmWave communications [37], and emerging communications paradigms such as task-oriented communications [38, 39] and semantic communications [40]. When adversarial attacks are launched against spectrum sensing, they can be utilized to enable covert communications [41–43] such that eavesdroppers cannot detect ongoing transmissions.

In training time, poisoning attacks can manipulate the training process of wireless systems. These attacks can be launched against individual spectrum sensors in forms of spectrum data falsification attacks or against a group of spectrum sensors such as used in cooperative spectrum sensing [44–47].

Backdoor (Trojan) attacks involve the insertion of hidden triggers (such as phase shifts) into the ML models of wireless systems during their training phase, enabling the models to produce malicious outputs when exposed to specific, predefined conditions in operational use. Backdoor attacks have been studied against spectrum sensing [48], task-oriented communications [38], and semantic communications [49].

From the privacy perspective, Membership Inference Attack (MIA) is a technique in which an adversary attempts to determine whether a specific data point was part of the training dataset of an ML model by leveraging the model's output responses. MIA can be effectively launched against wireless applications such as spectrum sensing [50, 51]. The sensitive information leaked by the MIA encompasses waveform, channel, and device attributes, and if disclosed, can be utilized by malicious actors to detect weaknesses in the underlying ML model (such as infiltrating authentication mechanisms).

Attackers can use ML for spoofing and hiding of wireless signals. Signal spoofing involves the intentional creation of deceptive signals to mimic legitimate communication sources, leading receivers to accept and act upon falsified information [52, 53] (for example with the use of generative adversarial networks (GANs) [54, 55], while signal hiding pertains to concealing wireless transmissions to prevent their detection or interception by unauthorized parties [56].

In wireless systems with distributed edge devices, federated learning finds applications in training common models by exchanging local models among edge devices. Malicious attacks such as jamming [57–59] or selfish behaviors such as free-riding [60] pose threats to the collaborative learning process in wireless systems.

3.5 Attacks on Reinforcement Learning for RAN Slicing

RL is an ML paradigm where an agent learns to make sequential decisions in an environment to maximize a cumulative reward signal through interactions and exploration. RL has been extensively used for wireless applications such as spectrum allocation, power control, and resource management to enhance network efficiency, quality of service, and user experience, while adapting to changing network traffic and spectrum conditions. To that end, RL has been effectively used for admission control and resource allocation in RAN slicing applications.

To serve RAN slices, communication resources (frequency-time blocks and transmit powers) and computational resources (processor usage) are assigned to stochastic arrivals of network slice requests, each with priority (weight), throughput, computational resource, and latency (deadline) requirements. Feasible requests are served by allocating available resources over their requested duration. Since each resource allocation decision affects future availability, a myopic approach optimizing current allocation proves inadequate for network slicing. Thus, RL can be applied to maximize network utility, measured by the total weight of granted slicing requests, within communication and computational limits over a time span. RL aims to sustain scalable performance with growing user numbers and efficiently allocates resources, even accommodating dynamic spectrum sharing with incumbent users in 5G networks [1, 61].

One option for attackers is to jam the spectrum, thereby manipulating how the reward is generated. Black-box attacks represent the most challenging scenario for attackers, where they have minimal or no information about the target model. In this case, attackers rely on techniques such as transferability or query-based attacks. Transferability exploits the observation that adversarial examples generated for one model can often fool a different but similar model. Query-based attacks involve iteratively querying the model to learn about its decision boundaries and optimize attacks. The adversaries can leverage spectrum observations to construct a surrogate RL-based model, which guides the selection of RBs for jamming while adhering to an energy budget. The goal is to maximize the count of failed requests resulting from jammed RBs. By disrupting the allocation of RBs, the adversary diminishes the RL algorithm's reward. Since this reward drives the RL algorithm's updates, the system's performance remains compromised even after the jamming ceases. To mitigate such attacks, reactive and proactive defensive approaches can be applied ranging from safeguarding the RL algorithm's updates to misleading the adversary's learning process [62–64].

In addition to jamming attacks, the adversaries can orchestrate the generation of fabricated (fake) network slicing requests, consuming valuable RAN resources that should have been allocated to legitimate requests. Through spectrum observations, the adversaries can construct a surrogate model via RL, which strategically formulates fake requests aimed at diminishing the reward obtained by genuine requests over time [65]. To that end, a novel flooding attack on RAN slicing was formulated in [65] and a framework was set up for detecting flooding attacks, where adversaries generate fake network slicing requests to consume the RAN resources that would be otherwise available to real requests. These resources include the virtual RBs for RAN. The adversary observes the spectrum and builds a surrogate model on the network slicing algorithm through an RL algorithm that decides how to craft fake requests to minimize the reward of real requests over time. This flooding attack is more effective than other benchmark attacks such as random fake requests and fake requests with the minimum resource requirement. To detect and mitigate these flooding attacks, the defense framework considered in [65] aims to distinguish normal user requests from adversarial requests.

3.6 Multi-task Learning for RAN Slicing Security

Network slices share the physical network infrastructure, and each slice requires a certain amount of resources (e.g., bandwidth and computing capacity). Authentication ensures that resources are allocated to authorized slices based on predefined policies, preventing one slice from consuming more than its fair share. Radio fingerprinting entails the extraction of distinctive patterns or features spanning the protocol stack including PHY layer, Medium Access Control (MAC) layer, and upper layers. Deep learning has been extensively used for device identification using Radio Frequency (RF) fingerprinting [66–68]. In this section, we leverage a multi-task classifier to enable authentication for network slicing by considering both waveform and device features (including device locations and hardware characteristics). To that end, we consider two tasks for RAN slicing. Task 1 is to classify the waveform and task 2 is to identify the device, namely the UE. We assume the channel induces path loss according to the Signal-to-Noise Ratio (SNR). There are two types of devices that have 5dB difference in received signals due to their locations and transmit powers, and each involves up to $\pi/20$ shift in phase. Transmissions are carried out either with Binary Phase Shift Keying (BPSK) or Quadrature Phase Shift Keying (QPSK) modulation.

We consider both *single-task learning* and *multi-task learning* for the two tasks of waveform classification and device identification. In multi-task learning, a single model is trained to perform multiple related tasks simultaneously. The goal of multi-task learning is to leverage the shared information or relationships among tasks to improve the overall performance of the model on all tasks. This approach contrasts with traditional single-task learning, where individual models are trained for each task separately. As multi-task learning involves training a common DNN instead of two different DNNs, one for each task, it also requires less memory and computation power to execute multiple tasks at the same time. Multi-task learning was used to enable semantic communications in [69, 70] and joint sensing and communications in [39, 71, 72].

For multi-task learning, let T be the total number of tasks. Let N be the total number of examples in the dataset. The dataset consists of pairs $(x_i, y_i^{(t)})$, where x_i is the input example and $y_i^{(t)}$ is the corresponding label for the ith example and the tth task. For each task t, the model has task-specific DNN parameters $\phi^{(t)}$. The shared model parameters are denoted as θ.

The goal of multi-task learning is to determine the shared parameters θ and task-specific parameters $\{\phi^{(t)}\}$ that optimize the joint loss function $L(\theta, \{\phi^{(t)}\})$:

$$L(\theta, \{\phi^{(t)}\}) = \sum_{t=1}^{T} w_t \sum_{i=1}^{N} L^{(t)}(\theta, \phi^{(t)}, x_i, y_i^{(t)}),$$

where $L^{(t)}(\theta, \phi^{(t)}, x_i, y_i^{(t)})$ is the loss function for the tth task and w_t is the weight for task t. The joint loss function L combines the individual task losses for all tasks. The model is trained to jointly minimize the combined loss, allowing it to leverage shared information and relationships between tasks. Note that by setting $w_j = 0$ for all $j \neq i$, the common model of multi-task learning is reduced to single-task model for task t.

During training, the parameters θ and $\{\phi^{(t)}\}$ are updated using gradient descent to minimize the joint loss function L. The shared representation captured by θ helps the model generalize across tasks, while the task-specific parameters $\{\phi^{(t)}\}$ allow the model to specialize for each task. After training, the model can make predictions for all tasks using the shared representation and task-specific layers. Given an input x, the model computes predictions $\{y^{(t)}_{\text{pred}}\}$ for all tasks.

The DNN of multi-task learning consists of a main branch and two sub-branches, one for each task. Categorical cross-entropy is used as the loss function and Adam (with learning rate 10^{-3}) is used as the optimizer. The DNN has a total of 71,156 parameters and its architecture is shown in Table 3.1. In case of single-task classifier, only main branch and one of the sub-branches are used to construct the DNN for the given task. 5,000 samples are generated with 80%-20% split for training and testing.

Figures 3.1 and 3.2 show the classifier accuracy for single-task learning and multi-task learning, respectively, when we vary the SNR. Both learning processes sustain similar levels of classification accuracy that improve with the SNR for both tasks.

Table 3.1 DNN for multi-task learning.

Network	Layer	Properties
Main branch	Input	size: 2 × 32
	Conv2D	filter size: 32, kernel size: (1,3) activation: ReLU
	Flatten	–
	Dense	size: 32, activation: ReLU
Sub-branch for Task 1	Input	size: 32
	Dense	size: 16, activation: ReLU
	Dropout	Dropout rate: 0.1
	Dense	size: 8, activation: ReLU
	Dropout	Dropout rate: 0.1
	Dense	size: 2, activation: Softmax
Sub-branch for Task 2	Input	size: 32
	Dense	size: 16, activation: ReLU
	Dropout	Dropout rate: 0.1
	Dense	size: 8, activation: ReLU
	Dropout	Dropout rate: 0.1
	Dense	size: 2, activation: Softmax

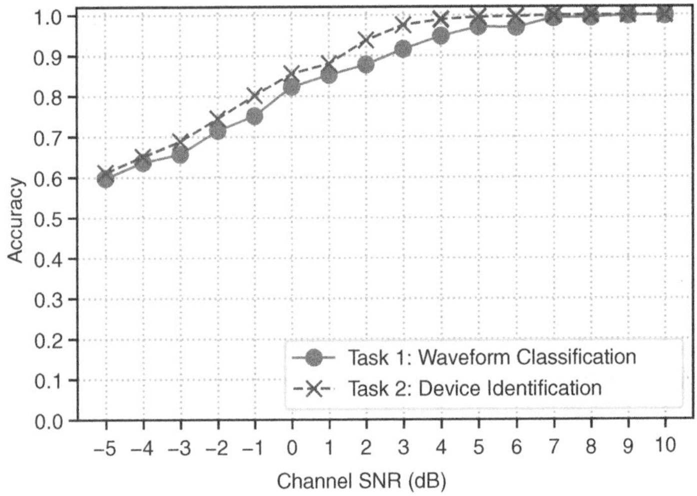

Figure 3.1 Single-task classifier accuracy.

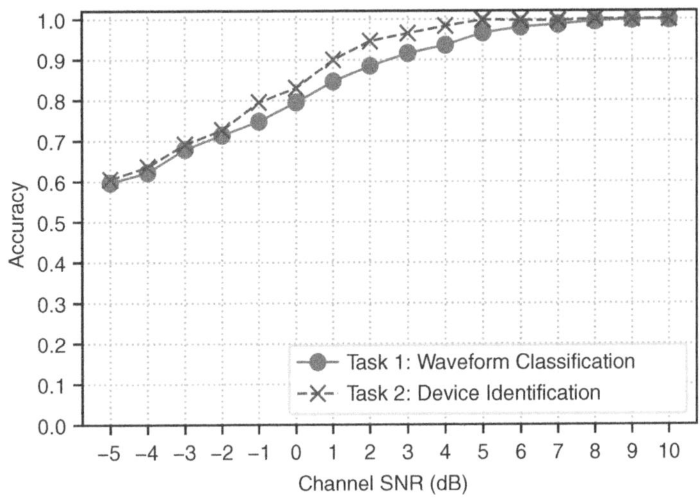

Figure 3.2 Multi-task classifier accuracy.

3.7 Adversarial Attacks on Single-task and Multi-task Learning for RAN Slicing

We use the *Fast Gradient Sign Method (FGSM)* as the adversarial attack technique to generate adversarial examples. The main idea behind FGSM is to perturb the input data by adding a small amount of noise in the direction that

(i) maximally increases the loss or error of the target model for the *untargeted attack* or (ii) maximally decreases the loss or error of the target model with respect to the target label for the *targeted attack*. This perturbation is determined by the gradient of the model's loss with respect to the input data. Let x be the original input data (e.g., an image), y_{true} be the true label of the original input, θ be the parameters of the target model, $L(\theta, x, y_{\text{true}})$ be the loss function of the target model for input x and true label y_{true}, ϵ be the small perturbation magnitude (often called the "step size").

The *untargeted attack* pursues the following steps:

1. Calculate the gradient of the loss function with respect to the input data: $\nabla_x L(\theta, x, y_{\text{true}})$.
2. Generate the adversarial perturbation by taking the sign of the gradient and scaling it by the perturbation magnitude ϵ: $\delta = \epsilon \cdot \text{sign}(\nabla_x L(\theta, x, y_{\text{true}}))$.
3. Create the adversarial example by adding the perturbation to the original input: $\tilde{x} = x + \delta$.
4. Feed the adversarial example into the target model and observe the model's prediction y_{pred} for the adversarial input.

The *targeted attack* pursues the following steps:

1. Calculate the gradient of the loss function with respect to the input data: $\nabla_x L(\theta, x, y_{\text{target}})$, where y_{target} is the target label for the adversarial example.
2. Generate the adversarial perturbation by taking the sign of the negative gradient and scaling it by the perturbation magnitude ϵ: $\delta = -\epsilon \cdot \text{sign}(\nabla_x L(\theta, x, y_{\text{target}}))$.
3. Create the adversarial example by adding the perturbation to the original input: $\tilde{x} = x + \delta$.
4. Feed the adversarial example into the target model and observe the model's prediction y_{pred} for the adversarial input.

The strength of adversarial attacks, namely ϵ is controlled by the *Perturbation-to-Noise Ratio (PNR)*, where perturbation is from the adversarial attack and noise refers to receiver noise. For numerical results, we vary the PNR and set SNR as 3 dB. Figures 3.3–3.5 show the attack success rates of the untargeted and targeted adversarial attacks, respectively, on single-task learning, when we vary the PNR. For the untargeted attack, the attack is successful when the predicted label is different from the original label for any task. For the targeted attack, the attack is successful when the predicted label is flipped from the original label to a target label of a given task (since original labels of the sample for different tasks may be different, the targeted attack needs to target the specific label of a specific task). In the targeted attack case, we show attack success results averaged over attacks that aim to flip original labels to different target labels. As shown in Figures 3.3–3.5, attack success increases for both classifiers as the PNR increases. Adversarial attack achieves higher attack success rate compared to using Gaussian noise as perturbation.

Figures 3.3–3.6 show the attack success rates of the untargeted and targeted adversarial attacks, respectively, on multi-task learning. Again, the attack success increases with the PNR. Compared with single-task learning, the adversarial attack remains less effective for the untargeted attack case, where it is

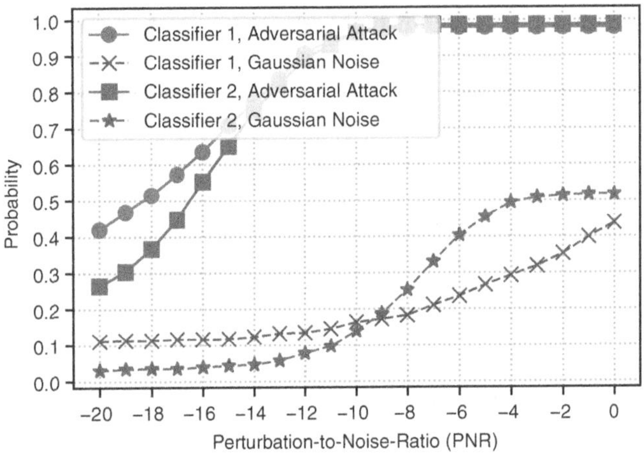

Figure 3.3 Untargeted attack success on single-task classifier.

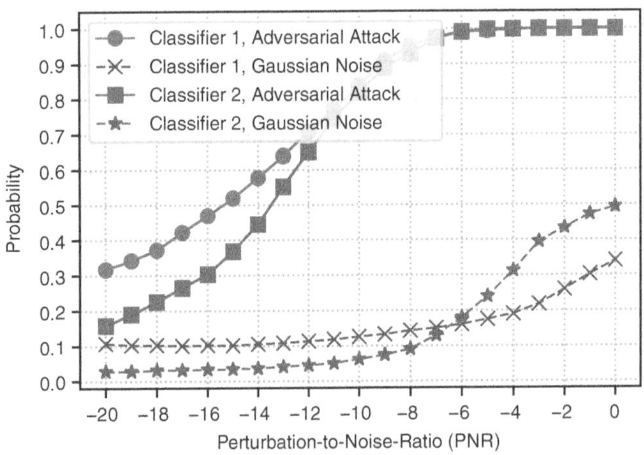

Figure 3.4 Targeted attack success on single-task classifier.

hard to cause errors on both classifiers at the same time by using a common perturbation. On the other hand, targeted attack is highly effective for multi-task learning, when the perturbation is crafted for a specific label of a specific classifier.

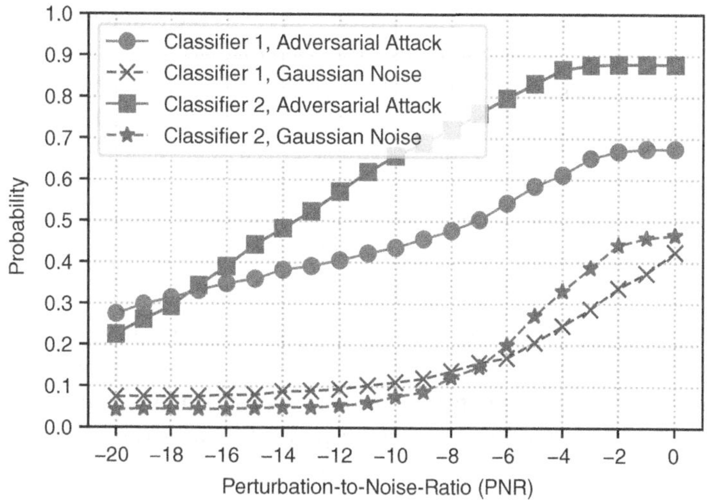

Figure 3.5 Untargeted attack success on multi-task classifier.

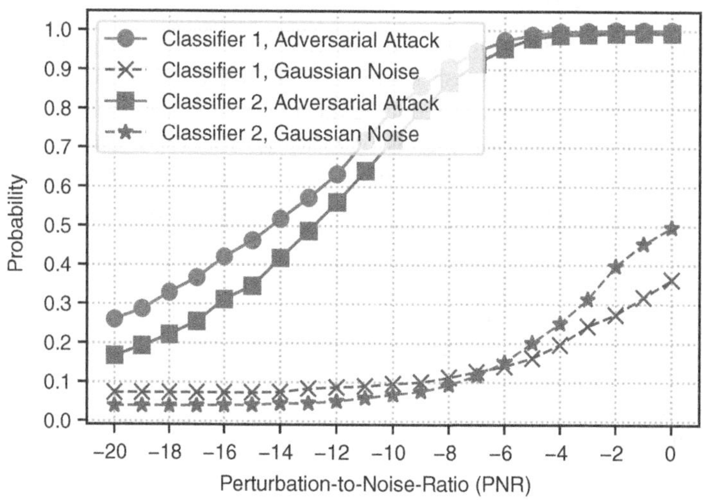

Figure 3.6 Targeted attack success on multi-task classifier.

3.8 Adversarial Training as the Defense Against Adversarial Attacks on RAN Slicing

We use adversarial training to improve the robustness of models against adversarial attacks. The main idea behind adversarial training is to train the model on a combination of regular (clean) examples and adversarial examples generated by attacking the model with FGSM (or other attacks). Let D be the training dataset containing pairs of clean inputs x and their corresponding labels y_{true}. Let N be the total number of examples in the dataset. The training dataset is represented as $D = \{(x_i, y_{\text{true},i})\}_{i=1}^{N}$. For each input x_i, the corresponding ground-truth label is denoted as $y_{\text{true},i}$. The model being trained is parameterized by θ. The loss function $L(\theta, x_i, y_{\text{true},i})$ measures the discrepancy between the model's prediction and the ground-truth label for the ith example. Adversarial examples x_i^{adv} are generated from the clean inputs x_i using an adversarial attack.

The adversarial training process can be summarized as follows:

1. Prepare the training dataset D with clean inputs and corresponding labels.
2. In each training iteration, sample a batch of examples $\{x_i\}$ from the dataset D.
3. For each x_i in the batch, compute its corresponding adversarial example x_i^{adv}.
4. Update the model's parameters θ using the gradient of the loss function for the adversarial examples: $\theta \leftarrow \theta - \eta \cdot \nabla_\theta \sum_i L(\theta, x_i^{\text{adv}}, y_{\text{true},i})$, where η is the learning rate.
5. After training, evaluate the model's performance on a separate validation or test set. Calculate the model's accuracy on clean examples x_i and its accuracy on adversarial examples x_i^{adv}.

Tables 3.2–3.4 show the performance of the defense against untargeted attacks on single-task learning and multi-task learning, respectively, when the PNR is set to 0dB. The defense is effective in reducing the attack success rate, while the classifier accuracy on clean samples remains high after defense. The defense works better against attacks on single-task classifiers, as it is more difficult to obtain adversarial examples and repeat the training process for the larger common model of multi-task learning.

Table 3.2 Defense against attacks on single-task classifier – attack on classifier 1.

Classifier accuracy before attack	0.923
Attack success (adversarial)	0.972
Attack success (Gaussian noise)	0.441
Clean sample accuracy after defense	0.911
Attack success (adversarial) after defense	0.160

Table 3.3 Defense against attacks on single-task classifier – attack on classifier 2.

Classifier accuracy before attack	0.973
Attack success (adversarial)	0.981
Attack success (Gaussian noise)	0.498
Clean sample accuracy after defense	0.913
Attack success (adversarial) after defense	0.363

Table 3.4 Defense against attacks on multi-task classifier.

Task 1 accuracy before attack	0.921
Task 2 accuracy before attack	0.971
Attack success for classifier 1 (adversarial)	0.761
Attack success for classifier 2 (adversarial)	0.855
Attack success for classifier 1 (Gaussian)	0.401
Attack success for classifier 2 (Gaussian)	0.504
Clean accuracy for classifier 1 after defense	0.847
Clean accuracy for classifier 2 after defense	0.850
Attack success for classifier 1 after defense	0.476
Attack success for classifier 2 after defense	0.517

3.9 Key Takeaways

Network slicing holds paramount significance in NextG communications, aligning communication networks with the diverse and dynamic requisites of contemporary applications and services. ML-based solutions have exhibited promise in facilitating this functionality, encompassing resource management and security aspects. With ML taking a central role in NextG communications, concerns are mounting about its susceptibility to adversarial impacts. In this chapter, we commenced by providing a synopsis of existing efforts in ML-driven resource management and security for RAN slicing. Subsequently, we delineated the various attack types that could be launched against these systems. We introduced a solution based on DNNs that leverages multi-task learning for network slicing authentication. Moreover, we designed adversarial attacks targeting this authentication mechanism and assessed their repercussions, gauging the decline in classifier accuracy for both single-task and multi-task learning scenarios. Employing adversarial training, we described the defense of RAN slices against adversarial attacks. Our results highlighted the vulnerabilities of RAN slicing to adversarial attacks and demonstrated how to secure RAN slices against malicious actors.

References

1 Shi, Y., Sagduyu, Y. E., and Erpek, T. (2020c). Reinforcement learning for dynamic resource optimization in 5G radio access network slicing. In *IEEE 25th international workshop on computer aided modeling and design of communication links and networks (CAMAD)*, pages 1–6. IEEE.

2 Dai, J., Li, L., Safavinejad, R., Mahboob, S., Chen, H., Ratnam, V. V., Wang, H., Zhang, J., and Liu, L. (2025). O-RAN-Enabled intelligent network slicing to meet service-level agreement (SLA). *IEEE Transactions on Mobile Computing*, 24: 890–906.

3 Zhou, H., Elsayed, M., and Erol-Kantarci, M. (2021). Ran resource slicing in 5G using multi-agent correlated q-learning. In *2021 IEEE 32nd Annual International Symposium on Personal, Indoor and Mobile Radio Communications (PIMRC)*, pages 1179–1184. IEEE.

4 Zhou, H. and Erol-Kantarci, M. (2022). Knowledge transfer based radio and computation resource allocation for 5G RAN slicing. In *2022 IEEE 19th Annual Consumer Communications & Networking Conference (CCNC)*, pages 617–623. IEEE.

5 Filali, A., Nour, B., Cherkaoui, S., and Kobbane, A. (2022b). Communication and computation O-RAN resource slicing for URLLC services using deep reinforcement learning. *arXiv preprint arXiv:2202.06439.*

6 Filali, A., Mlika, Z., Cherkaoui, S., and Kobbane, A. (2022a). Dynamic SDN-based radio access network slicing with deep reinforcement learning for URLLC and eMBB services. *IEEE Transactions on Network Science and Engineering*, 9(4): 2174–2187.

7 Zhou, H., Erol-Kantarci, M., and Vincent Poor, H. (2022). Learning from peers: Deep transfer reinforcement learning for joint radio and cache resource allocation in 5G RAN slicing. *IEEE Transactions on Cognitive Communications and Networking*, 8(4): 1925–1941.

8 Sulaiman, M., Moayyedi, A., Salahuddin, M. A., Boutaba, R., and Saleh, A. (2022). Multi-agent deep reinforcement learning for slicing and admission control in 5G C-RAN. In *NOMS 2022-2022 IEEE/IFIP Network Operations and Management Symposium*, pages 1–9.

9 Zhang, H., Zhou, H., and Erol-Kantarci, M. (2022). Federated deep reinforcement learning for resource allocation in O-RAN slicing. In *GLOBECOM 2022-2022 IEEE Global Communications Conference*, pages 958–963. IEEE.

10 Abouaomar, A., Taik, A., Filali, A., and Cherkaoui, S. (2022). Federated learning for RAN slicing in beyond 5G networks. *arXiv preprint arXiv:2206.11328.*

11 Shao, Y., Li, R., Hu, B., Wu, Y., Zhao, Z., and Zhang, H. (2021). Graph attention network-based multi-agent reinforcement learning for slicing resource management in dense cellular network. *IEEE Transactions on Vehicular Technology*, 70(10): 10792–10803.

12 Hu, T., Liao, Q., Liu, Q., Wellington, D., and Carle, G. (2022). Inter-cell slicing resource partitioning via coordinated multi-agent deep reinforcement learning. In *ICC 2022-IEEE International Conference on Communications*, pages 3202–3207. IEEE.

13 Wu, W., Zhou, C., Li, M., Wu, H., Zhou, H., Zhang, N., Shen, X. S., and Zhuang, W. (2022). AI-native network slicing for 6G networks. *IEEE Wireless Communications*, 29(1): 96–103.

14 Nouruzi, A., Rezaei, A., Khalili, A., Mokari, N., Javan, M. R., Jorswieck, E. A., and Yanikomeroglu, H. (2022). Toward a smart resource allocation policy via artificial intelligence in 6G networks: Centralized or decentralized? *arXiv preprint arXiv:2202.09093.*

15 Luu, Q.-T., Kerboeuf, S., and Kieffer, M. (2022). Admission control and resource reservation for prioritized slice requests with guaranteed SLA under uncertainties. *IEEE Transactions on Network and Service Management*, 19(3): 3136–3153.

16 Rose, S., Borchert, O., Mitchell, S., and Connelly, S. (2020). Zero trust architecture. Technical report, National Institute of Standards and Technology.

17 Hakiri, A., Gokhale, A. S., Barve, Y., Formicola, V., Shekhar, S., Mahmoudi, C., Rahman, M. A., Ghosh, U., Hasan, S. R., and Guo, T. (2022). Techniques for realizing secure, resilient and differentiated 5G operations. In *2022 14th IFIP Wireless and Mobile Networking Conference (WMNC)*, pages 113–117. IEEE.

18 Dangi, R., Jadhav, A., Choudhary, G., Dragoni, N., Mishra, M. K., and Lalwani, P. (2022). ML-based 5G network slicing security: A comprehensive survey. *Future Internet*, 14(4): 116.

19 Salahdine, F., Liu, Q., and Han, T. (2022). Towards secure and intelligent network slicing for 5G networks. *IEEE Open Journal of the Computer Society*, 3: 23–38.

20 Thantharate, A., Paropkari, R., Walunj, V., Beard, C., and Kankariya, P. (2020). Secure5G: A deep learning framework towards a secure network slicing in 5G and beyond. In *2020 10th annual computing and communication workshop and conference (CCWC)*, pages 0852–0857. IEEE.

21 Kuadey, N. A. E., Maale, G. T., Kwantwi, T., Sun, G., and Liu, G. (2021). Deepsecure: Detection of distributed denial of service attacks on 5G network slicing—deep learning approach. *IEEE Wireless Communications Letters*, 11(3): 488–492.

22 Sagduyu, Y. E., Shi, Y., Erpek, T., Headley, W., Flowers, B., Stantchev, G., Lu, Z., and Jalaian, B. (2023d). Adversarial machine learning: A new threat paradigm for next-generation wireless communications. In *AI, Machine Learning and Deep Learning: A Security Perspective*, pages 35–52. CRC Press.

23 Adesina, D., Hsieh, C.-C., Sagduyu, Y. E., and Qian, L. (2022). Adversarial machine learning in wireless communications using RF data: A review. *IEEE Communications Surveys & Tutorials*, 25(1): 77–100.

24 Sagduyu, Y. E., Shi, Y., Erpek, T., Headley, W., Flowers, B., Stantchev, G., and Lu, Z. (2020). When wireless security meets machine learning: Motivation, challenges, and research directions. *arXiv preprint arXiv:2001.08883.*

25 Erpek, T., O'Shea, T. J., Sagduyu, Y. E., Shi, Y., and Clancy, T. C. (2020). Deep learning for wireless communications. In Pedrycz, W., Chen, SM. (eds) *Development and Analysis of Deep Learning Architectures*. Studies in Computational Intelligence, vol 867, pages 223–266. Springer, Cham. https://doi.org/10.1007/978-3-030-31764-5_9

26 Sagduyu, Y. E., Erpek, T., and Shi, Y. (2021). Adversarial machine learning for 5G communications security. *Game Theory and Machine Learning for Cyber Security*, pages 270–288.

27 Luo, Z., Zhao, S., Lu, Z., Sagduyu, Y. E., and Xu, J. (2020a). Adversarial machine learning based partial-model attack in IoT. In *Proceedings of the 2nd ACM Workshop on Wireless Security and Machine Learning*, pages 13–18.

28 Sagduyu, Y. E., Shi, Y., and Erpek, T. (2019b). IoT network security from the perspective of adversarial deep learning. In *2019 16th Annual IEEE International Conference on Sensing, Communication, and Networking (SECON)*, pages 1–9. IEEE.

29 Sagduyu, Y. E. (2022a). Adversarial machine learning and defense game for NextG signal classification with deep learning. In *IEEE Military Communications Conference (MILCOM)*, pages 1076–1081. IEEE.

30 Erpek, T., Sagduyu, Y. E., and Shi, Y. (2018). Deep learning for launching and mitigating wireless jamming attacks. *IEEE Transactions on Cognitive Communications and Networking*, 5(1): 2–14.

31 Shi, Y., Sagduyu, Y. E., Erpek, T., Davaslioglu, K., Lu, Z., and Li, J. H. (2018b). Adversarial deep learning for cognitive radio security: Jamming attack and defense strategies. In *2018 IEEE international conference on communications workshops (ICC Workshops)*, pages 1–6. IEEE.

32 Kim, B., Sagduyu, Y. E., Erpek, T., Davaslioglu, K., and Ulukus, S. (2021c). Channel effects on surrogate models of adversarial attacks against wireless signal classifiers. In *ICC 2021-IEEE International Conference on Communications*, pages 1–6. IEEE.

33 Kim, B., Sagduyu, Y. E., Erpek, T., Davaslioglu, K., and Ulukus, S. (2020c). Adversarial attacks with multiple antennas against deep learning-based modulation classifiers. In *2020 IEEE Globecom Workshops (GC Wkshps*, pages 1–6. IEEE.

34 Kim, B., Sagduyu, Y. E., Davaslioglu, K., Erpek, T., and Ulukus, S. (2021b). Channel-aware adversarial attacks against deep learning-based wireless signal classifiers. *IEEE Transactions on Wireless Communications*, 21(6): 3868–3880.

35 Kim, B., Sagduyu, Y. E., Davaslioglu, K., Erpek, T., and Ulukus, S. (2020b). Over-the-air adversarial attacks on deep learning based modulation classifier over wireless channels. In *2020 54th Annual Conference on Information Sciences and Systems (CISS)*, pages 1–6. IEEE.

36 Kim, B., Shi, Y., Sagduyu, Y. E., Erpek, T., and Ulukus, S. (2021d). Adversarial attacks against deep learning based power control in wireless communications. In *2021 IEEE Globecom Workshops (GC Wkshps)*, pages 1–6. IEEE.

37 Kim, B., Sagduyu, Y., Erpek, T., and Ulukus, S. (2021a). Adversarial attacks on deep learning based mmWave beam prediction in 5G and beyond. In *2021 IEEE Statistical Signal Processing Workshop (SSP)*, pages 590–594. IEEE.

38 Sagduyu, Y. E., Ulukus, S., and Yener, A. (2023e). Task-oriented communications for NextG: End-to-end deep learning and AI security aspects. *IEEE Wireless Communications*, 30(3): 52–60.

39 Sagduyu, Y. E., Erpek, T., Yener, A., and Ulukus, S. (2024d). Will 6G be semantic communications? Opportunities and challenges from task oriented and secure communications to integrated sensing. In *IEEE Network*, volume 38, pages 72–80.

40 Sagduyu, Y. E., Erpek, T., Ulukus, S., and Yener, A. (2023a). Is semantic communications secure? A tale of multi-domain adversarial attacks. *IEEE Communications Magazine*, 61(11): 50–55.

41 Kim, B., Sagduyu, Y., Davaslioglu, K., Erpek, T., and Ulukus, S. (2022b). Adversarial machine learning for NextG covert communications using multiple antennas. *Entropy*, 24(8): 1047.

42 Kim, B., Erpek, T., Sagduyu, Y. E., and Ulukus, S. (2022a). Covert communications via adversarial machine learning and reconfigurable intelligent surfaces. In *2022 IEEE Wireless Communications and Networking Conference (WCNC)*, pages 411–416. IEEE.

43 Kim, B., Sagduyu, Y. E., Davaslioglu, K., Erpek, T., and Ulukus, S. (2020a). How to make 5G communications "invisible": Adversarial machine learning for wireless privacy. In *2020 54th Asilomar Conference on Signals, Systems, and Computers*, pages 763–767. IEEE.

44 Luo, Z., Zhao, S., Duan, R., Lu, Z., Sagduyu, Y. E., and Xu, J. (2021). Low-cost influence-limiting defense against adversarial machine learning attacks in cooperative spectrum sensing. In *Proceedings of the 3rd ACM Workshop on Wireless Security and Machine Learning*, pages 55–60.

45 Luo, Z., Zhao, S., Lu, Z., Xu, J., and Sagduyu, Y. E. (2020b). When attackers meet AI: Learning-empowered attacks in cooperative spectrum sensing. *IEEE Transactions on Mobile Computing*, 21(5): 1892–1908.

46 Sagduyu, Y. E., Shi, Y., and Erpek, T. (2019a). Adversarial deep learning for over-the-air spectrum poisoning attacks. *IEEE Transactions on Mobile Computing*, 20(2): 306–319.

47 Shi, Y., Erpek, T., Sagduyu, Y. E., and Li, J. H. (2018a). Spectrum data poisoning with adversarial deep learning. In *IEEE Military Communications Conference (MILCOM)*, pages 407–412. IEEE.

48 Davaslioglu, K. and Sagduyu, Y. E. (2019). Trojan attacks on wireless signal classification with adversarial machine learning. In *IEEE International Symposium on Dynamic Spectrum Access Networks (DySPAN)*, pages 1–6. IEEE.

49 Sagduyu, Y. E., Erpek, T., Ulukus, S., and Yener, A. (2023b). Vulnerabilities of deep learning-driven semantic communications to backdoor (Trojan) attacks. In *2023 57th Annual Conference on Information Sciences and Systems (CISS)*, pages 1–6. IEEE.

50 Shi, Y. and Sagduyu, Y. (2022a). Membership inference attack and defense for wireless signal classifiers with deep learning. *IEEE Transactions on Mobile Computing*.

51 Shi, Y., Davaslioglu, K., and Sagduyu, Y. E. (2020b). Over-the-air membership inference attacks as privacy threats for deep learning-based wireless signal classifiers. In *Proceedings of the 2nd ACM Workshop on Wireless Security and Machine Learning*, pages 61–66.

52 Shi, Y., Davaslioglu, K., and Sagduyu, Y. E. (2019). Generative adversarial network for wireless signal spoofing. In *Proceedings of the ACM Workshop on Wireless Security and Machine Learning*, pages 55–60.

53 Shi, Y., Davaslioglu, K., and Sagduyu, Y. E. (2020a). Generative adversarial network in the air: Deep adversarial learning for wireless signal spoofing. *IEEE Transactions on Cognitive Communications and Networking*, 7(1): 294–303.

54 Davaslioglu, K. and Sagduyu, Y. E. (2018). Generative adversarial learning for spectrum sensing. In *IEEE international conference on communications (ICC)*, pages 1–6. IEEE.

55 Ayanoglu, E., Davaslioglu, K., and Sagduyu, Y. E. (2022). Machine learning in NextG networks via generative adversarial networks. *IEEE Transactions on Cognitive Communications and Networking*, 8(2): 480–501.

56 Hou, T., Wang, T., Lu, Z., Liu, Y., and Sagduyu, Y. (2021). IotGAN: GAN powered camouflage against machine learning based IoT device identification. In *2021 IEEE International Symposium on Dynamic Spectrum Access Networks (DySPAN)*, pages 280–287. IEEE.

57 Shi, Y., Sagduyu, Y. E., and Erpek, T. (2023b). Jamming attacks on decentralized federated learning in general multi-hop wireless networks. *arXiv preprint arXiv:2301.05250*.

58 Shi, Y. and Sagduyu, Y. E. (2022b). How to launch jamming attacks on federated learning in NextG wireless networks. In *IEEE Globecom Workshops (GC Wkshps)*, pages 945–950. IEEE.

59 Shi, Y. and Sagduyu, Y. E. (2022c). Jamming attacks on federated learning in wireless networks. *arXiv preprint arXiv:2201.05172*.

60 Sagduyu, Y. E. (2022b). Free-rider games for federated learning with selfish clients in NextG wireless networks. In *IEEE Conference on Communications and Network Security (CNS)*, pages 365–370. IEEE.

61 Shi, Y., Costa, M., Erpek, T., and Sagduyu, Y. E. (2023a). Deep reinforcement learning for NextG radio access network slicing with spectrum coexistence. *IEEE Networking Letters*.

62 Shi, Y., Sagduyu, Y. E., Erpek, T., and Gursoy, M. C. (2022a). How to attack and defend NextG radio access network slicing with reinforcement learning. *IEEE Open Journal of Vehicular Technology*, 4: 181–192.

63 Shi, Y., Sagduyu, Y. E., Erpek, T., and Gursoy, M. C. (2022b). Jamming attacks on NextG radio access network slicing with reinforcement learning. In *IEEE Future Networks World Forum (FNWF)*, pages 397–402. IEEE.

64 Wang, F., Gursoy, M. C., Velipasalar, S., and Sagduyu, Y. E. (2022). Robust deep reinforcement learning based network slicing under adversarial jamming attacks. In *IEEE 33rd Annual International Symposium on Personal, Indoor and Mobile Radio Communications (PIMRC)*, pages 752–757. IEEE.

65 Shi, Y. and Sagduyu, Y. E. (2021). Adversarial machine learning for flooding attacks on 5G radio access network slicing. In *2021 IEEE International Conference on Communications Workshops (ICC Workshops)*, pages 1–6. IEEE.

66 Sankhe, K., Belgiovine, M., Zhou, F., Riyaz, S., Ioannidis, S., and Chowdhury, K. (2019). ORACLE: Optimized Radio clAssification through Convolutional neuraL nEtworks. In *IEEE INFOCOM - IEEE Conference on Computer Communications*, pages 370–378.

67 Sagduyu, Y. E. and Erpek, T. (2025). Adversarial attack and defense for LoRa device identification and authentication via deep learning. In *IEEE Internet of Things Journal*, volume 12, pages 20261–20271.

68 Sagduyu, Y. E. and Erpek, T. (2023). Adversarial attacks on LoRa device identification and rogue signal detection with deep learning. In *IEEE Military Communications Conference (MILCOM)*, pages 385–390.

69 Sagduyu, Y. E., Erpek, T., Yener, A., and Ulukus, S. (2024c). Low-latency task-oriented communications with multi-round, multi-task deep learning. In *Proceedings of the 30th Annual International Conference on Mobile Computing and Networking*, ACM MobiCom '24, page 2365–2370. Association for Computing Machinery.

70 Sagduyu, Y. E., Erpek, T., Yener, A., and Ulukus, S. (2023c). Multi-receiver task-oriented communications via multi-task deep learning. In *IEEE Future Networks World Forum (FNWF)*, pages 1–6.

71 Sagduyu, Y. E., Erpek, T., Yener, A., and Ulukus, S. (2024a). Joint sensing and semantic communications with multi-task deep learning. volume 62, pages 74–81.

72 Sagduyu, Y. E., Erpek, T., Yener, A., and Ulukus, S. (2024b). Joint sensing and task-oriented communications with image and wireless data modalities for dynamic spectrum access. In *IEEE International Symposium on Dynamic Spectrum Access Networks (DySPAN)*, pages 57–62.

4

Securing Management with Blockchain-based Federated Learning

Rui Ning[1], Hongzhi Guo[2] and Chunsheng Xin[3]

[1]*Computer Science, Old Dominion University, USA*
[2]*Engineering Department, Norfolk State University, USA*
[3]*Department of Computer Science, Iowa State University, USA*

4.1 Introduction

Deep Learning (DL) has quickly become a cornerstone of the artificial intelligence space, revealing its strength across a wide array of domains and applications. From self-driven cars to medical imaging and from text to computer vision, the ability of DL to automatically discover complex patterns through data has rendered it a necessary component of several forms of intelligence systems. But the nature of the strength itself is such that it becomes data-driven. That is, for a DL system to become effective and reach state-of-the-art performance, it requires a massive amount of data to train upon.

In the traditional centralized learning framework, the common method comprises pooling data from disparate sources and sending it to a central server for model training. While it might seem sensical computationally, it does pose a whole raft of problems, specifically with user privacy. As data becomes diverse, ranging rom individual data to medical records, centralizing such sensitive information inevitably breeds problems. Data breaches, misuse, and unauthorized access become a significant real possibility. Recognizing this challenge, the research community has been working hard to find ways to train models without putting user privacy at risk, knowing that this is a problem [1–4].

6G Security: New Frontiers in Mobile Network Technologies and Verticals, First Edition.
Edited by Luiz A. DaSilva.

Federated Learning (FL) is one of the many proposed methods and frameworks that stands out. It is not only innovative, but it is also becoming more popular in industrial applications [4]. FL changes the way we think about training by decentralizing it instead of centralizing data. In this system, each device or node that takes part keeps its own copy of the DL model and uses its own data to train. These devices don't share raw data; instead, they compute and share only the model updates, which are usually in the form of gradients. Stochastic Gradient Descent (SGD) or one of its variations [5] are then used to combine these gradients on a central server and update the global model. This method protects user data by keeping the raw data on the local device, which eases many of the privacy concerns that come with traditional centralized training.

But, as with all maturing technology, FL does not lack a host of issues. Among the major concerns, the excessive dependence on a central server, which can become a point of failure, can have serious consequences. Most importantly, the central nature, though helpful when it comes to pooling the increments, can be exploited. An exploited server can impart biases, tamper with the model weights to promote select malicious tasks, or even add backdoors that can later be exploited. Secondly, the FL framework as it exists today lacks proper motivation to encourage the participants to continuously contribute their data and their computing power, as a result of the lack of a concrete incentive mechanism.

To address these challenges, researchers are more and more interested in adding blockchain technology to the FL framework to help solve these problems. [6–9]. Blockchains are ledgers that are decentralized and can't be changed, which makes them a good way to fix some of FL's centralization issues. In a FL system that uses a blockchain, updates to models from nodes that are part of the system are seen as transactions. After that, these transactions are grouped into blocks. Then, consensus protocols like Proof-of-Work (PoW) or Proof-of-Stake (PoS) check these blocks and add them to the blockchain. This method makes sure that the system is open, can't be changed, and is decentralized, with everyone having an equal say in how it is validated. Also, by keeping model updates on a distributed network, the risks that come with having a single central server are much lower.

4.1.1 Challenges and Contributions

Combining blockchain technology with Federated Learning systems is a huge step toward solving some of the most important problems with centralized learning systems. The fact that blockchains are decentralized and FL can keep your information private could change the way we think about machine learning in distributed settings. But, like any new combination, this integration between blockchain and FL comes with its own problems:

1. The most important thing to think about is how well the system can grow. Traditional blockchain methods use raw model weights to make blocks by putting them together. Once these blocks are examined, they need to be sent out to everyone on the network and stored in their own ledgers. This broadcasting process creates a lot of network traffic, especially in big networks with millions of users. Also, storing these blocks with a lot of weight is very difficult, which makes it impractical in places where storage space is limited.
2. Privacy, which is very important to FL, could be at risk in a system based on blockchain. Even though the raw data isn't shared, everyone can see the model weights that are uploaded to the blockchain. If a smart enemy knows a lot about machine learning, they might be able to reverse-engineer or analyze these weights to figure out things about user data. This risk is significantly higher when the data is private, valuable, or safety-critical.
3. Blockchains work best when they are decentralized. This decentralization, on the other hand, also makes it easier for attacks to happen, especially for model poisoning attacks. Recent research has shown that there have been times when bad actors have added poisoned data to the global model in order to make it work worse. Even worse, these people could put in backdoors that would let them change how the model works to fit their evil plans.

Recognizing these challenges, we introduce BlockFed, a state-of-the-art solution designed to alleviate these concerns and pave the way for a truly scalable, privacy-preserving, and secure blockchain-based FL system. Our key contributions in this domain are:

First, the main goal of BlockFed is to encourage sparse local training. This method makes sure that only a small number of weights are changed during training. We use a new importance ranking algorithm to filter and compress these weights even more. This cuts down on the amount of data that needs to be sent during the broadcasting phase by a huge amount. By only looking at important data, we automatically lower the chance that private information will get out. In addition, our system uses differential privacy mechanisms [10] to add noise, which makes user data even more private.

Second, to deal with the problem of block size, especially when multiple clients send their updates at once, we use the power of Singular Value Decomposition (SVD). SVD lets us store data in a smaller space while still getting the main idea across. Our real-world results show that this method works, as the block size dropped by a huge 200 times in a network with 100 participants.

Third, any distributed learning system must be able to protect itself from poisoning attacks. BlockFed has a strong way to find and get rid of poisonous weights. Using the built-in features of the SVD algorithm, our system breaks down weight submissions and carefully removes parts that are unnecessary or could be harmful.

The next parts of this chapter dive into more detail about the complicated parts of blockchain-based FL systems, which will help build the BlockFed framework.

We back up our claims with detailed experimental results and end with some thoughts on what this field might look like in the future.

4.2 Background

4.2.1 Blockchain

Blockchain is a digital ledger technology that keeps track of transactions on many computers in a way that makes it impossible to change them after they have been recorded. This basic technology is the basis for many cryptocurrencies, especially Bitcoin. It is known for being secure, open, and difficulty to change.

The blockchain consists of a number of blocks, and each one has a list of transactions. Cryptographic hashes connect these blocks in the order they were created, making a chain. There is a timestamp, a link to the previous block, and transaction data in each block. This makes sure that transactions are in the right order and are safe.

Overall, key features of blockchain are:

- **Decentralization:** Blockchain works on a peer-to-peer network, which means that control is spread out across many nodes instead of being held by one central entity. This decentralization makes things safer and lowers the chance of a single point of failure.
- **Openness:** Everyone who is part of a blockchain can see all of its transactions. This openness makes it possible for everyone to check the validity of transactions, which builds trust and responsibility.
- **Immutability:** Once a transaction is on the blockchain, it can't be changed or taken off. This unchangeability makes it possible to keep a reliable audit trail and stops fraud and tampering.
- **Consensus Mechanisms:** Blockchain uses consensus mechanisms like PoW and PoS to check that transactions are real. These systems make sure that everyone agrees on the validity of transactions, which keeps the ledger safe.

Blockchain can be used in many areas besides cryptocurrencies, such as healthcare, finance, supply chain management, and federated learning, as discussed. Blockchain technology is a key innovation in the digital age because it solves challenging problems that traditional systems cannot.

4.2.2 Federated Learning

Federated Learning (FL) is a decentralized machine learning method that helps various clients (e.g., organizations or mobile devices) jointly train a model while their data stays decentralized. Unlike the normative centralized machine approaches, which involve bringing data together to a single server, FL overcomes data security and privacy issues by enabling data to stay decentralized on the local devices.

Key features of federated learning are:

- **Decentralized Training:** For FL, the client trains the model at their local location with their own data set and sends the updates of the model (the gradients) to the central server, rather than the raw data. Thus, it reduces the risk of privacy.
- **Model Aggregation:** The central server combines the model updates of all the clients involved to create a global model. The combined model then gets returned to the clients, and the process repeats iteratively until the process converges.
- **Privacy Preservation:** Preserving the local data and distributing model updates enables FL to improve privacy substantially. As an additional measure of privacy-preservation, other approaches such as differential privacy and Secure Multi-Party Computation (SMC) could be integrated to better secure sensitive data.
- **Scalability:** FL can scale to millions of devices and thus is fine for mass-scale applications. That it is a decentralized system allows parallel processing as well as efficient use of the computing resources.

Federated Learning becomes highly effective when data is decentralized and cannot be brought to a central location because of privacy, regulatory, or bandwidth issues. FL uses have applications in a wide array of areas such as healthcare, fintech, and the (IoT), through which it facilitates the building of strong models without the leaking of data privacy.

4.2.3 Blockchain-empowered FL

It is the combination of blockchain technology and FL, which has piqued the interest of the research fraternity, giving rise to a wide array of pioneering studies [7–9, 11–13]. The sole motivation behind these explorations comes as a result of the intrinsic limitations encountered by the classical FL framework, which frequently experiences centralization problems as well as vulnerabilities.

The core strength of blockchain lies in its decentralized architecture, enabling secure collaboration and data sharing across a network of untrusted parties. Such a setup is especially pivotal for deep learning processes that necessitate the aggregation of model weights from multiple nodes. In the conventional FL setup, this aggregation typically relies on a central server, making it susceptible to many-to-one traffic flows and centralized control dependence. Blockchain's decentralized nature alleviates these concerns, providing a safeguard against potential malicious server interventions.

Also, the use of an immutable distributed ledger implements another security feature. An immutable distributed ledger records and stores data from all nodes involved in cryptographically secured blocks. Such a system guarantees that any server-side attack of poisoning leaves a trace, which forms a strong defense for such attacks.

The consensus protocols of the blockchain, such as PoW and PoS, along with the use of incentive mechanisms, establish a cooperative setting. Such mechanisms are used to ensure participants are encouraged to train models and distribute their local gradients, thereby increasing the efficiency and the integrity of the training process as a whole.

4.2.4 Critical Limitations of Blockchain-based FL

This combination of FL and blockchain is indeed revolutionary, yet it comes with issues. While the benefits are real, a number of practical challenges have hindered the wide adoption of FL systems based on blockchain.

Overhead. Communication and storage overhead are still major issues today in the state-of-the-art blockchain-powered FL setting. Current methodologies oblige clients to broadcast their locally trained models, which include all weight increments, for others to validate. Communication costs are thus incurred through the process of broadcasting. Validators then collectively put these weights provided by various clients into new blocks, which are again broadcast and stored on all ledgers. This exacerbates the network, as well as storage, overhead. Penalties are even higher in the case of large-scale blockchain systems, wherein the incremental growth of block size exponentially increases the storage overhead over longer periods of training. BlockFed, our intended solution, seeks to mitigate these issues through 1) sparse local training, which diminishes the number of weights submitted, and 2) SVD of multiple client submissions, leaving behind the vital components.

Security and privacy concerns. Blockchain-based FL's decentralized nature, though a worthwhile benefit, opens new fronts for cyberattacks. Since participants use arbitrary data to update common models without outside judging, bad actors can take advantage. They can add poisoned data, which makes the summed model incorrectly classify all (untargeted attack) or selectively (following a set, predetermined structure) (targeted attack) the inputs. BlockFed solves the problem efficiently, eliminating malicious submissions found as the les significant components of the SVD decomposition process. BlockFed thus provides a simplified solution that is much faster than scrutinizing and cleaning each submission individually.

Privacy continues to take center stage in any distributed learning setup. For the case of blockchain-driven FL, the confidentiality of the clients' private local data becomes a must. Studies have shown astute hackers, given the information of new gradients, can possibly infer, and even rebuild the very first training data [14]. BlockFed implements safeguards to mitigate such a risk. By restricting the data uploaded to just the essential and highly abstracted weights, BlockFed reduces information leak possibilities. Moreover, our system intertwines perfectly with the existing differential privacy methods, strengthening the system's privacy requirements even further.

4.3 BlockFed Framework

The BlockFed framework represents a breakthrough paradigm for federated learning by leveraging the power of the blockchain. By leveraging the decentralized and tamper-proof nature of the blockchain with the cooperative nature of the federated learning, BlockFed attains a secure, efficient, and scalable distributed deep-learning system. We detail the BlockFed system in the next section, defining the process of how it works and the new methods utilized to enhance the performance, security, and efficiency.

4.3.1 System Overview

Figure 4.1 gives a visual description of the BlockFed system, which explains how it operates as well as interactions between different entities, i.e., the nodes (or clients) as well as the validators (or miners). Steps of the system can be detailed as:

(a) **Initialization and Sparse Local Training:** During initiation, all nodes of the network fetch the current model residing on the blockchain. This current model becomes an initial template for nodes to undergo sparse local training. Note that the very first block, which is referred to as Block 0, contains a model of randomized initialization, thus paving the way for the next cycles of training.

(b) **Weight Importance Ranking:** After local training, the new weights need to be prioritized and curated. For this, clients use our new Algorithm 4.2 to rank the weights by their importance. By communicating only the top-level weights, BlockFed dramatically reduces the communication overhead, which renders the model update broadcast process lean and efficient.

(c) **Validation and Compression:** During this stage, the validators or miners intervene to validate the new weights. They play a central role in their integrity and effectiveness. After validation, all submissions that pass undergo compression through the use of the SVD algorithm, facilitating efficient storage and bandwidth usage.

(d) Malicious Data Elimination: With security at its core, BlockFed introduces a robust mechanism to weed out potentially malicious and superfluous data. By leveraging the properties of SVD, validators can systematically discard components with minuscule singular values, effectively purging data that's deemed redundant or harmful. Following this purification process, the consensus protocol, specifically the PoW, is executed on the block to achieve network-wide agreement.

(e) Reward Allocation: Identification and rewarding participation are key factors for the achievement and longevity of any cooperative framework. Within BlockFed, data contributing nodes and nodes that participate in training, as well as validators that provide services of validation, are properly rewarded, which helps ensure a forward-thinking and cooperative environment.

(f) Blockchain Update Broadcast: Bringing the cycle to a conclusion, the newly updated blockchain, containing the new model, is broadcasted to the full network. The newly created model now becomes the standard for the next training cycles, thus keeping the full network synchronized and current.

The design of BlockFed ensures that it addresses the challenges typically associated with federated learning, offering a solution that's secure, efficient, and poised for scalability.

4.3.2 Key Steps in BlockFed

Sparse local training. At the beginning of each local training cycle, a client retrieves the current weight updates from the blockchain and applies them to

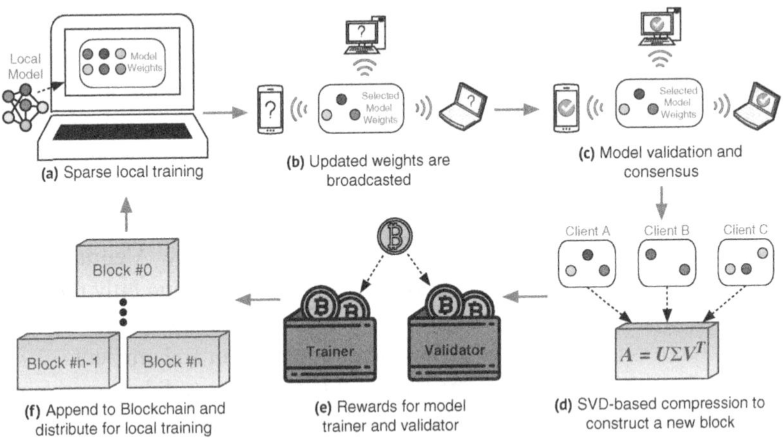

Figure 4.1 An overview of the proposed blockchain-based fully distributed federated learning (BlockFed) framework, showcasing the sequence of operations and the collaborative synergy between nodes and validators.

update its local model θ. Afterward, it performs training using locally stored data for a certain number of iterations to derive an updated model θ'. The weight changes between both versions, $\Delta\theta = \theta' - \theta$, are then broadcast to inform other clients. Our goal is to decrease the size of $\Delta\theta$ (i.e., the number of changed weights) and thus reduce the communication overhead. To decrease the number of changed weights, we adopt an $\mathcal{L}1$ regularization during training to encourage sparse updates (i.e., fewer updated weights) on the model's weights. Formally, we adopt the following objective function:

$$L_{local} = \mathcal{L}(f(x), y) + \alpha \parallel \Delta\theta \parallel_1, \tag{4.1}$$

where f denotes the model, $\mathcal{L}$ is a metric (e.g., cross-entropy) used to calculate the error between the prediction of input x and the target label y, $\parallel \cdot \parallel 1$ is the L_1 norm, and α is a hyperparameter balancing the two loss terms. The first term maximizes the predictive power of f, and the second term increases the sparsity of its updated weights. The L_1 norm term in the loss function, if used alone (named "LASSO"), generates a sparse solution that has been widely studied and found to be highly effective in various fields [15, 16].

To gain insights into sparse local training, we conduct a series of preliminary experiments to compare the differences between $\Delta\theta$ derived from regular training and sparse training. More specifically, we perform regular and sparse training on a LeNet [17] using the MNIST [18] dataset for a typical local training cycle of 3 epochs. We then calculate the $\Delta\theta$ based on the updated models and plot their distribution density map in Figure 4.2. We observe that the values of $\Delta\theta$ from sparse training are noticeably more concentrated around 0, resulting in a significantly smaller number of weights to be broadcast.

Additionally, we compare the accuracy of the models between the two types of training (see Figure 4.3), where we observe negligible differences. This can be explained by the over-parameterized nature of (DNNs), where recent studies have shown that regularized models (using norms such as L_1 and L_2) usually achieve equal or even better performance during training.

Weight ranking and broadcasting. To further reduce the number of changed weights, we filter out less important ones. Specifically, we rank the importance of each changed weight from the previous step using a carefully designed score:

$$s_k = \underbrace{|\Delta w_k|}_{\text{weight change}} \cdot \overbrace{\frac{1}{N}\sum_{i}^{N} \left| \frac{\partial f(x_i)^y}{\partial w_k} \right|}^{\text{average impact on predictions}}, \tag{4.2}$$

where w_k is an individual weight with index k, and $f(x_i)^y$ is the logit output of the input x_i for the correct class y. A total of N data samples are used to calculate the average importance score. It is worth noting that sparse local training encourages

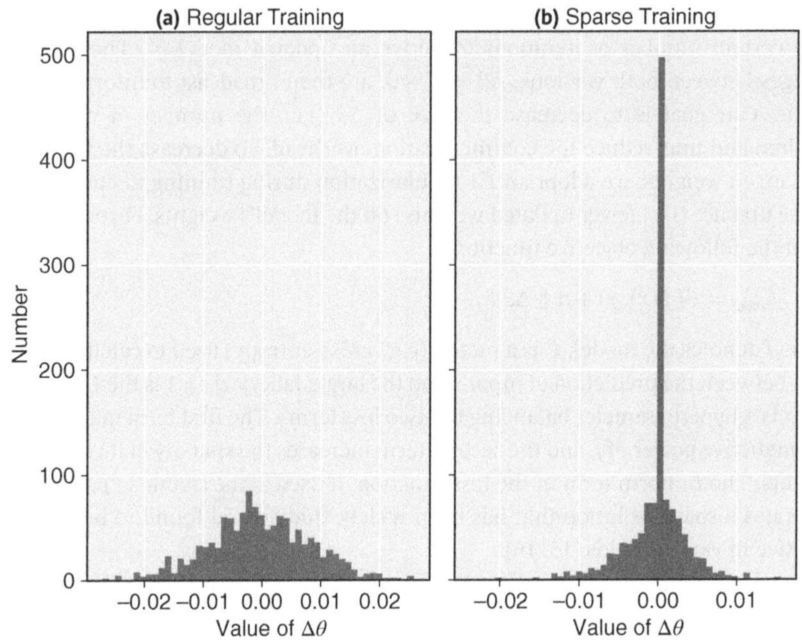

Figure 4.2 Compression of the distribution of $\Delta\theta$ between regular and sparse local training.

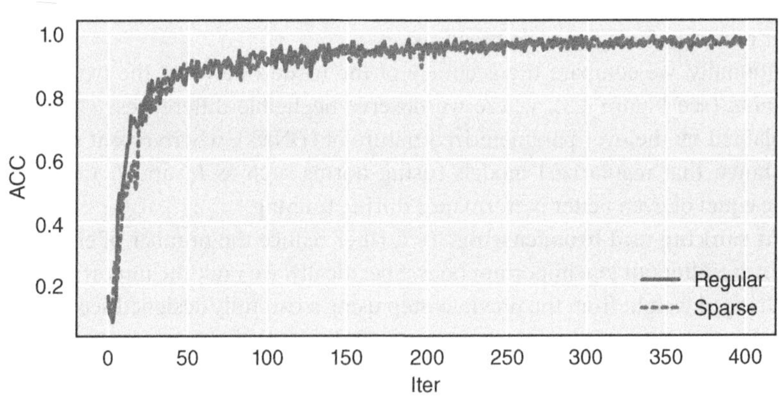

Figure 4.3 Comparison of the model's performance (accuracy) between regular and sparse local training.

a majority of weights with $\Delta\theta \approx 0$ (i.e., $\Delta w_k \approx 0$), yielding significantly lower s_k, which will be filtered out before uploading.

More specifically, we use an energy threshold to identify the most important m weights, formulated as follows:

$$\min_{M} \frac{\sum_{i=1}^{M} s_i^2}{\sum_{i=1}^{K} s_i^2} > T, \tag{4.3}$$

where K is the total number of ranked weights. To improve convergence speed, we propose dynamically adjusting the value of $T = \lambda \tanh(t)$, where $\lambda = 0.3$ is an empirically defined coefficient and t is the training time (number of rounds of uploading weights). Subsequently, the client broadcasts the highly ranked weights (with a total number of M) to all validators.

As shown in Table 4.1, we compare the number of uploaded weights for different models trained using the MNIST dataset between the original model and our approach (BlockFed). It is evident that BlockFed significantly reduces the number of uploaded weights across every model architecture by an average ratio of 4.5%.

It is worth mentioning that previous studies have shown that advanced attackers may reconstruct private training data if raw gradients (weight changes) are broadcast. However, in our approach, sparse training and weight filtering enforce clients to upload only highly abstracted and essential information, thereby naturally limiting information leakage. In our experiments in Section 4.4, we also demonstrate that our method can be seamlessly combined with differential privacy to enhance privacy protection, where artificial noises are injected into the updated weights.

SVD-based Weight Compression and Poison Removal

Weight Compression. On the validator's (miner's) side, it periodically packs the received submissions (weight changes $\Delta\theta$ from multiple clients) into a block (similar to submitted transactions in traditional blockchain) for consensus using protocols such as PoW and PoS. Here, we consider two critical challenges:

1. The number of received submissions may be massive, especially in a large-scale blockchain system, significantly increasing the block size and subsequently introducing overhead to the block's transmission and storage.

Table 4.1 Comparison of uploaded weights of models trained on MNIST dataset.

Architecture	LeNet	MobileNetV2	EfficientNetB0	ResNet18
Original	60K	3.4M	5.3M	31.7M
BlockFed	2.7K	0.2M	0.2M	1.5M

2. A small number of adversarial participants may upload malicious contributions to poison the shared weights; thus, efficiently removing them among all the submissions is crucial.

To address these issues, we propose using a sparse SVD scheme to (1) compress the combined matrix of all contributions, and (2) decompose contributions into orthogonal components and remove malicious contributions as less dominant components (based on the reasonable assumption that the majority of clients are benign).

More specifically, initially, a matrix A containing all received contributions is constructed by converting and concatenating each contribution as a 1-D column. Note that A has a size of $m \times n$ where n is the number of contributions and m is the maximum index of received weights. Then we decompose A using:

$$A = U\Sigma V^T, \tag{4.4}$$

where U consists of decomposed orthogonal components, Σ contains the singular values (importance scores) of each component in U, and V^T represents the composition of each component for the contributions. Previous studies have shown that 10% of the most important components effectively approximate the entire A. Therefore, we adopt an energy threshold $T_{block} = 0.1$ similar to Equation (4.3) to select corresponding components:

$$\min q \frac{\sum i = 1^q \sigma_i^2}{\sum_{i=1}^{p} \sigma_i^2} > T_{block}, \tag{4.5}$$

where p is the total number of singular values, and q is the number of retained ones. The compression rate can be derived as

$$\tau = \frac{mq + q^2 + nq}{mq}. \tag{4.6}$$

We conduct similar experiments to calculate the compression rate of models trained using the MNIST Dataset. As shown in Table 4.2, BlockFed achieves an average compression ratio of 7.97%. Combined with the average ratio (4.5%) of reduced submitted weights, BlockFed can approximately decrease the communication overhead to 0.3%.

Table 4.2 Centering compression rate of models trained on MNIST dataset.

Architecture	LeNet	MobileNetV2	EfficientNetB0	ResNet18
BlockFed	7.6%	8.2%	8.2%	7.9%

Malicious Contribution Removal. Another advantage of removing less important components during weight compression is the mitigation of poisoning attacks. In other words, malicious contributions are naturally identified as less important components during SVD decomposition as there are usually fewer attackers in the blockchain, thus can be naturally removed by weight compression. This can be evidenced by our experimental results in Table 4.4, as it can reduce the overall attack success rate to 5% under two advanced attacks.

Following this, the validator carries out the mining process (e.g., PoW and PoS) to verify the newly created block and achieve consensus among all the nodes. Once the mining is complete, the verified block is added to the blockchain and broadcast to all local devices. Both trainers and validators are rewarded accordingly.

4.4 Experimental Results

Delving into the experimental aspects of our study, we aim to rigorously assess the efficacy, efficiency, and resilience of the proposed BlockFed framework. Through an extensive set of experiments, we explore how BlockFed fares across various neural network architectures, datasets, and non-IID settings. Furthermore, in the wake of increasing concerns about data security and integrity in distributed frameworks, we also evaluate BlockFed's capability to withstand privacy breaches and potential poisoning attacks.

4.4.1 Experimental Setting

4.4.1.1 Blockchain

We base our experimental design on a prototyped hybrid blockchain framework. The framework, which simulates a real-world large-scale setting, hosts an array of virtual nodes over a single physical node. We use BlockFed implementation to train and aggregate the models with the aid of Pytorch 1.11.0, while FLASK 2.1.0 comes in for easy communication between nodes. For our experiments to reflect real-life conditions, we host them on ODU's state-of-the-art computing facility, the Wahab cluster. The facility, which serves as our workhorse, has 158 compute nodes and 18 GPU nodes, all with 40 cores, 2.4 GHz, and 368 GB RAM per compute node, and the GPU nodes are reinforced with four NVidia V100 GPUs, with 640 tensor cores per V100. Interconnectivity for nodes comes courtesy of a Mellanox InfiniBand EDR backbone, which enables fast data transfer with a 100 Gbps system throughput.

4.4.1.2 Dataset and Architecture

Any deep learning experiment's backbone is the dataset. For our analysis, we rely on well-established benchmark datasets: MNIST, the golden standard for handwritten digit classification [18], and CIFAR-10, a standard bearer in image recognition studies [19]. Alongside these datasets, we utilize two cutting-edge deep learning architectures: MobileNet-V2 [20], which shines for mobile vision applications, and ResNet18 [21], recognized for its deep residual learning prowess.

4.4.1.3 BlockFed Training

For a holistic evaluation, we adhere to the industry-standard setup. BlockFed is run until each participating node completes a total of 100 rounds of local training. Within each round, local training spans five epochs. The training regimen for the local models is grounded on SGD with an initial learning rate pegged at 0.01. Additionally, a weight decay factor of 1e-4 is applied with a batch size configuration of 128. Recognizing the complexities posed by non-IID data distributions, especially in federated settings, we simulate data imbalance using the Dirichlet distribution [22]. The degree of imbalance is controlled using a hyperparameter, $\alpha = 0.8$, where an increased value of α signifies a heightened level of imbalance.

4.4.1.4 Attack Configuration

Security against potential attacks comes first in the context of federated learning. Through our experiments, we seek to measure BlockFed's robustness against a variety of attacks. We, in particular, study the defense of BlockFed against a data recovery attack, known as DeepLeakage [14], and two backdoor poisoning attacks: MRA [23] and DBA [24]. We are interested in finding the robustness of BlockFed when bad actors seek to violate the integrity of the global model or have unauthorized knowledge about the sensitive data.

4.4.2 Performance Comparison

4.4.2.1 Performance Comparison

First, we demonstrate the superior performance of BlockFed in terms of average per-client communication costs (CC), average block size (BS), and performance loss (PL: decreased model accuracy from original FL). We compare BlockFed to three baseline approaches: original blockchain-based FL (where original model weights are broadcast and stored in the blockchain), quantization-based (where the precision of model weights is quantized to 8 bits), and Top-K (where only the top k most updated weights are broadcast; for a fair comparison, we adopt a k equal to the number of selected weights in BlockFed). We train the global model using a blockchain with a total of 100 participants until each client completes 100 rounds of local training.

As shown in Table 4.3 (row 2, columns 4-7), BlockFed can reduce per-client communication costs in each round of local training to 0.42 MB from an original size of 14.37 MB, while only introducing a negligible performance loss

Table 4.3 Performance comparison of per-client communication cost.

Arch	Dataset	Metric	Num of C = 100				C = 300
			Ori	Top-K	Quant	B-Fed	B-Fed
MobileV2	MNIST	CC (MB)	14.37	**0.42**	2.87	**0.42**	**0.42**
		BS (MB)	358.8	11.06	72.81	**1.81**	**1.93**
		PL (%)	0.00	2.57	2.14	**0.83**	**0.85**
	CIFAR10	CC (MB)	14.98	**0.44**	2.92	**0.44**	**0.45**
		BS (MB)	363.4	11.32	71.96	**1.90**	**1.98**
		PL (%)	0.00	3.02	2.71	**0.88**	**0.90**
Resnet18	MNIST	CC (MB)	47.82	**1.02**	5.15	**1.02**	**1.02**
		BS (MB)	1246.7	26.26	130.55	**3.13**	**3.20**
		PL (%)	0.00	2.13	2.07	**0.80**	**0.82**
	CIFAR10	CC (MB)	48.01	**1.14**	5.20	**1.14**	**1.16**
		BS (MB)	1294.1	27.33	132.62	**3.67**	**3.88**
		PL (%)	0.00	2.80	2.59	**0.89**	**0.93**

(0.83% of accuracy) to the derived global model. Additionally, comparable performance can be found across different architectures and datasets. The reason for this is that the sparse local training and carefully designed weight filtering scheme in BlockFed are extremely effective at 1) compressing the extracted information to a significantly smaller number of weights, and 2) accurately identifying those most important weights to eliminate redundant information prior to broadcasting. This can also be proven by comparing to the blockchain-based FL where only the Top-K most updated weights are broadcast. Observably, it can reduce communication costs to approximately the same level as BlockFed (with the same number of uploaded weights), but does so at the expense of nearly 3% in the model's precision. In addition to that, BlockFed can further minimize the storage cost by compressing a block to an average size of 1.81 MB (row 3) from its original size of 358.82 MB, which is the result of a combination of local weight filtering and SVD-based compression. The reason for this is that the updated models of many clients still share a substantial quantity of common knowledge (as they are training a global model cooperatively), which can be compressed and decomposed to a few essential components using SVD. This unique characteristic also renders BlockFed highly effective for large-scale training, where a huge number of updated models can be compressed to nearly the same number of key components. As shown in column 8 of Table 4.3, the average block size of BlockFed with 300 clients is roughly identical to that of the BlockFed with 100 users.

4.4.2.2 Comparison of Privacy Leakage

Another critical metric of blockchain-based FL (and other distributed learning frameworks) is its privacy level. More specifically, it is crucial to safeguard clients' private local data, as earlier research [14] has demonstrated that sophisticated attackers can infer or reconstruct the original training data based on the updated gradients. For example, DeepLeakage [14] is capable of reconstructing an image (Figure 4.4b) that is nearly identical to the original training sample (Figure 4.4a) using the broadcast weights in a blockchain-based FL system. As previously stated, we hypothesize that BlockFed can naturally limit information leakage by uploading only essential weights with highly abstracted knowledge. To prove this, we conduct similar experiments to reconstruct an image depicted in Figure 4.4e, which is much noisier than recovery from raw, quantized, and Top-K weights (Figure 4.4b–d, respectively). It also proves that BlockFed only uploads essential weights as DeepLeakage failed to reconstruct the background contents.

In addition to that, we also demonstrate that BlockFed can be easily integrated with existing differential privacy techniques to further enhance its privacy level. Differential privacy is a widely used technique to protect users' privacy by injecting artificial noise into the model weights before broadcasting to prevent information leakage. It has been embraced for modern FL and distributed learning systems. We conduct an experiment to reconstruct the original image on a BlockFed framework protected by a widely adopted differential privacy algorithm [10], where each client injects a certain amount of noise into its model weights prior to uploading them. As shown in Figure 4.4f, the recovered image contains no information from the original sample.

4.4.2.3 Resilience to Data Poisoning Attack

We also examine the resilience of BlockFed against two sophisticated attacks: DBA and MRA. Specifically, we conduct experiments with a BlockFed system consisting of 300 clients under both poisoning attacks, where 10 malicious clients attempt to

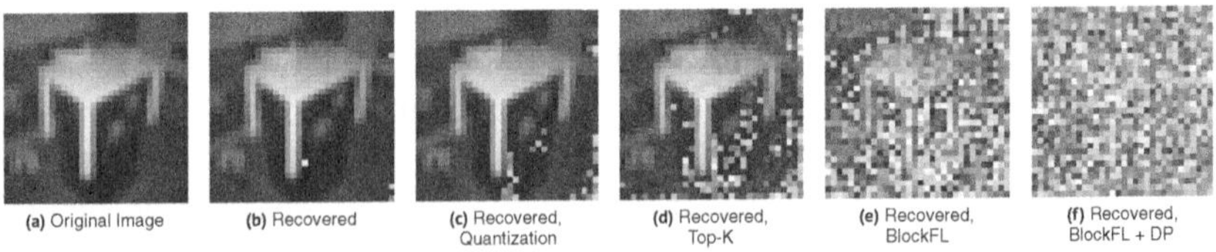

Figure 4.4 Comparison of recovered images using updated weights under the DeepLeakage attack. (a) Original image employed for local training. (b) Recovered image derived from raw updated weights. (c) Recovered image obtained from quantized weights. (d) Image restored using only the Top-K most updated weights. (e) Reconstructed image utilizing filtered weights within the BlockFed framework. (f) Reconstructed image generated from the DP-enhanced BlockFed.

Table 4.4 Resilience comparison of blockchain-based FL against two data poisoning attacks.

Attack	Gradient Clip	Krum	Median	BlockFed	Trimmed Mean
MRA	78.21%	63.21%	72.8%	**4.61%**	38.5%
DBA	82.01%	82.66%	73.19%	**5.48%**	42.13%

inject a backdoor into the global model. We then compare BlockFed with standard data-poisoning defenses (i.e., Gradient Clipping [25], Krum [26], Median [27], and Trimmed Mean [27]) in terms of the final global model's Attack Success Rate (ASR), which represents the proportion of malicious inputs misclassified to the target class. As demonstrated in column 5 of Table 4.4, BlockFed exhibits significantly enhanced robustness against both attacks, reducing the ASR to 5%. This is attributable to the fact that the malicious contributions from attackers are deemed less significant components (due to the typically lower number of attackers) and are consequently removed during the SVD-based weight compression process.

4.5 Key Takeaways

In recent years, the pairing of blockchain technology with Federated Learning (FL) has been of considerable interest. For the future, the pairing has the promise of addressing the inherent limitations seen in typical FL systems, i.e., the vulnerabilities of a centralized node and lack of genuine incentives for participants. By forsaking the use of the typical central parameter server in favor of a decentralized blockchain, the new system guarantees that model updates are not merely shared, but stored securely in an immutable distributed ledger. Such a design inherently increases the robustness of the system, decreasing the risks of individual node collapses and potential malicious uses.

But the integration with the blockchain-FL is not without challenges. The same character of the decentralization, while good for the most part, produces a set of roadblocks. Most importantly, the storage and communication overhead gets augmented. As all the nodes of the net are provided with access to model updates and are required to have them stored, the following overhead becomes a bottleneck as the deployments are scaled up. Such overhead, left unchecked, can become significant stumbling blocks to the feasible scalability and implementation of the FL-based blockchain systems.

Also, the decentralized setup, along with the dependence on local fine-tuning with local datasets, makes the system susceptible to potential poisoning attacks. Adversaries, through poisoned datasets, could bias the global model, instilling biases or even placing backdoors which could be exploited later.

Recognizing these challenges and the immense potential of blockchain-empowered FL, this chapter introduced *BlockFed*. This new framework is meticulously designed to address the aforementioned concerns. At its core, BlockFed aims to optimize communication and storage overhead, ensuring that only crucial model updates are broadcast and stored. Moreover, it introduces robust mechanisms to detect and counteract poisoning attacks, thereby safeguarding the integrity of the global model.

Truly, *BlockFed* is a reminder of the transformation of federated learning. It encapsulates the prospect of a future when distributed deep learning may safely, efficiently, and at scale take place, spur a wide variety of areas and applications forward.

References

1 Mohassel, P. and Zhang, Y. (2017). Secureml: A system for scalable privacy-preserving machine learning. In *Proceedings of the IEEE Symposium on Security and Privacy (SP)*, pages 19–38.

2 Bonawitz, K., Ivanov, V., Kreuter, B., Marcedone, A., McMahan, H. B., Patel, S., Ramage, D., Segal, A., and Seth, K. (2017). Practical secure aggregation for privacy-preserving machine learning. In *Proceedings of the ACM Conference on Computer and Communications Security (SIGSAC)*, pages 1175–1191.

3 Xu, K., Yue, H., Guo, L., Guo, Y., and Fang, Y. (2015). Privacy-preserving machine learning algorithms for big data systems. In *Proceedings of the IEEE International Conference on Distributed Computing Systems (ICDCS)*, pages 318–327.

4 Konečný, J., McMahan, H. B., Yu, F. X., Richtárik, P., Suresh, A. T., and Bacon, D. (2016). Federated learning: Strategies for improving communication efficiency. *arXiv preprint arXiv:1610.05492.*

5 Bottou, L. (2010). Large-scale machine learning with stochastic gradient descent. In *Proceedings of COMPSTAT'2010: 19th International Conference on Computational Statistics, Paris, France, August 22–27, 2010 Keynote, Invited and Contributed Papers*, pages 177–186. Springer.

6 Kim, H., Park, J., Bennis, M., and Kim, S.-L. (2018). On-device federated learning via blockchain and its latency analysis. *arXiv preprint arXiv:1808.03949.*

7 Li, Y., Chen, C., Liu, N., Huang, H., Zheng, Z., and Yan, Q. (2020). A blockchain-based decentralized federated learning framework with committee consensus. *IEEE Network*, 35(1): 234–241.

8 Pokhrel, S. R. and Choi, J. (2020). Federated learning with blockchain for autonomous vehicles: Analysis and design challenges. *IEEE Transactions on Communications*, 68(8): 4734–4746.

9 Bao, X., Su, C., Xiong, Y., Huang, W., and Hu, Y. (2019). FLchain: A blockchain for auditable federated learning with trust and incentive. In *Proceedings of the International Conference on Big Data Computing and Communications (BIGCOM)*, pages 151–159.

10 McMahan, H. B., Ramage, D., Talwar, K., and Zhang, L. (2017). Learning differentially private recurrent language models. *arXiv preprint arXiv:1710.06963*.

11 Nguyen, D. C., Ding, M., Pham, Q.-V., Pathirana, P. N., Le, L. B., Seneviratne, A., Li, J., Niyato, D., and Poor, H. V. (2021). Federated learning meets blockchain in edge computing: Opportunities and challenges. *IEEE Internet of Things Journal*. 8(16): 12806–12825.

12 Majeed, U. and Hong, C. S. (2019). FLchain: Federated learning via MEC-enabled blockchain network. In *Proceedings of the Asia-Pacific Network Operations and Management Symposium (APNOMS)*, pages 1–4.

13 Zhou, S., Huang, H., Chen, W., Zhou, P., Zheng, Z., and Guo, S. (2020). Pirate: A blockchain-based secure framework of distributed machine learning in 5g networks. *IEEE Network*, 34(6): 84–91.

14 Zhu, L., Liu, Z., and Han, S. (2019). Deep leakage from gradients. *Advances in Neural Information Processing Systems*, 32.

15 Zhang, H., Wang, J., Sun, Z., Zurada, J. M., and Pal, N. R. (2019). Feature selection for neural networks using group lasso regularization. *IEEE Transactions on Knowledge and Data Engineering*, 32(4): 659–673.

16 Lim, M. and Hastie, T. (2015). Learning interactions via hierarchical group-lasso regularization. *Journal of Computational and Graphical Statistics*, 24(3): 627–654.

17 LeCun, Y. et al. (2015). Lenet-5, convolutional neural networks. *URL: http://yann.lecun.com/exdb/lenet*, 20(5): 14.

18 Yann, L. (2010). MNIST handwritten digit database. *ATT Labs.*

19 Krizhevsky, A. et al. (2009). Learning multiple layers of features from tiny images. Toronto, ON, Canada.

20 Sandler, M., Howard, A., Zhu, M., Zhmoginov, A., and Chen, L.-C. (2018). Mobilenetv2: Inverted residuals and linear bottlenecks. In *Proceedings of IEEE Conference on Computer Vision and Pattern Recognition (CVPR)*, pages 4510–4520.

21 He, K., Zhang, X., Ren, S., and Sun, J. (2016). Deep residual learning for image recognition. In *Proceedings of the IEEE Conference on Computer Vision and Pattern Recognition (CVPR)*, pages 770–778.

22 Minka, T. (2000). Estimating a dirichlet distribution. Technical report, MIT.

23 Bagdasaryan, E., Veit, A., Hua, Y., Estrin, D., and Shmatikov, V. (2020). How to backdoor federated learning. In *Proceedings of the International Conference on Artificial Intelligence and Statistics (AISTATS)*, pages 2938–2948.

24 Xie, C., Huang, K., Chen, P.-Y., and Li, B. (2019). Dba: Distributed backdoor attacks against federated learning. In *Proceedings of the International Conference on Learning Representations (ICLR)*.

25 Sun, Z., Kairouz, P., Suresh, A. T., and McMahan, H. B. (2019). Can you really backdoor federated learning? *arXiv preprint arXiv:1911.07963.*

26 Blanchard, P., El Mhamdi, E. M., Guerraoui, R., and Stainer, J. (2017). Machine learning with adversaries: Byzantine tolerant gradient descent. In *Proceedings of Advances in Neural Information Processing Systems (NeurIPS)*, 30.

27 Yin, D., Chen, Y., Kannan, R., and Bartlett, P. (2018). Byzantine-robust distributed learning: Towards optimal statistical rates. In *Proceedings of International Conference on Machine Learning (ICML)*, pages 5650–5659.

5

Securing Non-terrestrial Networks

Rahul Varma Chintalapati, Paul Masur, Nishith D. Tripathi and Jeffrey H. Reed

Electrical and Computer Engineering, Virginia Tech, USA

5.1 Introduction

Non-terrestrial Networks (NTNs) encompass efforts by the 3rd Generation Partnership Project (3GPP) to expand cellular communication services beyond the traditional terrestrial domain. An NTN harnesses spaceborne platforms such as satellites or airborne platforms, or High-Altitude Platform Stations (HAPSs), in either transparent or regenerative fashion, to provide services to user equipment (UEs) in geographically remote or rural areas, which are typically underserved by traditional cellular communication systems. However, providing cellular communications services from these altitudes present certain challenges to adapt principles and network operations of a traditional 5th Generation (5G) New Radio (NR) network into a new paradigm. In this chapter, the identified challenges to implementing an NTN are described. In addition, the solutions to these problems, as they are available in the 3GPP 5G NTN specifications, are additionally discussed. NTN represents an ongoing work project by the 3GPP, as the NTN specifications are enhanced to design a better NTN.

5.1.1 Non-terrestrial Network: The Definition

There is no universally accepted definition of an NTN. According to the 3GPP, NTNs are networks or parts of a network in which RF resources like base stations

6G Security: New Frontiers in Mobile Network Technologies and Verticals, First Edition.
Edited by Luiz A. DaSilva.

(i.e., gNodeBs in 5G) or relay nodes are carried by space borne vehicles (e.g., Low Earth Orbit (LEO), Medium Earth Orbit (MEO), and Geo-Synchronous Earth Orbit (GEO) satellites) or airborne vehicles (e.g., HAPS and Unmanned Aircraft Systems (UASs)). Generally, a non-terrestrial network utilizes one or more NTN-Gateways connecting communication payloads or NTN payloads residing on spaceborne or airborne vehicles to the communication infrastructure on the ground. An NTN includes several of the following components: GEO satellites, non-GEO satellites, a radio link between the NTN-Gateway and the NTN payload called a feeder link, a radio link between the UE and the NTN payload called a service link or access link satellite, a satellite or a UAS platform implementing transparent or regenerative NTN payloads (discussed later in this chapter), inter satellite links per NTN design, and the NTN UE. Examples of NTN platforms are LEO satellites with the altitude ranging from (30 to 1500 km), MEO satellites with the altitude ranging from 7000 to 25000 km, GEO satellites at the altitude of – 35786 km, and UASs with the altitude ranging from 8 to 50 km including HAPSs. GEO satellites and HAPSs appear stationary to a fixed point on the earth's surface.

Some works define UASs as being part of an NTN. 3GPP [1] defines an aerial as an airborne vehicle embarking a bent pipe payload or a regenerative payload telecommunication transmitter, typically at an altitude between 8 and 50 km. A UAS utilizes an aerial vehicle or airborne platform. Airborne vehicles encompass Tethered UAS, Lighter than Air UAS, and Heavier than Air UAS, all operating at altitudes typically between 8 and 50 km. A terrestrial network operates with the support of only the infrastructure built on the ground whereas, an NTN operates with the support of both on ground infrastructure (e.g., an NTN-Gateway) and non-terrestrial infrastructure (e.g., satellites or HAPS).

5.1.2 Motivation for the NTN

The motivation for using NTNs is based on use cases of three categories: service continuity, service ubiquity, and service scalability. Service continuity is the combined utilization of both terrestrial and non-terrestrial networks in situations where 5G services cannot be offered by terrestrial networks, like in an airplane or on a ship. In these situations, service continuity enables a better user experience using both a TN and an NTN, with the UE using one type of network at a time. Service ubiquity enables services in geographical areas where terrestrial networks are unavailable. Additionally, in scenarios where a terrestrial network is unable to serve a geographical area (e.g., due to natural disasters), NTNs can assist in solving connectivity issues. The use of Starlink in Ukraine in the early 2020s to establish connectivity when the terrestrial networks were down due to war is an example of service ubiquity. Service scalability leverages the large coverage area of NTNs to multicast or broadcast similar content over a large geographical area. Simultaneous use of a terrestrial network and an NTN in a future version of 3GPP-based NTN or a custom version of the 3GPP-defined NTN increases resiliency of communications, because information can be duplicated in two types of networks.

5.1.3 NTNs – A Historical Perspective and Emerging Industry Trends

Intelsat is an organization that owns and manages a constellation of communication satellites to provide international telecommunication and broadcasting services. Intelsat has launched more than a hundred satellites. Example services provided by Intelsat include backhaul for cellular networks, broadcasting services, connectivity for airplanes and business jets, and direct satellite connectivity for suitable devices. The oldest active Intelsat is the ninth generation Intelsat 901 satellite, which has been active since June 2001 and provides direct-to-home and internet services. Intelsat 46 (also called Amazonas Nexus), a geostationary satellite, which covers the Americas continent, North Atlantic corridor, and Greenland, was launched in February 2023 on the SpaceX Falcon 9 Block 5 launch vehicle. Intelsat provides global connectivity services through its fleet of GEO, LEO, and MEO satellites, using the C, Ku, and Ka bands; wide beams; spot beams, and frequency reuse technology. Their terrestrial network, Intelsat One, works concurrently with the satellite fleet to provide a hybrid satellite – fiber architecture for resilient cost-effective connectivity. Intelsat plans to deploy software defined network services globally.

EchoStar provides internet services and satellite communications all over the world through Hughes Networks systems. EchoStar has launched nine GEO satellites operating in the Ku (12–18 GHz), Ka (26.5–40 GHz), and S (2–4 GHz) bands, with the earliest satellite launched in 2003. EchoStar launched a satellite in 2017 using the Falcon 9 space vehicle. They provide broadcasting and internet services throughout the world. EchoStar provides enterprise services like data and video transmission for disaster recovery, satellite systems engineering for satellite design, procurement and launch, spectrum management, government services, mobile satellite services, and broadcasting services.

GlobalStar is an organization that maintains Low Earth Orbit communication satellite constellation for low speed data communications and satellite phone communication services. They provide services like asset tracking, emergency and remote communications, one way SMS messaging, geolocation services, and voice call services. In 2022, GlobalStar partnered with Apple to allow iPhone 14 to send SOS messages over satellites. In June 2022, GlobalStar launched its satellite using the Falcon 9 space vehicle.

Inmarsat is a British organization that provides telephone and data services using 14 geostationary satellites which communicate with ground stations.

In May 2023, Inmarsat was acquired by another satellite company named ViaSat. Inmarsat/ViaSat provides aviation, IoT, and maritime connectivity services across the globe. Inmarsat launched Inmarsat-6 satellite in February 2023 on the Falcon 9 rocket.

Starlink is a satellite constellation owned by SpaceX, and it provides internet services with a latency of around 25 ms to over 56 countries. As of early 2024, there were over 4000 small LEO satellites in the Starlink network. Starlink provides a variety of services for residential, business, portable, maritime, aviation, and IoT applications.

OneWeb is an organization that aims to provide satellite broadband services. They planned to have a total of 648 LEO satellites, out of which 544 were launched and 542 were functional as of early 2024. One web provides carrier, enterprise, government, maritime, aviation, and mobility solutions to users.

Amazon Kuiper is a satellite constellation of 3236 LEO satellites to provide high speed and affordable broadband services all over the world. This Amazon project started in 2018 and was granted the FCC license to operate in 2020. They aim to launch 2 prototype satellites by mid-2023 and start serving customers by the end of 2024. In addition, they aim to deploy and operate at least half of their constellation by 2026. Various NTN initiatives are shown in Figure 5.1.

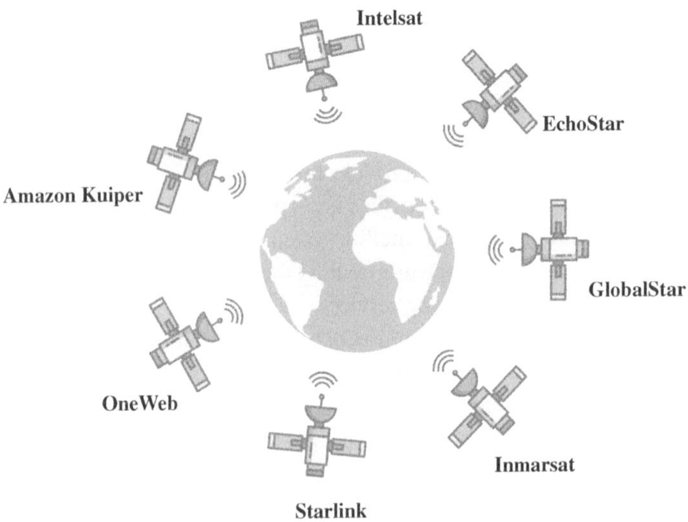

Figure 5.1 Various NTN initiatives.

5.1.4 3GPP NTN Roadmap

The NTN standardization started in the 3GPP Release 15 in 2017, where the study item focused on deployment scenarios and channel models. After the initial study was conducted, the 3GPP system aspects work group focused its efforts in identifying the use cases for NTN, which served as the motivation for non-terrestrial networks. The use case categories were service continuity, ubiquity, and scalability [2].

In the 3GPP Release 16, the RAN working group worked on identifying the requirements to support the NTN in 5G NR, specifically LEO and GEO NTN platforms. Some identified requirements were 5G NR Frequency Division Duplexing (FDD), UEs with Global Navigation Satellite System (GNSS) capabilities, and support for transparent payloads.

The working groups in RAN, WG1 and WG2, discussed challenges like propagation delays, Doppler effects, and non-stationary cells. The solutions to these challenges were identified to be appropriate timing relationships, Hybrid Automatic Repeat Request (HARQ) and uplink synchronization. The RAN WG2 (RAN2) discussed enhancements in the user plane like increasing the timer values, the Random Access Channel (RACH) procedure, and uplink scheduling considering the longer propagation delays in the NTN. In the control plane, mobility management enhancements were discussed. The RAN WG3 (RAN3) focused on architectural enhancements like the feeder link switch, automatic neighbor relation, UE registration, and UE paging. The RAN WG4 (RAN4) focused on RF requirements and radio resource management. The Service and System Aspects (SA) WG1 (SA1) focused on requirements for NTN use cases and NTN service requirements for 5G.

The SA WG2 (SA2) worked on the architecture aspects for NTN platform access in 5G networks. The SA2 identified key challenges for a satellite based NTN in 5G and came up with potential solutions for both direct satellite access and satellite backhaul networks. The work included a study on mobility management, satellite delays, QoS aspects with direct satellite access networks and satellite backhaul networks, and non-GEO regenerative payload, among others.

The SA WG5 (SA5) identified management and orchestration aspects of NTN. The Core Network and Terminals work group studied the 5GC aspects related to NTN. Some work included Public Land Mobile Network (PLMN) selection, narrowband IoT for NTN, and enhanced machine type communications (eMTC) for the NTN [2].

The 3GPP also studied the frequency bands for operation taking into consideration scenarios like operating frequency bands for a GEO-based NTN, LEO based NTN with steerable beams, LEO-based NTN with fixed beams, and MEO based NTN with fixed beams [3].

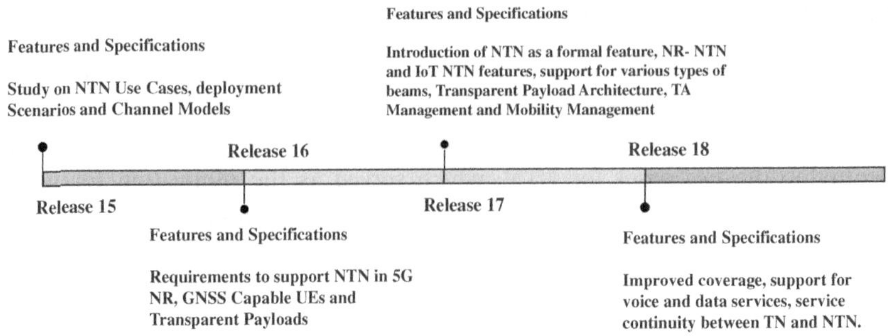

Figure 5.2 NTN advances through multiple releases.

The 3GPP Release 17 introduces Non-Terrestrial Networks as a formal feature to the standard. Release 17 talks about transparent payload architecture and integration of narrowband IoT with NTNs. It also introduces different types of beams supported on NTN Platforms, tracking area management and mobility management in NTNs. Release 17 makes assumptions such as GNSS capable UE, support of IoT features up to Release 16, and support of single carrier and multi carrier operations in narrowband IoT.

In 3GPP Release 18, NTN enhancements include improved coverage and support of voice and data services, enabling service continuity when UE switches between terrestrial networks and NTNs. Release 18 NR-NTN includes support for operating frequencies beyond 10 GHz and UE location verification by the network. Some other IoT-NTN enhancements include coverage, mobility, and performance enhancements compared to Release 17. An overview of NTN evolution throughout the releases is shown in Figure 5.2. Work for Release 19 is expected to start from summer 2024 and continue until late 2025 or early 2026.

5.2 3GPP NTN Architecture

This section discusses the main elements of the NTN systems. Section 5.2.1 talks about the different kinds of NTN platforms, Section 5.2.2 talks about the types of beams used by the NTN platforms, and Section 5.2.3 explains various NTN architectures defined in the 3GPP standard.

5.2.1 Types of NTN Platforms

The 3GPP's vision of the NTN includes utilizing NTN platforms such as GEO satellites, Non-Geosynchronous Orbit (NGSO) satellites, and HAPS. GEO satellites operate at the standard geosynchronous altitude of 35,786 km. NGSO satellites include LEO, MEO, and Highly Elliptical Earth Orbit (HEO) satellites, each with their own altitude. Lastly, HAPS systems are seen as airborne platforms that operate at an atmospheric altitude such as 20 km [1, 2]. The 3GPP-defined NTN also supports air-to-ground systems, where the radio network is on the ground but UEs being served are in the air (e.g., airplanes).

The 3GPP's design approach of the NTN includes several core objectives. First, terrestrial cellular communications systems are susceptible to natural disasters that lead to partial damage or even complete destruction of infrastructure. Terrestrial cellular communications systems are furthermore prone to terrorist attacks and armed conflicts. Second, a cellular system that could better support a fast-moving UE on a high-speed train or an airplane is particularly attractive. Third, most of Earth's surface is water. Expanding coverage beyond Earth's landmass and coastal areas is highly beneficial. Fourth, a satellite-based system can serve as a backup for terrestrial systems, particularly in times of emergency or for critical communications. Lastly, an additional layer of 5G services allows for the network as a whole to accommodate and provide further services [1].

5.2.2 Types of Beams

The NTN includes three types of beams: Earth-fixed beams, Earth-moving beams, and quasi-Earth-fixed beams. Figure 5.3 provides an illustrative example of an Earth fixed beam. The service satellite type is GEO. The GEO orbit maintains its

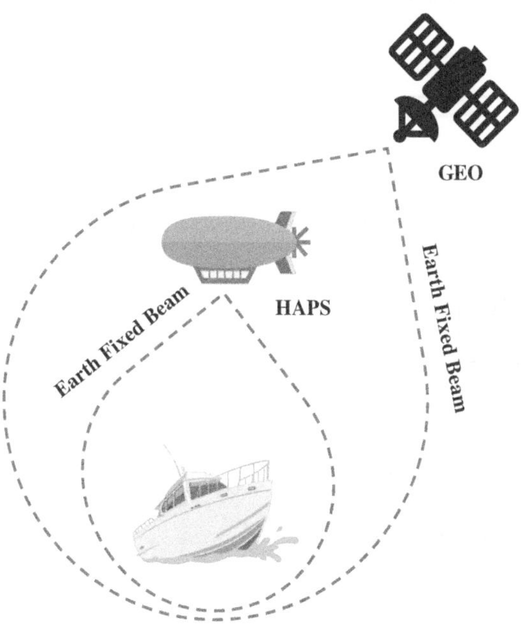

Figure 5.3 Visual representation of an earth-fixed beam.

position over the same portion of Earth, and its beam illuminates the same area of the Earth's surface. A fixed position of the NTN platform (e.g., a GEO satellite or HAPS) enables the NTN platform to serve the same portion of the Earth at all times, which is particularly attractive in remote areas [4].

The second NTN beam type is Earth-moving, which can be seen in the time-lapse of Figure 5.4. The 3GPP standardization efforts have focused on LEO satellites and GEO satellites as two extreme cases. In this example, the beam direction from the LEO satellite is fixed. As the LEO satellite follows its non-stationary orbit, the service beam continuously sweeps across the Earth's surface through time. In other words, the beam orientation remains the same, and the beam illuminates different geographic areas from one instant to the very next instant. Information regarding the movement of the satellite and its position is known as ephemeris information, and a UE may use this information for various network operations, including mobility [4, 5].

The third type of beam in the NTN is quasi-Earth fixed, whose time-lapse representation is shown in Figure 5.5. This scenario is encountered in LEO or non-GEO satellites. In this scenario, the satellite is able to control its beam direction to extend the coverage time of a particular geographic region on the surface. As can be seen, the LEO satellite moves across the sky through time and continues to adjust its beam direction to cover the same geographic region on the surface for a certain duration. In other words, the NTN platform covers the same geographic area for a certain time period, such as the sum of three periods shown in Figure 5.5.

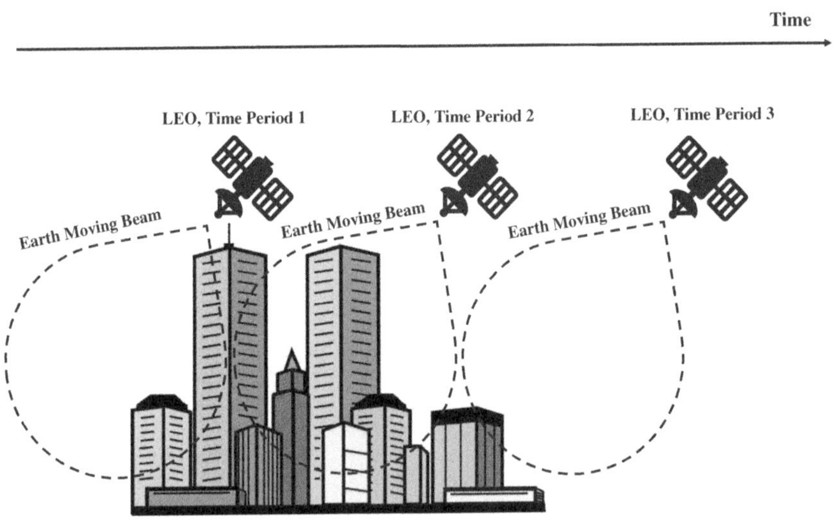

Figure 5.4 Visual representation of an earth-moving beam.

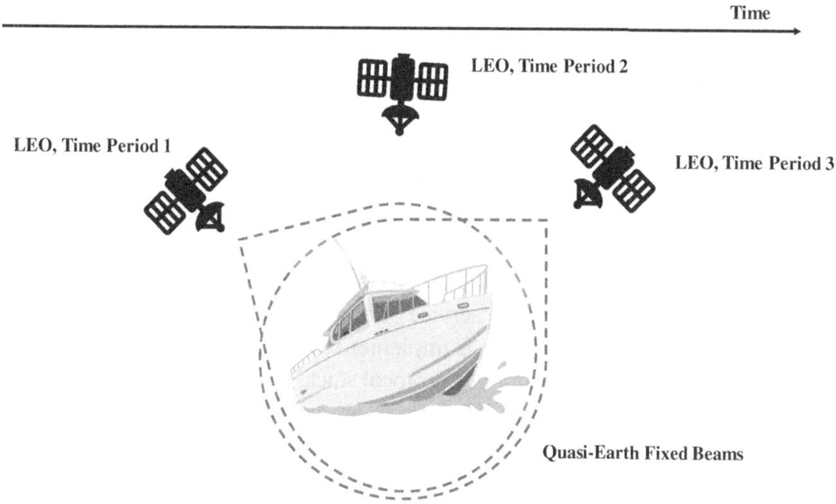

Figure 5.5 Visual representation of a quasi-earth-fixed beam.

The NTN platform can cover a completely different geographic area, similarly in another time window. As with Earth-moving beams, ephemeris information can be used to predict satellite movement and the effective life of tracking areas [4, 5].

5.2.3 NTN Architectures

3GPP defines two types of NTN architectures, transparent (also known as bent pipe) payload and regenerative payload architectures. A transparent NTN payload does not process the baseband signal; it processes RF signals by carrying out functions such as RF filtering, frequency conversion, and amplification. A regenerative NTN payload, in addition to having all the transparent payload capabilities, carries out baseband processing including modulation, demodulation, coding and decoding and can also perform gNB functions like switching, routing, and inter-gNB signaling. The regenerative NTN payload implies a full gNB or part of a gNB on board an NTN platform. The 3GPP studied various types of NTN architectures, including transparent and regenerative payloads, prior to Release 17. As of Release 18, only the transparent payload architecture is supported for the NTN. Section 5.2.3.1 describes the NTN architecture with a transparent payload, while Section 5.2.3.2 illustrates examples of NTN architectures with regenerative NTN payloads.

5.2.3.1 Transparent Payload Architecture

An end-to-end 5G NTN-based system with a transparent NTN payload consists of NTN-capable UEs, transparent NTN payloads residing on NTN platforms (e.g., satellites or HAPSs), terrestrial NTN-Gateways, non-NTN infrastructure gNBs on the ground, the 5G Core network, and the data network. An end-to-end 4G NTN-based system typically uses LTE eNBs and the 4G Evolved Packet Core (EPC). The UE and the NTN payload exchange 5G NR signals using the NR-Uu interface through the service link or the access link. The NTN-Gateway and the NTN payload residing on the NTN platform communicate through the feeder link as shown in Figure 5.6. While the service or access link utilizes the NR air interface protocol stack, the feeder link is implementation-specific and hence follows the operator-determined air interface protocol stack.

The transparent NTN payload functions as an analog RF repeater, a frequency converter, and an amplifier. The user traffic flow in the uplink (i.e., UE-to-network), as shown in Figure 5.6, starts with the UE, which sends information to the transparent NTN payload through the service link. The transparent payload then performs necessary RF processing to the received signal such as RF filtering and frequency conversion and amplification and relays the RF signal to the NTN-Gateway through the feeder link. The NTN-Gateway sends the signal to the non-NTN infrastructure portion of the gNB through an implementation-specific wireline or wireless interface. The gNB transports the

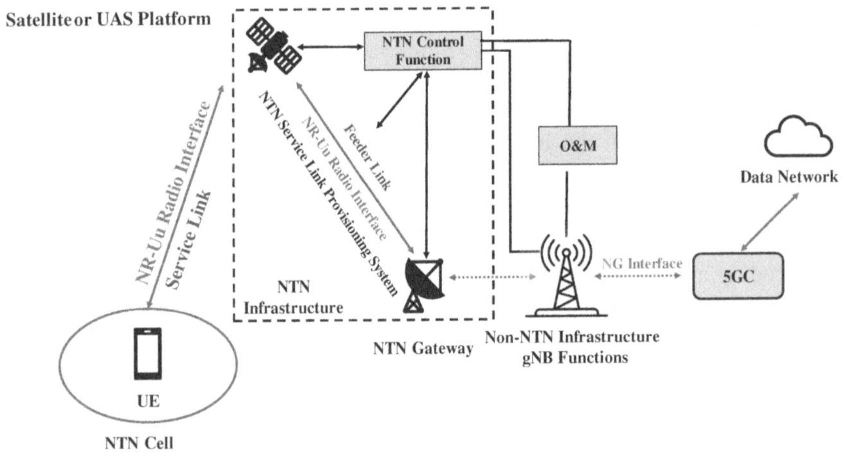

Figure 5.6 An end-to-end NTN-based system architecture with a transparent NTN payload.

UE information such as IP packets to the UPF in the 5GC via the N3 interface. The Non-Access Stratum (NAS) Signaling (e.g., signaling related to session management and mobility management) in the uplink undergoes similar processing. The NAS signaling originates at the UE and is sent on the NR Uu-based service link. The NTN payload relays the received RF signal to the NTN-Gateway after suitable RF processing. The NTN-Gateway sends the signal to the non-NTN infrastructure portion of the gNB through an implementation-specific interface. The gNB conveys the UE's NAS signaling to the AMF using the traditional Next Generation Application Protocol (NGAP) signaling on the N2 interface. Information such as IP packets to and finally to the UPF in the 5GC.

In an example realization of the NTN-based NG-RAN, the NTN gNB is divided into the NTN infrastructure and Non-NTN gNB functions as shown in Figure 5.6 [4]. The NTN infrastructure, part of the NTN gNB, includes the NTN Service Link Provisioning System. The NTN Service Link Provisioning System consists of one or more transparent NTN payloads residing on NTN platforms and one or more NTN-Gateways. The NTN Service Link Provisioning System communicates with the UE over the NR-Uu interface. The NTN Service Link Provisioning System performs mapping between (i) the NR-Uu radio protocol and (ii) the radio resources of the NTN infrastructure such as beams, radio channels, and transmit power.

The NTN Control Function controls the NTN payloads and the radio resources of the entire NTN infrastructure and provides control data such as ephemeris to the non-NTN infrastructure portion of the gNB. The O&M element of the architecture provides different NTN parameters to the gNB based on the type of beams. In case of Earth-fixed beams, the O&M provides to the gNB the cell identifier mapped to the beam and the reference location of the cell, such as the cell center and the range. The O&M provides information such as cell identifier mapped to the beam, cell reference location, time window for feeder link, and service link switch overs. It also provides identifiers and time windows of serving NTN platforms (e.g., satellites) and NTN-Gateways to the gNB in case of quasi-Earth-fixed beams. In case of Earth-moving beams, the O&M provides the gNB with information such as the cell identifier mapped to the beam, beam's direction and motion of the beam's footprint on Earth, elevation/orientation of the beam relative to the NTN payload, schedule of successive serving NTN-Gateways, and schedule of successive feeder link and service link switch overs.

The NTN Service Link Provisioning System and the NTN Control Function constitute the NTN infrastructure. The NTN payload resides on an NTN platform such as a spaceborne or airborne vehicle, which provides the framework of a physical structure, power, commanding, telemetry, attitude control, and a thermal environment including radiation shielding. Figures 5.7 and 5.8 describe the User Plane and the Control Plane of 5G NR NTN protocol architecture. The radio protocol stacks for both the Control Plane and User Plane are similar to 5G-NR.

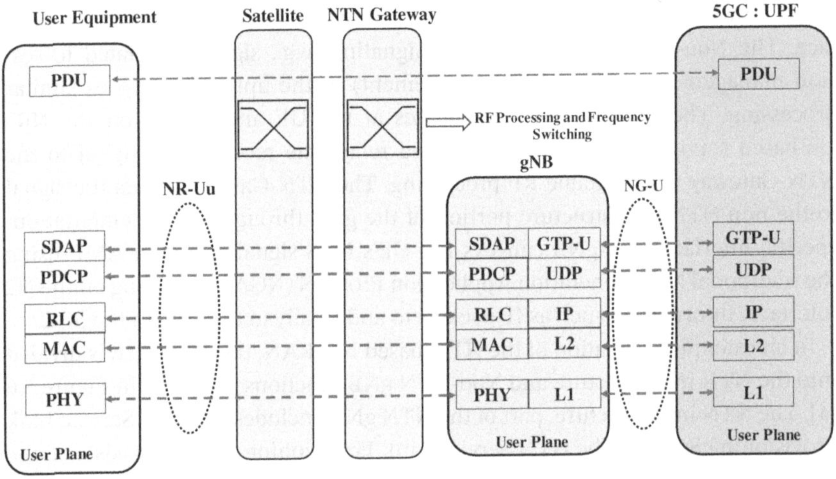

Figure 5.7 User Plane protocol stacks in the transparent NTN payload architecture.

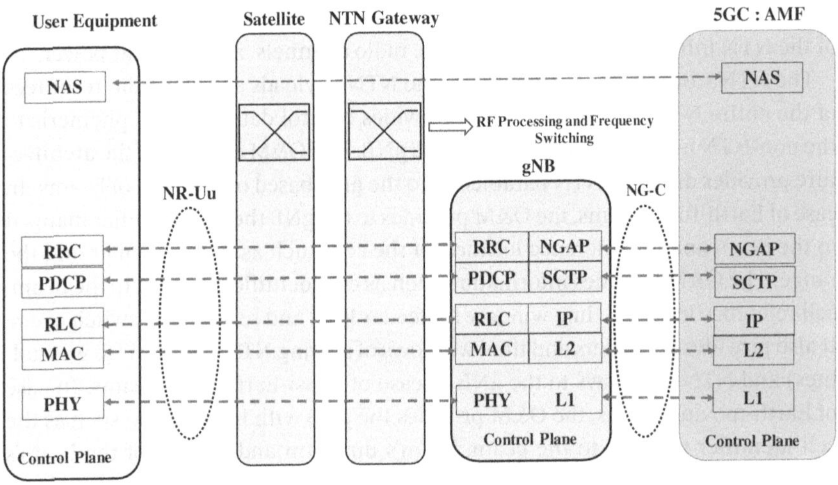

Figure 5.8 Control Plane protocol stacks in the transparent NTN payload architecture.

5.2.3.2 Regenerative NTN Payload Architectures

The end-to-end NTN architecture with a regenerative NTN payload consists of NTN-capable UEs, regenerative NTN payload on an NTN platform, NTN-Gateway, and the core network. The UE and the regenerative NTN payload communicate through the service link via the RF signal associated with the NR Uu interface. The NTN platform and the NTN-Gateway communicate through the feeder link

(NG over Satellite Radio Interface (SRI)). Depending upon the type of the regenerative payload and the overall NTN architecture, two regenerative NTN payloads may communicate with each other through the Inter Satellite Link (ISL) which may either be a radio interface or an optical interface, as shown in the Figure 5.10. For example, when the regenerative payload is a gNB, two gNBs may communicate with each other via the ISL [5]. Two regenerative payload architectures are shown in Figures 5.9 and 5.10.

In Figure 5.9, the regenerative payload functions as a gNB that has the capabilities to perform all the tasks and functions of a gNB. The general uplink traffic flow in a regenerative payload architecture, as shown in Figure 5.10, starts with the UE, which transmits information to the NTN gNB, where the information is processed and transmitted to the NTN-Gateway via the SRI interface on the feeder link. The NTN-Gateway sends the information to the 5GC via a suitable interface that reflects the NG-U interface. If an NTN platform hosts more than one gNB,

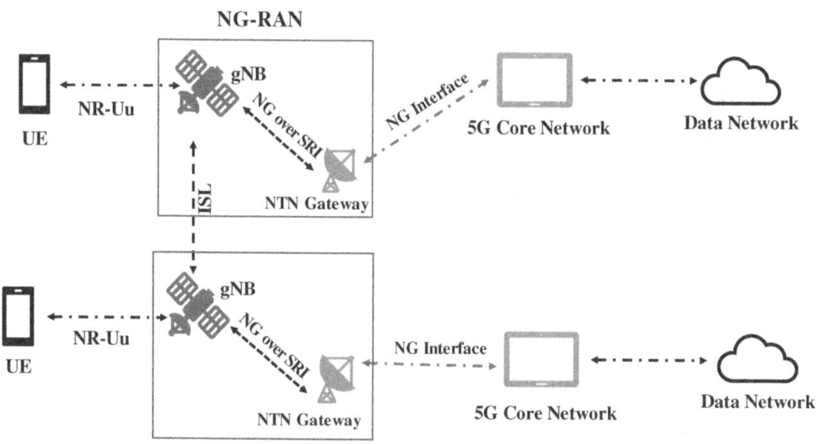

Figure 5.9 NTN architecture with a gNB as the regenerative payload on the NTN platform.

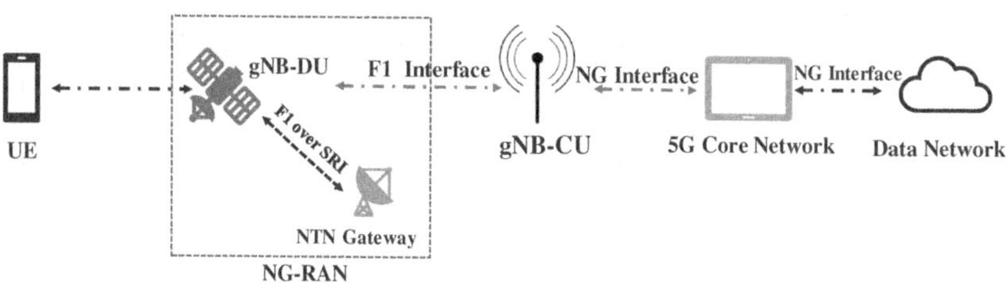

Figure 5.10 NTN architecture with a gNB-DU as the regenerative payload on the NTN platform.

the same NG over SRI interface is used to transmit information to the 5GC via the NTN-Gateway. The User Plane traffic is also transported between the NTN platform and the NTN-Gateway through the SRI interface. The Protocol Data Units (PDUs) are transported between the NTN payload (i.e., the gNB) and the 5GC over the GTP-U tunnels via the NTN-Gateway, similar to the terrestrial 5G network.

There is a possibility of a disaggregated gNB in the NTN architecture with a regenerative payload, where the gNB-DU is on board the NTN platform and the gNB-CU is on the ground, as shown in Figure 5.10. The gNB-DU consists of the lower layers of the NR radio protocol stack, and the gNB-CU contains the upper layer of the protocol stack. Information is transmitted between the UE and the NTN payload/gNB-DU over the NR-Uu interface, and the communications between the NTN payload/gNB-DU and the NTN-Gateway occur via the F1 interface over the SRI. Note that the gNB-DU and the gNB-CU communicate through the F1 interface. In this architecture, more than one gNB-DU onboard an NTN platform can be connected to a single CU on the ground and all the relevant information can be transmitted over SRI with multiple F1 instances.

5.3 Unique NTN Challenges and Associated Candidate Solutions

This section focuses on NTN challenges and various 3GPP solutions to the challenges. Section 5.3.1 discusses unique NTN challenges like delay, coverage, mobility, Doppler Shift among others. Section 5.3.2 discusses 3GPP enhancements to overcome the mentioned challenges and support NTN communications.

5.3.1 NTN-specific Challenges

The architecture and deployment types of the NTN make the NTN quite distinct from traditional terrestrial networks. The differences between the TN and the NTN introduce a number of distinct design challenges that must be addressed to successfully implement an NTN. Figure 5.11 lists the challenges unique to an NTN.

These challenges include long propagation delays due to high NTN platform altitudes, variable propagation delays as NTN platforms move, high and variable Doppler shifts as non-GEO NTN platforms move and the NTN platform elevations change, large coverage areas, mobility of cells that move with a NTN platform, different beam types such as Earth-fixed, quasi-Earth-fixed, or Earth-moving, and the NTN payload type such as regenerative or transparent [1]. Each of these issues is discussed in the following subsections.

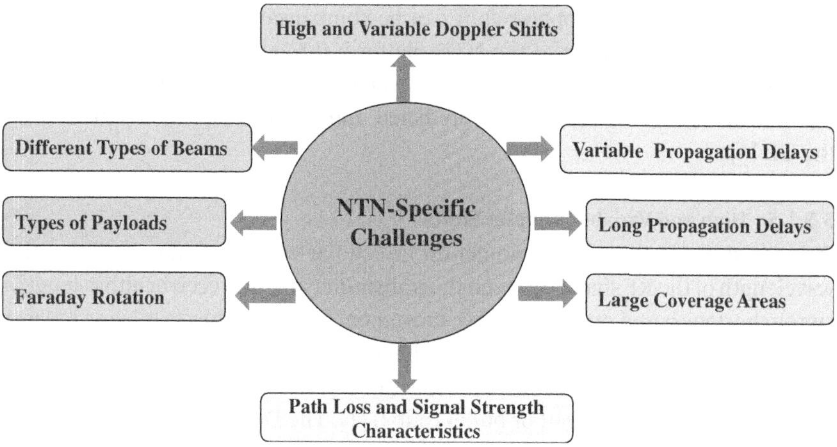

Figure 5.11 NTN-specific challenges.

5.3.1.1 Long Propagation Delays

Foremost among the challenges of implementing an NTN network is the RF propagation delay due to long distances between the Earth's surface and NTN platforms. Since the speed of light is constant, so a satellite based NTN communication network simply cannot compete with a terrestrial network in the overall communication latency. The target latency on the air interface in 5G NR is 1 ms. In contrast, LEO satellites range 30–50 ms, MEO satellites range 125–325 ms, and GEO satellites range 600–800 ms [2]. This delay needs to be accounted for in RACH operations, HARQ procedures, and discontinuous reception procedures, to name a few. Largely, the solution implemented is to include a Common Time Advance into previously defined 5G procedures and account for its value [4].

5.3.1.2 Variable Propagation Delay

When the NTN platform is not stationary, the propagation delay between the NTN UE and the gNB can vary significantly depending on the elevation angle of the NTN platform. If an NTN cell is not Earth-fixed (i.e., quasi-Earth-fixed and Earth-moving), the NTN platform such as a satellite responsible for a given NTN cell communicating with an NTN UE will have different elevation angles toward such UE. Simple trigonometry dictates that the distance between the UE and the NTN platform will change based on the angle between them, which is based on the NTN platform's position in the orbit. Therefore, the propagation delay varies based on the movement of the NTN platform [1]. This delay may exceed one TTI, which means significant timing advance adjustment, is needed

[1, 4]. Fortunately, the Doppler shift can be compensated by the UE by obtaining the speed and position of the NTN platform through the UE's own observations and retrieval of system information via a System Information Block (SIB). Hence, the variable delay may be predicted and subsequently compensated by the UE [6].

5.3.1.3 High and Variable Doppler Shifts

Doppler shift is a known phenomenon in cellular communications, where the wavelength of the RF signal between the transmitter and the receiver either lengthens or shortens based on their relative movement (e.g., two objects moving toward or away from each other). NTN transmissions are also subject to this phenomenon, particularly for non-GEO NTN platforms (and to a less extent with fast-moving UEs such as those on aircraft or bullet trains) [1]. The Doppler shift (or f) is influenced by the relative velocity between the UE and the angle (θ) between the UE and the NTN platform. As with variable propagation delays, if the variables are known in advance, the expected Doppler shift may be precomputed and appropriately pre-compensated.

5.3.1.4 Large Coverage Areas

Through directional antenna beams and high altitudes, NTN platforms can provide radio coverage to large geographic areas on the Earth's surface. There are certain positives to providing large coverage areas. First, larger and sparser regions may be covered without a huge infrastructure costs of radio, transport, and core networks, especially when there is not a viable business case for a terrestrial network providing coverage. Second, for Earth-moving beams and quasi-Earth-fixed beams, mobility signaling requirements may be reduced with a larger coverage area [5]. For moving IoT devices, reducing signaling needs is an important consideration for battery management.

Yet with any design choice, with positives come negatives. With larger coverage areas, paging capacity can be a challenge [5]. For densely populated areas, a large corresponding coverage area would be a poor business choice because the resources available per user would be less, especially when the available radio channel bandwidth is small. Furthermore, if the NTN is meant to provide emergency communications, congestion problems are not a welcome network feature. Additionally, in case of large quasi-Earth-fixed or Earth-moving cells, there would be a need to move a large number of users from the outgoing cell to the incoming cells. A large NTN cell implies more variations in the propagation delays for cell-center UEs and cell-edge UEs, requiring additional adjustments in timing synchronization.

5.3.1.5 Mobility of Cells

Moving NTN cells are a major departure from traditional, terrestrial networks. Suddenly, with Earth-moving and quasi-Earth fixed beams, the serving cell itself will change, even if the UE has not moved! This new paradigm requires design changes to mobility management procedures. Ephemeris information, including NTN platform positioning and trajectory information, needs to be indicated to the UE. Furthermore, neighboring cell information should be shared, for handover based on timing and known NTN platform trajectories. Adjustment to measurement gaps is also required, and RRC periods should be managed based on expected timing of coverage changes [4, 6].

5.3.1.6 Different Types of Beams

With three beam types (see Figures 5.3–5.5) and moving cells, beam management is an important topic within the NTN. Initial efforts aim to use Rel-15/Rel-16 operations. Ephemeris information along with the UE-determined position and velocity of the UE will provide the needed variables for the terminal to compensate for the radio environment's effects based on the beam type [5]. Different types of beams have different effects on mobility challenges. For Earth-fixed beams in a GEO-based NTN, NTN cell mobility is irrelevant and signal strength-based handover criterion would be inadequate. The location-based handover would be more significant for a GEO-based NTN. For quasi-Earth-fixed beams and Earth-moving beams, there would be a need to consider beam or cell mobility for effective mobility management. Furthermore, a large number of UEs would need to be moved from one cell to another. In a terrestrial network cell, Tracking Areas (TAs) are fixed and have fixed relationships with cell IDs. However, if a TA is associated with an NTN cell, mobility of cells in quasi-Earth-fixed and Earth-moving beams or cells would result in a TA change, leading to a tsunami of TA update related signaling.

5.3.1.7 Types of Payloads

As illustrated earlier, the NTN supports two different platform types: transparent (or bent pipe) and regenerative. Each payload type offers different levels of functionality and poses different challenges. The gNB is located on the ground in case of the NTN with a transparent NTN payload. Hence, the gNB-to-5GC connectivity in an NTN can be designed by reusing the TN architectural principles. Since an NTN platform may lose the sight of an NTN-Gateway due to the mobility of the NTN platform, a change in the feeder link is required A transparent NTN payload performs RF filtering, frequency conversion, and amplification and can be viewed as just a repeater at a very high altitude [5]. In some ways, the transparent NTN payload can simplify the design and operations. For example, ISLs are not needed or critical for the transparent payload. The NTN platform design prioritizes cost-efficiency by minimizing components susceptible to technical issues in the absence of on-site technicians, while ensuring the continuity of coverage. Keeping the gNB functionality farther away from the UE only adds delays for time-sensitive processing (e.g., HARQ and Automatic Repeat Request (ARQ) retransmissions and scheduling based on channel conditions for HAPS and LEOs), as a signal needs to travel from surface to surface through the NTN platform and other entities.

In contrast, the regenerative payload functionality can utilize a gNB on the NTN platform. RAN functionalities including modulation, coding, ARQ, HARQ, switching, and routing are brought closer to the UE [5]. This approach reduces timing requirements for network functions and control loops. Since the gNB (or gNB-DU) is placed on the NTN platform, the regenerative NTN payload is more complicated, expensive, and difficult to maintain compared to the transparent NTN payload. ISLs would be an important part of the NTN with a regenerative platform because multi-hop NTN platform links with ISLs can be used to provide coverage (e.g., middle-of-the-ocean coverage) even when land mass is unavailable to locate an NTN-Gateway on the ground.

5.3.1.8 Path Loss and Signal Strength Characteristics

While the propagation path loss in an NTN is typically free-space path loss (corresponding to the propagation path loss exponent n = 2 instead of n = 3.5 or so in a terrestrial network), long distances result in huge propagation path losses. Table 5.1 summarizes examples of path losses for different scenarios. The terms max and min are used to describe the minimum and maximum satellite distances from the Earth's surface.

Such large path losses place a significant constraint on the achievable signal strength and Signal-to-Noise-and-Interference Ratio (SINR) at the NTN platform. Furthermore, the signal strength variations are different in a terrestrial network and an NTN. Figure 5.12 illustrates expected signal strengths at different distances

Table 5.1 Free space path loss of various NTN platforms.

NTN Platform	Distance between NTN Platform and Earth Surface (km)	Free Space Path Loss (dB)
HAPS-Max	50	133.27
LEO-min	600	154.85
LEO-max	1500	162.81
MEO-min	7000	176.19
MEO-max	20000	185.31
GEO	35786	190.36

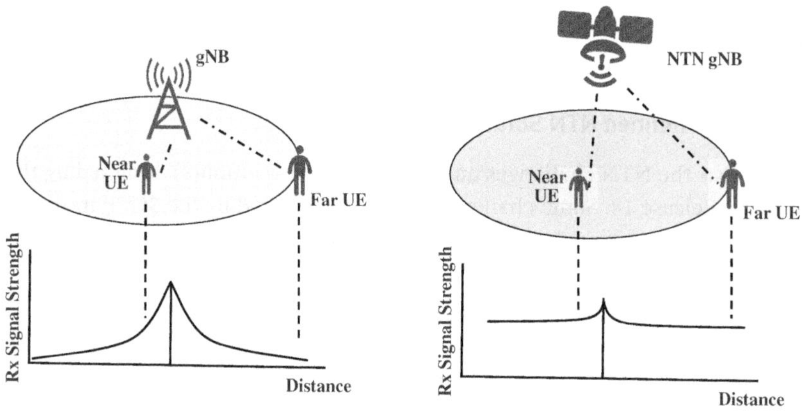

Figure 5.12 Signal strength characteristics.

from the center of an NTN cell. The left portion of the figure depicts a terrestrial network and the variations in the received signal strength as a user moves further away from the gNB. The user closer to the gNB has a very high received signal strength, and the user at the edge of the cell has a very low received signal strength. The right portion of the figure depicts an NTN system and the signal strength variations experienced by users at different positions in the cell. The user in the center of the cell will experience a similar signal strength as the user at the edge of the cell.

While a terrestrial network cell has wide variations in signal strength near the cell center and the cell edge, an NTN cell has only small variations in signal strength, even though NTN cells are quite large. Such lack of signal strength distinction between the cell center and the cell edge in an NTN requires non-signal strength criteria for effective mobility management mechanisms such as cell re-selection and handover.

5.3.1.9 Faraday Effect

The RF signals in NTN systems travel between the terrestrial components (UE or NTN-Gateway) and NTN Platforms. During this journey, the RF signal passes through different layers of the atmosphere. The ionosphere contains electron clouds because of it being ionized by solar radiation. These electron clouds will have deteriorating effects on the RF signal, and if the polarization of the RF signal changes due to the ionospheric effects, it is called Faraday Rotation.

The 3GPP defines Faraday Rotation as the rotation of the polarization of an RF signal when it interacts with the ionic medium in the Earth's magnetic field in its path [1]. Generally, the frequency of the RF signal is inversely proportional to Faraday Rotation. RF signals with higher frequencies are less prone to these ionospheric effects.

5.3.2 3GPP-defined NTN Solutions

Table 5.2 lists the NTN challenges and the associated solution(s) designed by the 3GPP as of Release 18. Some challenges may be addressed by the 3GPP in future releases. The solutions are briefly discussed in the following subsections.

Table 5.2 NTN challenges and 3GPP 5G NTN solutions.

Challenge ID	Challenge	Potential 3GPP NTN Solution(s)
1	Long propagation delays	Synchronization enhancements, timing advance enhancements, protocol timer adjustments, HARQ enhancements, timing relationship enhancements, QoS enhancements
2	Large variations in propagation delays	Synchronization enhancements, timing advance enhancements, protocol timer adjustments, HARQ enhancements, timing relationship enhancements, QoS enhancements
3	Large distances between the device and the NTN platform	Implementation-specific configurations of suitable high-gain antennas and high power amplifiers on the NTN platform
4	Large Doppler shifts and Doppler rates	Enhanced System Information, GNSS-capable UEs, Frequency pre-compensation
5	Large cells or coverage areas (Large differential delays for different UEs)	RACH enhancements, User location reporting
6	NTN cell mobility (handover and cell reselection needed even for stationary users)	Enhanced System Information, enhanced mobility triggers, enhanced Tracking Area management, feeder link switch management
7	3 Types of Beams	Enhanced System Information, enhanced mobility triggers, enhanced Tracking Area management
8	Types of Payloads	Support for transparent payloads in R17 and R18; potential support for regenerative payloads in a future 3GPP release
9	Faraday Rotation	Circular polarization

5.3.2.1 Synchronization Enhancements

Due to the long propagation delays in NTNs when compared to terrestrial networks, enhancements in synchronization procedure is required. The NTN uses the same synchronization channels and signals as terrestrial networks. To help UE identify the radio frame boundary and the cell ID, synchronization signals like Primary Synchronization Signal (PSS) and Secondary Synchronization Signal (SSS) are used. Every radio cell in 5G is identified by a cell ID which is arranged into 336 cell ID groups and 3 identities per cell group, making it a total of 1008 cell IDs. A given cell or sector is identified by the UE through the PSS ID and the group ID associated with the SSS. The PSS and SSS are packed together with the Physical Broadcast Channel (PBCH) and the entire block is called a Synchronization Signal Block (SSB).

Initially, to search for a cell, the UE uses PSS and SSS for frequency and time offset estimation and correction. The UE first decodes the PSS to identify the cell ID sector (i.e., 0, 1, or 2) and then decodes the SSS to identify the cell group ID (i.e., 0 to 335). Finally, it computes the serving cell ID (N_{ID}^{Cell}) in the following equation [7]:

$$N_{ID}^{Cell} = 3.N_{ID}^{(1)} + N_{ID}^{(2)} \tag{5.1}$$

where $N_{ID}^{(1)}$ is the ID of the cell group and $N_{ID}^{(2)}$ is the ID of the cell in the group.

The gNB periodically transmits an SSB burst consisting of multiple SSBs by using beam sweeping, with each SSB transmitted through a specific beam and at a predefined time. After the UE identifies the serving cell based on the PSS and the SSS in a given beam, it learns about the serving beam index corresponding to the SSB. The Demodulation Reference Signal (DMRS) is a function of the SSB index and when the UE decodes the SSB including the associated PBCH, it retrieves the master information block (MIB). The SSS and the PBCH DMRS can be used by the UE to determine the Reference Signal Received Power (RSRP) and Reference Signal Receive Quality (RSRQ) of the SSB and finally select a suitable serving beam [8]. While the NTN reuses synchronization signals defined for the terrestrial network, the NTN UE receiver would need to enhance its synchronization signal search procedure to be able to detect such signals in the NTN.

Currently, 5G uses Channel State Information-Reference Signal (CSI-RS) and SSB-based measurements for handover. SS/PBCH Measurement Timing Configuration (SMTC) is an SSB/PBCH Block based Radio Resource Management (RRM) measurement timing configuration window. It is configured by the network for the UE based on the expected propagation delays between the device and the neighbor cells (e.g., cells of an incoming satellite). The SMTC configurations associated with a frequency can be configured and mapped to a group of cells in the 3GPP Release 17.

5.3.2.2 Timing Advance Enhancements

Timing advance ensures all uplink transmissions from various UEs in a cell are in sync when they are received by the base station. As shown in Figure 5.13, a UE closer to the base station will have shorter propagation delay and thus a smaller timing advance. A UE that is away from the base station would have a longer propagation delay and thus a larger timing advance. The timing advance for every active UE is calculated by the gNB, initially aligned during the RACH procedure, and adjusted as needed during the active radio connection. Specifically, after the initial radio connection setup, the timing advance is monitored and updated by the MAC layer as the need arises; the gNB sends a MAC layer timing advance command to the UE. These differing timing advances for the UEs will ensure that the uplink radio frames from UEs in the cell have an identical time alignment at the base station. When the timing advance is calculated for a UE, the propagation delay for both the uplink and downlink is taken into consideration by the gNB. There is an additional (optional) parameter called Toffset, which can be used by the gNB to establish the timing relationship between the downlink radio frame and the uplink radio frame. Toffset can thus be viewed as the time difference between the uplink frame and its corresponding downlink frame.

The long propagation times in NTNs affect the timing advance procedures, specifically the RACH procedure, during which the timing advance values are determined for UEs and would need appropriate adjustments. Due to these longer delays, there is a high probability that the random access procedure attempts would fail in an NTN with current terrestrial delay limits. NTN systems would

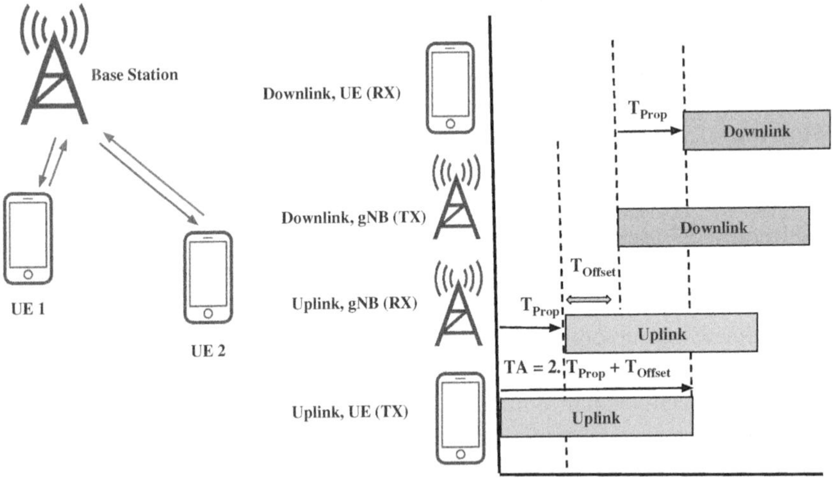

Figure 5.13 Timing advance in terrestrial networks.

need to support longer timing advance values to tackle longer delays. According to the 3GPP standard [9],

$$Total_{timingadvance} = N_{TA} + N_{TA,offset} + N_{common(TA,adj)} + N_{UE(TA,adj)} \tag{5.2}$$

where the parameters N_{TA} is 0 for PRACH and updated by base station during RACH procedure, N_{UE} is the a priori estimation of TA by the UE, N_{common} is the common network timing advance and N_{offset} is the fixed offset delay used in 5G terrestrial networks.

Broadly, the steps that occur during the TA procedure start with the UE establishing its position, the NT platform broadcasting information like its orbital position and its velocity, among others, through SIB. The UE then calculates the propagation delay with the information it has from the previous two steps, and finally the UE deriving the initial timing advance value for initial radio access.

5.3.2.3 Timing Relationship Enhancements

To address the long Round Trip Time (RTT) challenge in NTN, there are two optional parameters defined to solve the misalignment issues in uplink and downlink channels, namely Koffset and KMAC . The Koffset parameter is used to tackle the delay aspects in NTN, and it has gNB-specific and UE-specific components. The K offset timer is used to tackle delays in Physical Uplink Shared Channel (PUSCH) Downlink Control Information (DCI) scheduling, acfRAR procedure in RACH, and HARQ feedback in PUSCH. KMAC is a timing alignment parameter to address the misalignment issue in the uplink and downlink. KMAC refers to a reaction time, where the transmitter sends a command to be executed, the receiver needs to react within a certain time frame.

To maintain a successful radio link between the UE and the gNB, the UE needs to report control information like RSRP, Channel Quality Indicator (CQI), Channel State Information (CSI), or rank indicator to support efficient functioning of mechanisms like Multiple-Input Multiple-Output (MIMO), beamforming, and modulation and coding schemes. To support the UE in long-delay NTN situations, the network provides the UE with an additional timing offset parameter (K_{offset}).

Similar to the UE and the RAN, the 5GC network functions will need to cope with the long delays in NTNs. As a result, the session management and mobility management function timers would need longer times as compared to the present values. To make sure the changes to 5G NR is kept at a minimum, a new Radio Access Technology (RAT) type called the NTN RAT was introduced to the AMF. Extended timers are defined in the core functions and NTN-UE in this RAT type to handle longer delays.

Table 5.3 explains the current timer values and suggested minimum values of the timers in the worst case scenarios (worst-case round trip time) in these core network related timers.

Table 5.3 5GC-related timer enhancements for the NTN.

Timer Name	Timer Number	RTT Propagation Delay	Minimum Suggested Increase
5GMM –UE Timer	T3510	5 RTT	5 WCRTT
5GMM –UE Timer	T3517	3 RTT	3 WCRTT
5G SM UE Timer	T3580	4 RTT	4 WCRTT
5G SM UE Timer	T3581	1 RTT	1 WCRTT
5G SM UE Timer	T3582	1 RTT	1 WCRTT

Table 5.4 NR protocol timer durations in the NTN.

Protocol Layer	Timer Name	Duration
RLC	RLC t-PollRetransmit Timer	4000 ms
RLC	RLC t-Reassembly Timer	2200 ms
PDCP	PDCP discard timer	3000 ms

5.3.2.4 Adjustments in Protocol Timers

As a result of increased round trip time delays in an NTN, enhancements in Radio Link Control (RLC) and Packet Data Convergence Protocol (PDCP) layers are required, especially RLC and PDCP layer associated timers, to extend support to NTNs while simultaneously supporting 5G terrestrial networks. Table 5.4 shows the updated protocol timer durations for NTN systems.

Radio Link Control (RLC) layer in 5G supports three transport modes, transparent mode, unacknowledged mode, and acknowledged mode. These transport modes transport upper layer service data units (SDU) to the MAC layer. The MAC layer furthermore transports the Service Data Unit (SDU) to the physical layer for transmission. Unacknowledged and acknowledged modes both add an RLC header to the packet received from the upper layer.

One of the fields in the RLC header is the polling field, which is used to trigger a status report from the RLC peer. The *RLC t-PollRetransmit Timer* is used by the RLC transmitter for retransmitting the poll. The timer starts once the poll is transmitted and once the timer expires, the poll is retransmitted. The maximum configurable time is 4000 ms, which is sufficient for RTT in GEO NTNs. Segmentation indicator is another field in the RLC header which indicates if a specific PDU is a part of an RLC header or not and if the PDU is a segment, it indicates if it is at the beginning, the middle, or at the end of a packet. The *RLC t-Reassembly Timer* is used by the RLC (acknowledged mode) at the receiver to detect loss of RLC PDUs at lower layers. The timer begins once the PDU is received from the lower layer (MAC), placed in the buffer (at the receiver) and at least one byte is missing from the received PDU.

Additionally, the RLC protocol checks if the timer has already started running before it begins. This timer is configured to account for the arrival of individual segments of a PDU at the receiver after segmentation or HARQ. Since HARQ is supported in NTN, the *RLC t-Reassembly timer* should now at least support maximum time for HARQ transmissions and retransmissions, which is larger than the RTT in NTNs. Considering the maximum RTT of NTNs, 3GPP extends the timer from 200 ms to 2200 ms [10].

The PDCP layer maps upper layer protocol packets (e.g., IP packets or Ethernet frames) into the packets suitable for being transmitted over the air. Some other key functionalities of the PDCPlayer include header compression, duplicate data detection and authentication among others. Similar to the RLC layer timers, the PDCP timers also expire due to the long RTT in NTN. The *PDCP discard timer* starts at the transmitter when the SDU is received at the receiver PDCP. When the timer expires, or when the SDU is successfully delivered, the transmitter discards the PDCP SDU. In terrestrial networks, the *PDCP discard timer* can be set between 0.5 ms and 1500 ms or can optionally be switched off and be set to infinity based on specific applications and services. To support the NTN, this timer was extended to 2000 ms. The *PDCP t-reordering timer* is used to detect loss of data PDU, and this timer starts or resets when the SDU is delivered to the upper layers with a maximum expiration time of 3000 ms.

5.3.2.5 HARQ Enhancements

Procedures like ARQ, Forward Error Correction (FEC), and HARQ are used for error detection and correction at the receiver during transmission. In ARQ, the receiver detects the error in a data packet, discards it, and requests retransmission of the packet from the receiver. Forward error correction detects and corrects the error in the packet. The HARQ procedure detects the erroneous packet, stores the erroneous packet in the buffer for combining with future retransmissions, and requests retransmission of the erroneous packet. Then, the HARQ uses all the versions of the transmitted packet to correct any error(s) in the packet.

Generally, when the transmitter transmits information bits, it adds Cyclic Redundancy Check (CRC) bits at the end of the information bit stream at the PHY layer to help the receiver detect errors. The entire information bit stream along with the CRC bits are encoded by the FEC encoder. When the receiver receives this data, it first decodes the information bit stream, followed by the CRC check, which would allow the receiver to detect if there are any errors in the data stream.

If the received data stream has errors, the error data is stored in the buffer and a retransmission is requested. If the retransmitted data also has errors, it is also stored in the buffer along with previously received transmissions of the same packet, and another retransmission request is made. In HARQ, all the related erroneous data stored in the buffer are used to decode the original data sent by the transmitter accurately. Two popular HARQ combining techniques include chase combining HARQ and incremental redundancy HARQ. In chase combining HARQ, the transmitter sends identical information to the receiver for each retransmission request. Thus, the receiver will have *n* copies of the same information to decode the erroneous data accurately. In incremental redundancy HARQ, when an error is detected in the original data, and a retransmission request is sent, the transmitter would only send a fraction of the code symbols in addition to the original CRC bits. If the receiver is unable to decode the data accurately, in the second retransmission the transmitter includes a different set of code symbols (and CRC bits) and sends it to the receiver. This process happens until the receiver is able to accurately decode the originally transmitted data without errors or when the maximum number of configured retransmissions is reached. This procedure allows the receiver to have more diverse information for decoding and correcting the errors in the data. 5G uses adaptive and asynchronous HARQ for error detection and correction and supports up to 16 parallel HARQ processes. Since HARQ is inherently a stop and wait protocol, there is often underutilization of radio resources. To prevent this underutilization of resources, multiple HARQ processes are carried out in parallel, with one process responsible for transmission of one packet and another process responsible for another packet in a given period.

Due to the increase in delays for NTNs, enhancements to the pre-NTN HARQ parameters are required. Examples of HARQ enhancements for the NTN include (i) increasing the amount of parallel HARQ process per UE from 16 to 32, (ii) configuring the RLC layer to optionally disable the HARQ feedback process, which would allow the transmitter to transmit data without waiting for feedback, and (iii) enabling blind retransmissions [10].

5.3.2.6 QoS Enhancements

5G QoS model is based on QoS flows which either have a guaranteed bit rate or a non-guaranteed bit rate. A QoS Flow ID identifies the QoS flow in a 5G system. User traffic with the same flow ID in a PDU session have identical traffic routing and forwarding procedures. The QoS flow is controlled by the SMF in the 5GC, and the QoS flow is either pre-configured at the SMF, created during establishing the PDU session or modified during the PDU session. A QoS flow is characterized by a QoS profile, one or more QoS rules, QoS parameters and one or more packet detection rules defined for uplink and downlink by the SMF.

The QoS profile for a specific QoS flow would include a number of QoS parameters like the 5G QoS identifier and allocation and retention priority. The 5G QoS identifier is a scalar value that maps to a list of 5G QoS characteristics such as the resource type (i.e., guaranteed bit rate, non-guaranteed bit rate, and delay-critical guaranteed bit rate), default priority level. These QoS characteristics relate to prioritizing scheduling resources among various QoS flows, and the lowest priority level would equate to having the highest priority. The packet delay budget provides the maximum acceptable delay between the UE and the UPF. The Packet Error Rate (PER) provides the maximum acceptable rate of PDUs which were processed by the link layer of the sender but could not be delivered by the receiver to the upper layer. Default maximum data burst volume, and default averaging window defines the duration in which the guaranteed flow bit rate and maximum flow bit rate can be calculated.

3GPP reuses the existing terrestrial 5G QoS service model in NTN and introduces an additional 5G QoS indicator. In case of GEO satellites, where the packet delay is significantly larger, a new 5G is introduced with a packet delay budget of 1100 ms. The 3GPP, up to at least Release 18, expects non-GEO satellites and HAPS to reuse 5G QoS indicators defined for terrestrial networks.

5.3.2.7 Enhanced System Information

5G NR system information consists of the Master Information Block (MIB) and several System Information Blocks (SIBs). The MIB is critical system information that is broadcast by the gNB periodically, and the UE obtains the MIB information while decoding the PBCH as discussed earlier in the initial cell selection procedure . SIBs contain information which help the UE perform various tasks like establishing a radio connection with a base station, performing cell- reselection procedures, and inter-RAT cell selection procedures among others. The system information is divided into minimum system information and other system information. The minimum system information contains MIB and SIB1, which carry the necessary information for initial access. Other system information contains all the SIBs that are not broadcast in the minimum system information. The other system information may or may not be broadcast in 5G NR. The other system information can be provided on demand by the gNB to the UE whenever the UE requests specific SIBs.

The SIB1 contains an additional information field pertaining to NTN in addition to 5G NR system information. The *cellBarredNTN* field has two values, *barred* or *notBarred*. The value of *Barred* means that the UE is barred from accessing this cell to access the network and services via an NTN. The value of *notBarred* means that the UE is allowed to connect to this cell that is associated with the NTN. If the information corresponding to the field is not available, the UE considers that this cell is not connected to the NTN. The parameter*cellBarredNTN* is applicable to NTN UEs.

SIB 19 focuses on conveying NTN related assistance information to NTN supported UEs. It consists of information fields like (i) *distanceThresh*, which indicates the distance of the UE from the serving cell reference position, (ii) *ntn-config*, which contains required parameters for the UE to access 5G via NTN platforms, (iii) *ntn-NeighCellConfigList*, which contains a list of neighboring cells and their ntn-config parameters, carrier frequency and PhyCellId which is the cell ID of the base station, (iv) *referenceLocation*, which provides the reference location of the serving cell and (v) *t-Service*, which mentions the information when a serving cell stops serving the area served by the NTN platform. It also contains parameters like timing advance, K_{offset}, and ephemeris data, among others [11].

5.3.2.8 RACH Enhancements

A major challenge in the RACH procedure in an NTN is long round trip times. These long trip times have a negative impact on both the base station and the UE. Figure 5.14 shows a generic RACH procedure in 5G NR. From the base station perspective, the gNB broadcasts time windows of when the UE can send Physical Random Access Channel (PRACH) sequences. However, due to the non-stationary characteristic of the NTN platforms with respect to the UE, the reception of PRACH sequences at the gNB varies from one UE to another UE. This effect is visible at the gNB as a floating RACH time window. As a result, the gNB will have to increase the PRACH reception window time. Additionally, since the NTN platforms have a large coverage area and serve a large number of UEs, it may lead to RACH collisions and congestion of the channel.

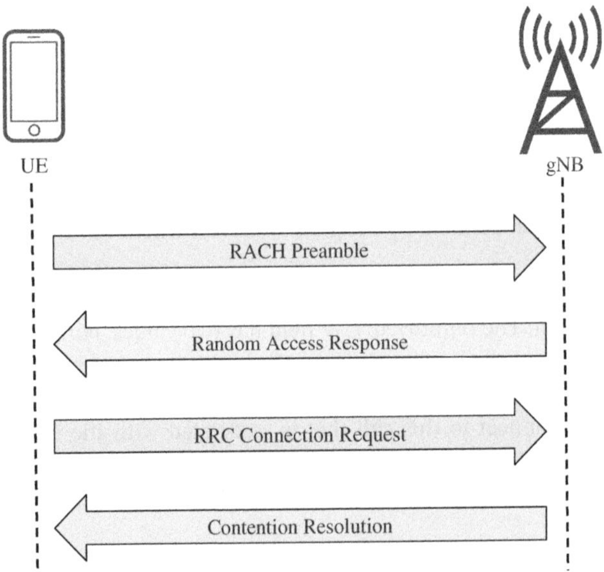

Figure 5.14 RACH procedure in 5G NR.

The second challenge from the gNB perspective is the PRACH ambiguity. Due to large propagation delays for different UEs in the cell, the PRACH preamble ambiguity can occur at the gNB.

From a UE perspective, after the UE transmits the PRACH preamble, it waits for the gNB to send a Random Access Response (RAR) message within a preconfigured time window. However, in certain scenarios, the UE may not receive the RAR within that specific window because of the long RTT in NTN and hence may retransmit the PRACH preamble. TS 38.213 proposes that the RAR reception time window begin after an additional ($TA + KMAC$) milliseconds time frame, where TA is the timing advance parameter and $KMAC$ is an optional parameter defined in the higher layers of the protocol stack.

The second challenge from a UE perspective is the contention resolution procedure after the UE sends the Radio Resource Control (RRC) connection request. After the RRC connection request, the UE monitors the contention resolution message for a specific time frame via a timer called the *ra-ContentionResolutionTimer*. Due to the long RTT, the timer may expire before the UE receives the contention resolution message from the base station and may end up not completing the procedure. The proposed solution to this challenge is either adding an offset before the beginning of the timer or extending the timer duration [10].

5.3.2.9 Feeder Link Switch Management

As mentioned earlier, the radio link between the NTN-Gateway and the NTN platform is called a feeder link. There are scenarios like moving NTN platforms, traffic offloading, capacity management, and maintenance where there is a need to switch between feeder links. There are many challenges that arise due to the feeder link switch, such as (i) transferring all the connected users from one NTN platform to another, (ii) transitioning between one gNB to another, and (iii) changing the serving AMFs. To solve this challenge, the 3GPP introduced some procedures, namely hard and soft feeder link switch.

In a soft feeder link switch, the NTN is temporarily connected to more than one NTN-Gateway at the same time when it transitions between one feeder link to another [4]. During this time, the upper layers of the core network would take care of issues like packet loss and duplication.

In a hard feeder link switch, the NTN is only connected to one terrestrial NTN-Gateway at a time. This would lead to interruption in radio links during the NTN platform's transition from one feeder link to the next. However, with network assistance, where the network provides UE with the supporting information and helps in RRC reestablishment similar to network assisted cell change procedure in terrestrial networks, it is possible to reduce connectivity interruption times [10].

5.3.2.10 GNSS-capable UEs

According to 3GPP, a UE which supports NTN services is required to be GNSS-capable and has the ability to know its own location [4]. The GNSS capability enables the UE to pre-compensate for propagation delays and Doppler shifts to align itself to the NTN platform. The UE the knowledge of its own location along with system information block (SIB19, *ntn-config field*), which provides the broadcast NTN platform ephemeris to determine its frequency and time pre-compensation. Having GNSS capability facilitates radio link synchronization. The GNSS-based location provided by the UE to the network can be used by the network to determine the correct PLMN based on the mobile country code of the county the UE is currently located in. If the UE's registration request is rejected by the 5GC because of its location, the UE can attempt to register to another PLMN.

5.3.2.11 Inter-country Mobility Management, Coarse User Location Reporting, and UE Location Verification

Terrestrial networks are designed to serve users in the country they are located in, thereby adhering to the compliance requirements. NTNs may cover multiple areas or regions at the same time. One area of concern would be when an NTN platform serves multiple countries and the NTN platform could connect to users from either countries. From a user perspective, a user from country X could connect and have access to services provided in country Y. Now, the question arises on how to follow compliance requirements of all the regions that the NTN platform is serving.

One proposed solution is using the AMF in the core network as a verification anchor. The AMF would check if the UE is located in the area it is allowed to serve. In this scenario, the RAN is connected to various AMF's which serve their respective regions and would assist the AMF's in verification and connection procedures. In case the NG-RAN detects that a user from country X is connected to an AMF serving country, it uses the NAS node selection function. Using the NAS node selection function, AMF determines the appropriate AMF to serve a specific user by considering information like slicing information, onboarding indication and UE location information. This procedure is executed by the RAN as it is pre-configured to make sure the serving AMF of the user is situated in the region of the user.

In Release 17, the UE reports its coarse GNSS-based location to the network to address the privacy concerns associated with a high-resolution UE location. One of the Release 18 studies focuses on the network verification of the UE-reported location [12].

5.3.2.12 Enhanced Mobility Triggers

In terrestrial networks, handover between two gNBs occur based on measurements such as RSRP and RSRQ. In a typical 5G NR commercial implementation, handover between two cells in a terrestrial network essentially occurs when the neighbor cell RSRP exceeds the RSRP of the serving cell by a certain amount (which is a function of several parameters, including a threshold). Such handover is based on so-called an A3 event trigger. Since RSRP measurements are primarily a function of the free-space path loss in an NTN. Since the RSRP difference between two NTN platforms as observed by a cell-edge UE may not be significant, RSRP-based handover may not work well in an NTN. Hence, RSRP cannot be the sole indicator for RRM mobility procedure for seamless mobility [10].

This RSRP related challenge serves as a motivation for introducing additional measurement triggers beyond signal strength-based measurements. While these new triggers can be used with traditional handover and conditional handover, the 3GPP focused more on conditional handover in Release 17. When conditional handover was initially introduced for a terrestrial network, the set of conditions for triggering the handover were A3 (neighbor cell RSRP better than the combined RSRP of serving cell and hysteresis) and A5 (serving cell RSRP becomes worse than threshold and neighboring cell RSRP becomes better than threshold) events.

For the NTN, new events with new trigger conditions are introduced: (i) Event D1 (which is a distance based trigger event) and (ii) CondEvent T1 (which is a time based trigger event). For a distance based handover event, if the distance between the UE and the NTN platform exceeds a certain maximum threshold, it would mean that the serving NTN cell is moving away from the UE and the handover is triggered. According to 3GPP, Event D1 is triggered based on two entering conditions and two leaving conditions. The event is triggered when the UE fulfills either both the entering conditions or at least one of the leaving conditions [11].

While evaluating the entering conditions, the evaluation is done with respect to two terrestrial reference points, a specific margin of error (hysteresis) and the distance between the UE, and its respective reference point based on specific condition. All the parameters follow the same unit of measurement (e.g., meters). The entering conditions are as follows:

- (Distance between the UE and the reference point 1) – hysteresis > (a threshold distance with respect to reference point 1);
- (Distance between the UE and the reference point 2 + hysteresis) < (Threshold distance with respect to reference point 2).

In an example implementation, Reference Point is the serving cell, and Reference Point 2 is an incoming neighbor cell. Similarly, while evaluating the leaving conditions, the evaluation is done based on similar parameters as the entering conditions. The leaving conditions are as follows :

- (Distance between the UE and the reference point 1 + hysteresis) < (a threshold distance W.R.T reference point 1);
- (Distance between the UE and the reference point 2 – hysteresis) > (Threshold distance WRT reference point 2).

The time-based handover event is suitable in scenarios where an NTN platform serves a region for a specific amount of time (e.g., quasi-Earth fixed cells). According to 3GPP TS 38.331, CondEvent T1 is triggered based on one entering condition and one leaving condition. The event is triggered when the UE fulfills either the entering condition or the leaving condition. While evaluating the entering conditions, the evaluation is done with respect to a time measured at the UE, the duration for which the NTN platform serves the region and a threshold duration after which a specific NTN platform serves a region. All the parameters follow the same unit of measurement (milliseconds). The conditions for entering and leaving are given below, respectively:

- Entering Condition: Time (at UE) > threshold time;
- Leaving Condition: Time (at UE) > threshold time + duration of service.

5.3.2.13 Enhanced Tracking Area Management

Mobility management in 5G terrestrial networks is similar to its predecessor LTE. The UE updates its location to the network regularly. It would not be efficient to track every UE in its idle mode whenever it moves into a different cell, and thus to make the process more efficient, a set of cells are grouped together under one umbrella, called tracking areas. A UE is assigned a registration area, which is made up of one or more tracking areas, which would allow the network to locate a UE and also makes it simpler and cost-efficient for the UE to update its location to the network. The UE sends a NAS message (e.g., Registration request in 5G) to the 5GC whenever it enters a Tracking Area that is not part of the Registration Area that the AMF has defined for the UE in the most recent Registration Accept message. A given cell in a terrestrial network is associated with a single tracking area [5].

Conversely, when a network needs to find an idle UE, it performs a paging procedure and establishes a signaling connection. When the UPF receives downlink traffic for a specific UE, and it does not have the tunneling address of the UE, a paging procedure is carried out to find out the location of the UE.

In NTNs, this process of tracking or registration area update is inefficient in case of moving NTN platforms. When an NTN platform moves, the NTN cell and the associated TA changes, even for a stationary UE. Such NTN platform mobility would require a large number of UEs to carry out TA/RA updates, significantly consuming precious resources. To solve this challenge for the NTN, the 3GPP introduced the concept of earth fixed tracking area, where the gNB broadcasts a set of Earth-fixed TAs in each cell based on the geographic area that it is serving at a given instant. Hence, a stationary UE observes the same tracking areas, even when an NTN platform moves.

5.3.2.14 Support for High-Gain Antennas and Suitable High-Power Amplifiers on the NTN Platform

The NTN reuses adaptive modulation and coding schemes of 5G NR. Both the NTN UE and the NTN gNB would support QPSK, 16-QAM and 64-QAM modulation schemes. However, due to weak signals resulting from large propagation path losses, high-order modulation schemes may not be feasible in several NTN configurations (e.g., GEO). Furthermore, to ensure adequate signal strength to meet the receiver sensitivity requirement even when propagation path losses are large, large EIRPs may be needed. See Section 5.4.2 for a high-level analysis of required EIRPs for different types of NTN payloads. A large EIRP can be realized in practice by using a suitable High-Power Amplifier (HPA) and a high-gain antenna.

5.3.2.15 Circular Polarization

Polarization is the orientation of electric field of a wave with respect to a specific axis. There are generally three types of polarization, (i). Linear polarization, where the electric field varies only with respect to one axis, (ii). Circular polarization, where the electric field varies between two axes at different times, and (ii). Elliptical polarization, where the magnitude of the electric field varies between 2 axes at different times.

Circular polarization occurs when the tip of an electric field vector of an EM wave traverses a circle on a plane normal to the direction of propagation. This happens when the electric field has two perpendicular components with equal magnitude and are 90 degrees out of phase with each other. The equation of a circularly polarized wave can be written as

$$\bar{E} = E_0 \sin(\omega t - kx)\vec{a}_y + E_0 \sin(\omega t - kx + \pi/2)\vec{a}_z. \tag{5.3}$$

Where $\bar{E}$ is the polarization vector, E0 is the magnitude of electric field, and ω is the angular frequency. Terrestrial communications use linear polarization. However, in NTNs , circular polarization is widely used. Circular polarization are furthermore classified as right-handed and left-handed circular polarization. One of the main concerns of circular polarization implementation is the Tx and Rx design. If both the UE and the gNB support circular polarization, there will be no polarization loss. However, if the NTN payload utilizes circular polarization but the UE uses linear polarization, there is a 3 dB polarization mismatch loss. Additionally, if circular polarization is applied in the opposite direction, the mismatch loss is close to infinity. In practice, a 3 dB polarization mismatch is accepted in favor of simpler NTN UE implementation. The NTN UE can also use multiple receive antennas to achieve diversity gain, overcoming the polarization mismatch loss.

5.4 Security Vulnerabilities in the NTN

Various kinds of NTN specific interfaces are discussed in this section. Section 5.4.1 discusses the vulnerabilities of wireless interfaces in NTN systems, and jamming attack scenarios in these wireless interfaces are described. A quantitative analysis for jamming attack scenarios is done in Section 5.4.2. Section 5.4.3 mentions the current security mechanisms deployed in 5G NR and describes candidate strategies for NTNs.

5.4.1 Overview of NTN-specific Interfaces

Table 5.5 summarizes the NTN interfaces that experience unique challenges compared to a TN for Transparent NTN Payload (TNP) and a Regenerative NTN Payload (RNP).

The radio link between the UE and the NTN payload (transparent or regenerative) is the service link, and transmission occurs over the NR-Uu interface. For a transparent payload architecture, the uplink and downlink communication occurs between the UE and the payload and for the regenerative payload architecture, the

Table 5.5 NTN-specific interface for security considerations.

NTN Interface	Interface Type
UE-to-NTN Payload Service/Access Link	Wireless
NTN Payload-to-UE Service/Access Link	Wireless
NTN Payload-to-NTN Gateway Feeder Link	Wireless
NTN Gateway-to-NTN Payload Feeder Link	Wireless
NTN Control Function-NTN Gateway	Wireline
NTN Control Function-NTN Gateway-NTN Payload	Hybrid Wireline and Wireless
NTN Control Function-Non-NTN Infrastructure gNB Functions	Wireline
NTN Gateway-to-Non-NTN Infrastructure gNB Functions	Wireline
NTN Control Function-O&M	Wireline
Non-NTN Infrastructure gNB Functions-O&M	Wireline

uplink and downlink communication occurs between the UE and the gNB or the gNB-DU, depending on the type of the NTN architecture deployed.

The radio link between the NTN payload (transparent and regenerative) and the NTN-Gateway is the feeder link, and the transmission occurs over the NG interface or the F1 interface via implementation-specific radio protocol stack (i.e., 3GPP-defined or non-3GPP-defined radio protocol stack). For a transparent payload architecture, the uplink and downlink communications occur between the NTN payload and the NTN-Gateway and for a regenerative payload architecture, the uplink and downlink communication occur between the gNB/gNB-DU and the NTN-Gateway.

As of 3GPP Release 18, the NTN Control Function feature is defined for the transparent NTN payload architecture. The uplink and downlink communication between the NTN Control Function and the NTN-Gateway is expected to be over wireline access, but wireless access can be implemented if the wireline infrastructure is infeasible or unavailable. The interface between the NTN Control Function and the NTN-Gateway is beyond the scope of 3GPP specifications. Additionally, uplink and downlink communications between the NTN control function and the non NTN infrastructure gNB functions is also expected to occur over wireline access, but wireless access can be implemented if the wireline infrastructure is infeasible or unavailable. There is a hybrid wireline-wireless interface between the NTN control function, NTN-Gateway and the NTN payload, with the interface between the NTN-Gateway and the control function being wireline and the interface between the NTN-Gateway and the payload being wireless.

The uplink and downline communications between the control center and the O&M function occur over wireline infrastructure, and the interface between the O&M and the non-NTN infrastructure gNB functions also occurs over wireline access.

In general, security vulnerabilities over wireless interfaces can be significant and security vulnerabilities over wireline access are relatively low.

5.4.2 Analysis of Jamming Attacks on Wireless Interfaces of the NTN

Section 5.4.2.1 discuss jamming scenarios for the wireless interfaces in NTNs. Four scenarios are considered, the wireless interface between the NTN Gateway and NTN Payload and the wireless interface between the UE and NTN Payload. Section 5.4.2.2 provides a quantitative perspective on jamming attacks on the wireless interfaces.

5.4.2.1 Jamming Scenarios in an NTN

As summarized in Table 5.5, there are two wireless interfaces and four links (i.e., two links per interface). These NTN interfaces and links have unique characteristics compared to a TN.

Figure 5.15 depicts a scenario where the attacker tries to cause interference in the feeder link from the NTN-Gateway to the NTN payload. The attacker tries to jam portion of the NR channel bandwidth or the entire NR channel bandwidth. If the distance between the attacker and the NTN platform is large (e.g., a GEO satellite), the impact of the jamming attack is expected to be low because the interference may not be high relative to the thermal noise floor at the NTN payload receiver.

Figure 5.16 depicts the scenario where the attacker tries to cause interference to the feeder link from the NTN payload to the NTN-Gateway. The receiver in this case is the NTN-Gateway. In this scenario, the received signal from the NTN platform can be quite weak due to long distance between the NTN payload the NTN-Gateway. If the attacker is situated in proximity to the NTN-Gateway, the

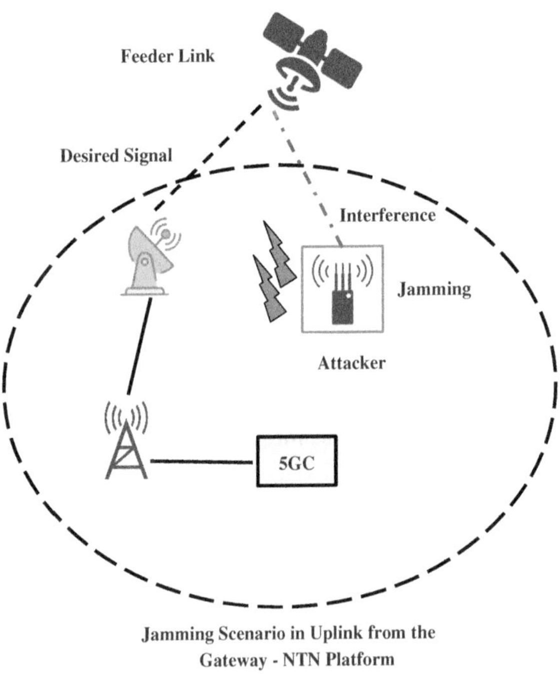

Figure 5.15 A jamming scenario in the feeder link from the NTN Gateway to the NTN Payload.

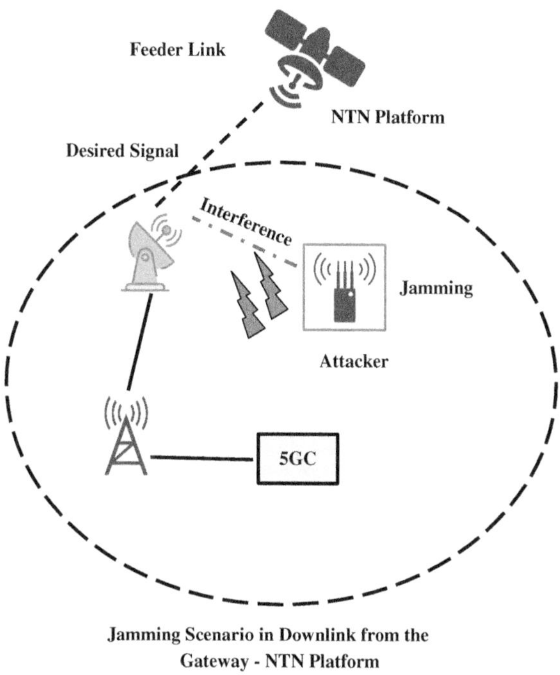

Figure 5.16 A jamming scenario in the feeder link from the NTN Payload to the NTN Gateway.

impact of such an attack could be quite severe, because the attacker can be located quite close to the NTN-Gateway receive antenna.

Figure 5.17 depicts the scenario where an attacker tries to cause interference in the uplink service link from the UE to the NTN payload. In this scenario, the attacker tries to jam the communication channel toward the receiver, which in this case is the NTN platform. If the distance between the attacker and the NTN payload receiver is large, the impact of this attack is expected to be low.

Figure 5.18 depicts a scenario where the attacker tries to cause interference in the service link from the NTN platform to the UE. In this scenario, since the attacker can be relatively in close proximity to the UEs compared to the NTN payload, the impact of a jamming attack can be high.

5.4.2.2 Quantitative Analysis of Jamming Attacks in an NTN

The levels of jamming interference at the receiver and the resulting Signal-to-Interference-plus-Noise Ratio (SINR) levels in various deployment scenarios are quantified below. Table 5.6 specifies numerical values of most relevant

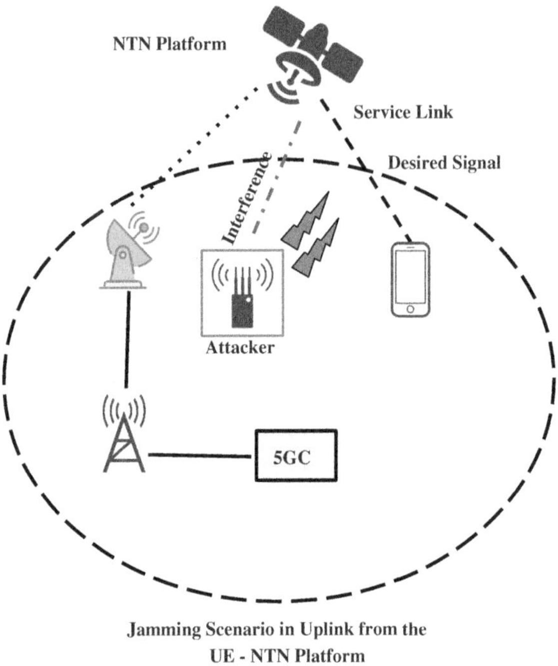

Figure 5.17 A jamming scenario in the service link from the UE to the NTN Payload.

parameters for the analysis. The path loss for all these scenarios is assumed to be Free Space Path Loss (FSPL), which is defined as [13]

$$FSPL(dB) = 10 * log_{10}(\lambda/(4\pi R))2 \tag{5.4}$$

where λ is the wavelength at an operating frequency fc and R is the distance between the NTN-Gateway and the NTN Payload. The analysis is carried out for the S frequency band (i.e., 2.2 GHz).

The received power at the receive antenna is the difference between the transmitter Effective Isotropic Radiated Power (EIRP) ($EIRP_{Transmitter}$) and the free space path loss and is calculated as:

$$RX(dBm) = EIRP_{Transmitter}(dBm) - FSPL(dB) \tag{5.5}$$

The received power after including antenna gain, Rx_{Gain} , is calculated as:

$$Rx_{Gain}(dBm) = Rx(dBm) + Rx_{AntennaGainSignal} \tag{5.6}$$

where $Rx_{AntennaGainSignal}$ is the gain of the receive antenna toward the desired signal.

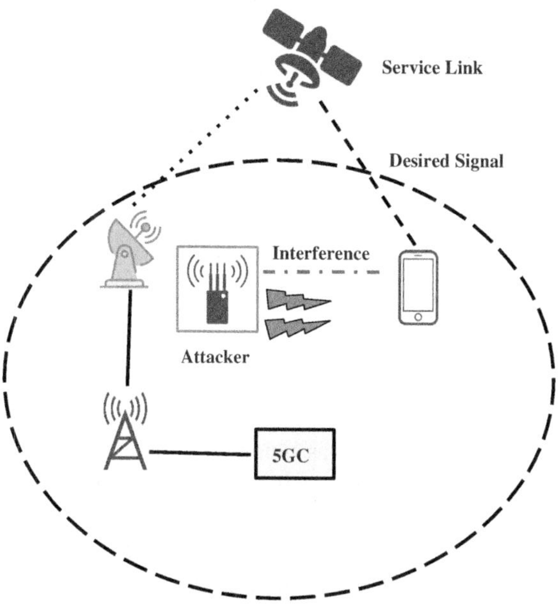

Figure 5.18 A jamming scenario in the service link from the NTN Payload to the UE.

Table 5.6 Parameters for numerical results.

Parameter	Value
Operating Frequency (fc)	2.2 GHz
Channel Bandwidth	10 MHz
Power Spectral Density	−174 dBm/Hz
Number of subcarriers	624
Noise Figure	8 dB
Number of Physical Resource Blocks	52
Subcarrier Spacing	15000 Hz

The Effective Interference caused by the attacker is the difference between the EIRP of the attacker ($EIRP_{Attacker}$) and the path loss experienced by the interfering signal ($FSPL_{interferingSignal}$) and is calculated as:

$$EffectiveInterference(EI)(dBm) = EIRP_{Attacker}(dBm) + Rx_{AntennaGainInterference} - FSPL_{interferingSignal}dB) \tag{5.7}$$

where $Rx_{AntennaGainInterference}$ is the gain of the receive antenna toward the interfering signal.

While estimating the noise (thermal noise floor) in the system, parameters like thermal noise Power Spectral Density (PSD) (in dBm/Hz), noise figure at the UE (NF), the number of Physical Resource Blocks (NPRB) in the channel bandwidth, the number of subcarriers per PRB (NSC), and Subcarrier Spacing (SCS) (in Hz) are considered.

$$ThermalNoiseFloor(TNF)(dBm) = PSD(dBm/Hz) + NF(dB) + \\ 10 * log_{10}(NPRB * NSC * SCS)(dB - Hz) \quad (5.8)$$

Finally, the SINR is calculated as :

$$SINR(dB) = 10 * log_{10}[Rx_{Gain}(mW)/(EI(mW) + TNF(mW))] \quad (5.9)$$

Table 5.7 describes the receive antenna gains assumed for various NTN platforms during the analysis. Their receive antenna gains ensure RSRP of at least −77 dBm.

Finally, we define an SINR threshold of (−3 dB) below which the feeder and service links are assumed to become vulnerable to jamming attacks.

5.4.2.3 Numerical Results

This section provides a numerical perspective of the vulnerabilities mentioned in Section 5.4.2. An analysis was carried out to estimate the impact of jamming interference when an attacker on the ground tries to disrupt the feeder link and/or the service link. Distance-based and attacker EIRP-based sensitivity analysis is performed, to understand the vulnerabilities of various links.

Figure 5.19 shows the SINR trends, as the distance between the UE and the attacker varies when the attacker EIRP remains constant at 50 dBm. As the distance between the UE and attacker increases, the SINR increases, which shows that as the attacker moves further away from the UE, the impact of attacker interference decreases. Therefore, the service link is less vulnerable to jamming attacks

Table 5.7 Receive antenna gain of different NTN platforms.

NTN Platform	Receive antenna gain of NTN platform (dB)
HAPS-Max	35
LEO-min	56
LEO-max	64
MEO-min	78
MEO-max	87
GEO	92

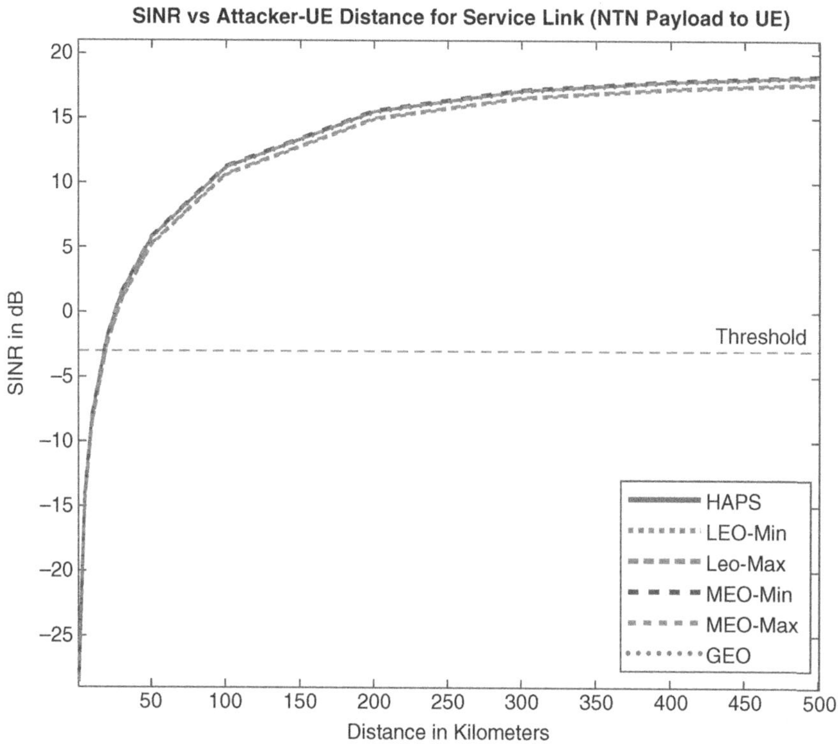

Figure 5.19 Attacker-UE receiver distance vs SINR for the service link.

from long distances. The vulnerable distance, according to the analysis, is below 50 km.

Figure 5.20 shows the SINR trends as the distance between the NTN-Gateway and the attacker varies. Since the attack is done in proximity, the feeder link between the NTN Gateway and the NTN payload is more vulnerable to attacks when the attacker is in proximity to the NTN Gateway.

Figure 5.21 shows the trends in the SINR at the UE receiver as the EIRP of the attacker varies and the distance between the attacker and the UE remains constant (75 Km), on the service link. As the attacker EIRP increases, the SINR decreases, and at high attacker EIRP (>65 dBm), the service link becomes vulnerable to jamming attacks.

Figure 5.22 shows the trends in the SINR at the NTN-Gateway receiver as the EIRP of the attacker varies at constant distance (100 m). The feeder link between the NTN platform and the NTN Gateway is highly vulnerable even at very low attacker EIRP.

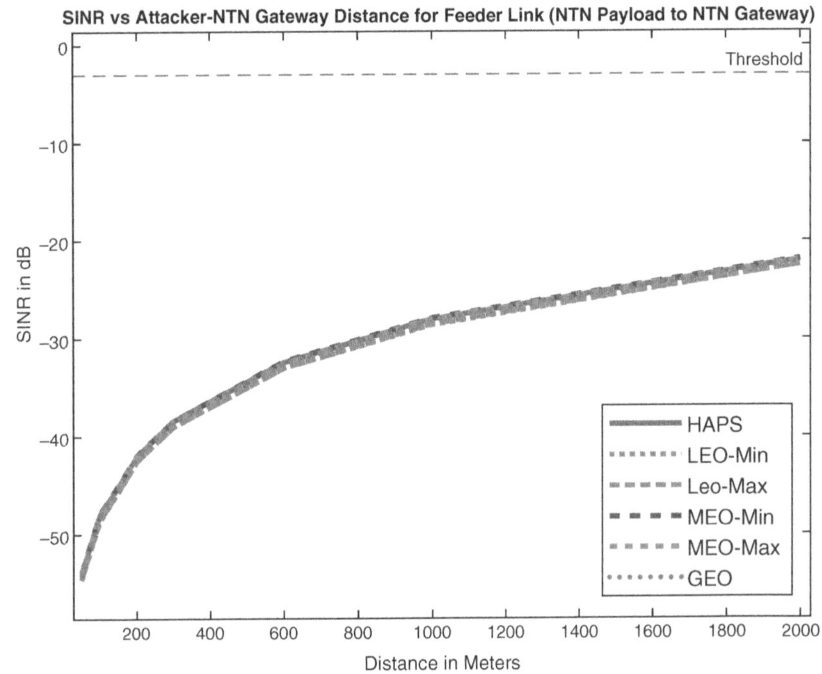

Figure 5.20 Attacker-NTN-Gateway receiver distance vs SINR for the feeder link.

It is evident from the above analysis that the feeder link is more vulnerable when the receiver is the NTN-Gateway and the service link is more vulnerable when the receiver is the UE. There is some degree of vulnerability when the receiver is the NTN Platform, but the degree of vulnerability is much lower when compared to the cases of the NTN-Gateway and the UE as receivers.

5.4.3 Candidate Strategies for Enhanced NTN Security

An overview of the current 5G security mechanisms mentioned in Figure 5.23 is given below, described. Furthermore, security mechanisms for wireline and wireless interfaces of the NTN are also discussed.

5.4.3.1 Security Apparatus in 5G: An Overview

Network access security enables a mobile user to authenticate and safely access services between (i) the Mobile Equipment (ME) and the Universal Subscriber Identity Module (USIM), (ii) the ME and the serving network (e.g., a visited or roaming network), (iii) the ME and a non-3GPP network, (iv) USIM and the serving network, and (v) the serving network and the home network.

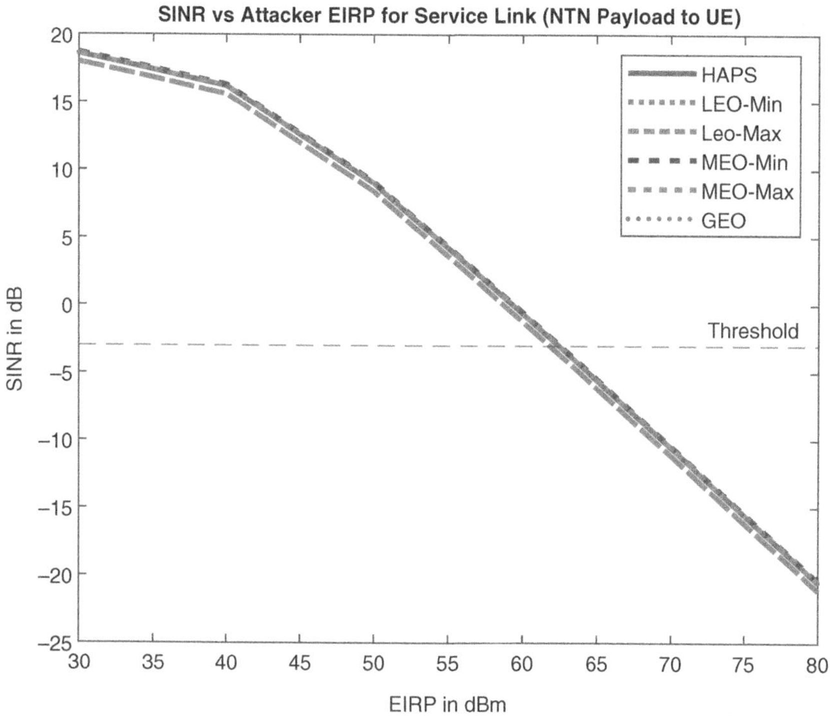

Figure 5.21 Attacker EIRP vs SINR at the UE receiver for the service link.

Network domain security enables network nodes to exchange signaling messages and user data securely, specifically between (i) the serving network and the home network and (ii) a non-3GPP network and the serving network. User domain security enables secure access between the ME and the USIM. Application domain security enables secure exchange of messages between applications on the user domain and the applications on the provider domain. The Service Based Architecture (SBA) domain architecture enables network functions to communicate securely with each other, specifically between the serving network and the home network. Visibility and configurability of security shows the user if a specific security mechanism is in operation.

5G provides various security mechanisms such as subscriber privacy, Control and User Plane integrity protection, intra-network security, and unified access control among others. Subscriber privacy is provided through identities such as 5G- Globally Unique Temporary Identifier (5G- GUTI) and Subscription Concealed Identifier (SUCI). 5G- GUTI is an 80-bit identifier allocated to the UE by the AMF. 5G-GUTI consists of a 24-bit Public Land Mobile Network ID

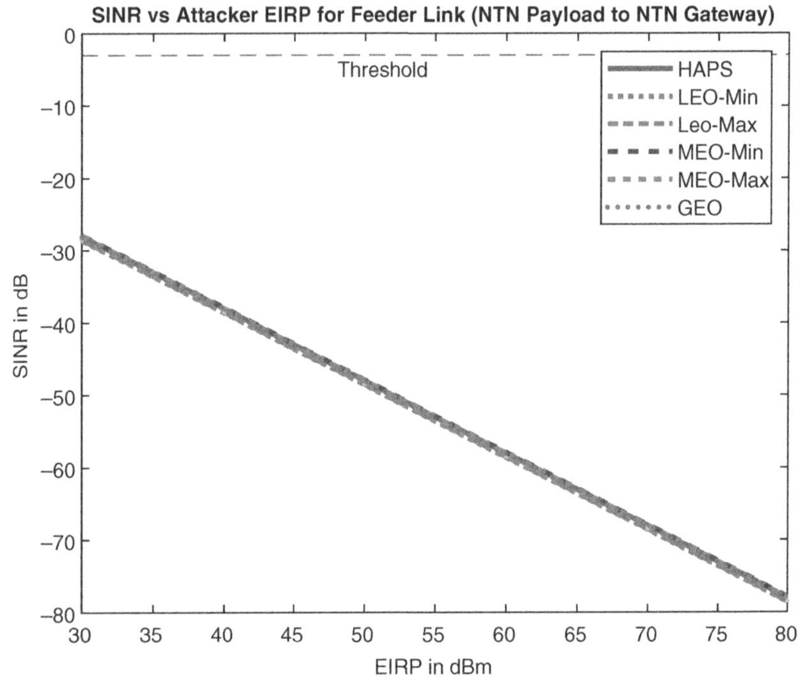

Figure 5.22 Attacker EIRP vs SINR at the NTN-Gateway receiver for the feeder link.

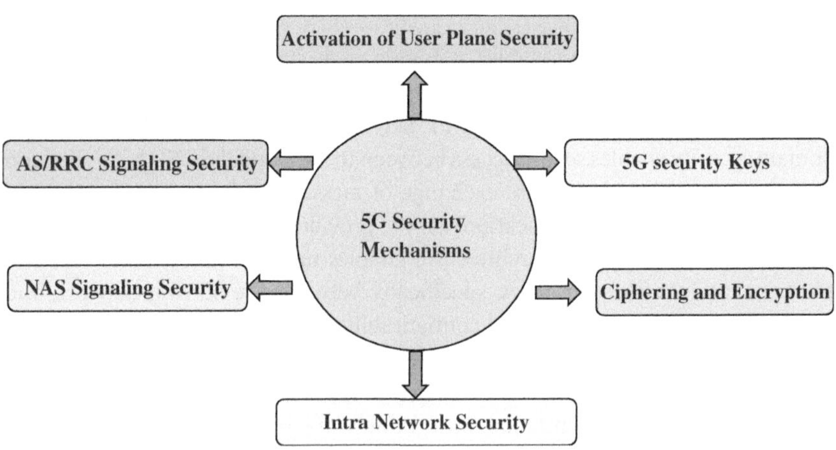

Figure 5.23 Types of 5G security mechanisms.

(PLMN ID) and a 24-bit AMF ID. This information allows for the identification of the AMF that allocated the GUTI. A shortened version of 5G-GUTI is a 32-bit 5G Temporary Mobile Subscriber Identity (5G TMSI) which identifies the specific UE in the AMF. 5G- GUTI is a temporary identifier and thus not a permanent identity unique to an individual or a mobile phone. The AMF provides a 5GGUTI to the UE when it receives a registration request message or a service request message from a 5G UE. SUCI or subscription concealed identifier is an identifier that protects the permanent identifier such as the International Mobile Subscriber Identity (IMSI) to protect the identity of the subscriber and prevent any rogue entity from acquiring the IMSI from the radio interface. Hence, SUCI rectifies this issue of IMSI catching in 2G, 3G and 4G networks, thereby enhancing the security and privacy of the subscriber.

The Control Plane is protected at various levels like (i) protection of Non-Access Stratum (NAS) signaling between the UE and the AMF, and (ii) protection of Access Stratum (AS) signaling between the UE and the gNB. Integrity protection of the signaling messages using a code called Message Authentication Code (MAC) to prevent altering of the NAS/AS signaling message. Confidentiality of the NAS/AS message by encrypting the message itself through ciphering or encryption mechanisms. The User Plane is also protected through encryption and integrity protection in 5G.

A unified procedure called Authentication and Key Agreement (AKA) is defined to enable the use of the same authentication framework for 3GPP access as well as non-3GPP access. The 5G AKA procedure is used for 3GPP networks, and EAP-AKA (Extensible Authentication Protocol) is used for non-3GPP networks. This AKA procedure enables mutual authentication, generates keying material, and supports unified access control for both 3GPP network elements like gNBs and non-3GPP elements like Wi-Fi Access Points.

5.4.3.2 Mutual Authentication

Mutual authentication is done during the registration process through a series of steps involving the AMF, the AUSF, and the UDM. Figure 5.24 illustrates the mutual authentication procedure in 5G.

The UE sends a registration request as a NAS signaling message to the AMF, and this message contains the UE identification information, such as SUCI or 5G-GUTI. The AMF sends a UE authentication request to the AUSF, and this authentication request consists of the UE ID and the Serving Network details. The AUSF sends this request and constituent information obtained from the AMF to the UDM for processing the information. The UDM, which consists of the Subscription Identifier De-Concealing Function (SIDF), is used to de-conceal the SUCI and obtain the SUPI, The UDM also consists of the Authentication Credential Repository and Processing Function (ARPF) to get SUPI-relevant information like the

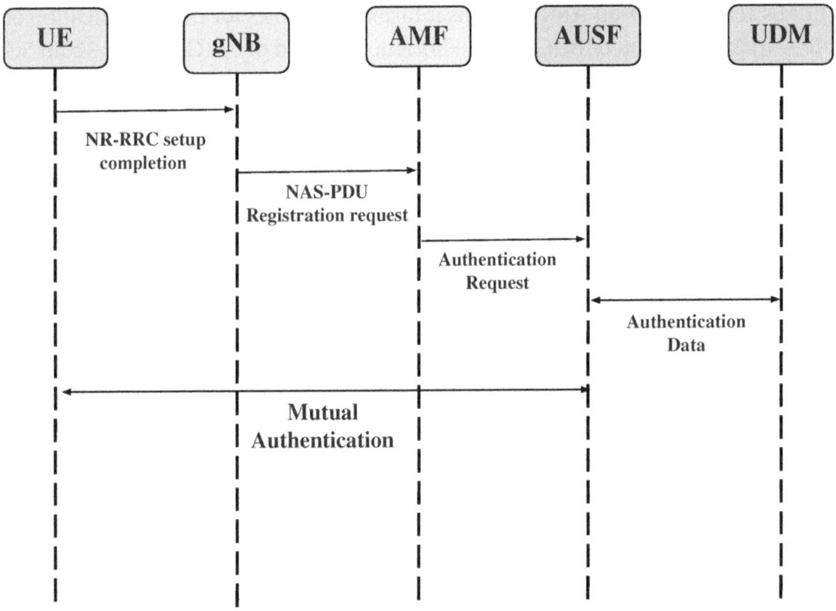

Figure 5.24 Mutual authentication mechanism in 5G.

IMSI and the authentication method (e.g., 5G-AKA or EAP-AKA) associated with the IMSI. The UDM then constructs the 5G Home Environment Authentication Vector (5G HE AV) consisting of a random number, authentication token, expected resonance and the K_{AUSF}.

The UDM then sends the authentication vector and SUPI back to the AUSF. The AUSF will calculate the 5G Authentication vector from the 5G HE AV, and change a few parameters like the expected response with the Hx Response and the K_{AUSF} with the K_{SEAF}. The AUSF then sends the random number, authentication token, Hx Response and the K_{SEAF} to the AMF in the form of a new vector called 5G Serving environment Authentication vector. The AMF sends the authentication-related parameters in an authentication request message through NAS signaling to the UE.

The UE evaluates the authentication request message and responds to the authentication request message by sending an authentication response message back to the AMF, which evaluates if the UE response is the same as the expected response. This step ensures that the authentication is successful from the serving network (SN) perspective. Once the SN authentication is successful, the AMF sends an authentication response to the AUSF where the UE response and the expected response are compared and if they match, the home network side authentication is also deemed successful. Once the home network authenticates the UE, it sends an authenticate response to the AMF in the serving network. In summary, there are two steps to mutual authentication. Firstly, the UE authenticates the serving network, and the serving network authenticates the UE. In the end, we generate the anchor key, which is used to create other security keys for securing communication between the network and the UE [14].

5.4.3.3 5G Security Keys

Once the AUSF generates the anchor key, K_{AUSF} , the AMF generates K_{AMF} . The K_{AMF} is now used to create two types of keys for NAS security, namely, K_{NASint} which is used for integrity protection of NAS signaling and K_{NASenc} which is used for encryption of NAS signaling. The gNB also generates four keys for radio interface security, K_{RRCint} and K_{RRCenc} for integrity protection and encryption of RRC signaling, respectively, and K_{Upint} and K_{Upenc} for integrity protection and encryption for the User Plane, respectively. All these four keys for the radio interface are used at the PDCP layer.

5.4.3.4 Security Algorithms for 5G

Over the air interface between the UE and the NG-RAN (i.e., the Uu interface) is protected using ciphering and encryption. There are four different types of ciphering or encryption algorithms, widely divided into ciphering and integrity protection algorithms. NEA0 is a null ciphering algorithm. NEA1 is a 128-bit SNOW 3G based algorithm. NEA2 is a 128-bit AES based algorithm. NEA3 is a 128-bit ZUC-based algorithm. There are 4 different types of integrity protection algorithms-NIA0, NIA1, NIA2, and NIA3, which are similar to the ciphering algorithms.

5.4.3.5 Activation of NAS Signaling Security

Activation of security in NAS signaling takes place in 2 steps. After the authentication process is completed, the AMF sends a (NAS) SECURITY MODE COMMAND message to the UE, which contains security related information like types of NAS security algorithms being used, ng-KSI indicating a security context to the UE and a NAS Message Authentication Code (MAC). Once the UE receives this message, it starts the ciphering and integrity protection procedures in the uplink and is ready to decipher the downlink messages. The AMF at this point will also start deciphering in the uplink once it sends the security command message. In the next step, the UE sends a security mode complete message, and the AMF will start ciphering in the downlink.

5.4.3.6 Activation of AS/RRC Signaling Security

The gNB sends an RRC Security Mode Command message to the UE. This message contains the security algorithm configuration, like the ciphering and integrity protection algorithms used for secure communications. Once the UE receives this message, it starts RRC integrity protection in the uplink and RRC deciphering in the downlink. At the same time, the gNB starts with RRC ciphering in the downlink. In the next step, the UE sends a Security Mode Complete message to the gNB as a response to the Security Mode Command message. Once the gNB receives the message, it starts RRC deciphering in the uplink. In addition, there is a replay protection activated along with integrity protection [15]. Replay protection ensures that the authentication messages are exchanged exactly once between the user and the gNB to prevent unauthorized entities/attackers from replaying these messages and gaining access to the UE.

5.4.3.7 Activation of User Plane Security

The gNB sends an RRC Reconfiguration Message, which consists of radio bearer configuration such as the ciphering and integrity protection algorithms. Once the UE receives this message, it starts the User Plane integrity protection and ciphering in the uplink. In the downlink, the UE starts User Plane integrity verification and deciphering. The gNB starts User Plane integrity verification and deciphering in the uplink and User Plane integrity protection and ciphering in the downlink. The UE then sends an RRC Reconfiguration Complete message as a response to the previous message.

5.4.3.8 Inter-network Function Intra-network Security

5G introduces a security mechanism for new service-based interfaces. 5G provides transport-layer security which include server-side and client-side certificates, network layer security like IPsec, mutual authentication between network repository function and the network function, authorization of network function service by NRF, and interSEPP (service edge protection proxy) communications between the serving network and the home network using transport layer security in case there is no IP exchange network between these networks. According to the 3GPP [15], the NRF and NF can mutually authenticate each other during the discovery and registration phase.

Mutual authentication occurs either through direct communication or indirect communication. In the direct communication scenario, if there is transport layer security, the authentication for transport layer protection procedure would be used for mutual authentication between the NRF and the NF. If there is no prior transport layer security, mutual authentication occurs either through NDS/IP or physical layer security procedures. In the indirect communication scenario, mutual authentication is done using the transport layer protection solution or the client credentials assertion-based authentication.

5G also defines security for non-service-based interfaces. For non-service-based interfaces, NDS/IP for IP based interfaces is available for use. For authentication, we use the IKEv2 certificate-based mechanism and for integrity protection we use IPsec ESP tunnel mode mechanism. Examples of non-service-based interfaces include the N2 interface between the gNB and the AMF and N3 interfaces between the gNB and the UPF, Xn-C and Xn-U interfaces between two gNBs, and the Rx interface between the PCF and the IMS [14].

5.4.3.9 Security Mechanisms for Wireline Interfaces of the NTN

Security mechanisms for wireline interfaces in 5GC are classified into service-based wireline interfaces and non-service based wireline interfaces. SBI security mechanisms include network layer security, transport layer security, inter SEPP communications, NRF authorization of NF services, and NRF-NF mutual authentication. In non-service-based interfaces, authentication is done through the IKEv2 certificate-based procedures and integrity, and confidentiality protection is carried out by the IPsec ESP procedure. Table 5.8 summarizes the candidate security mechanisms in wireline interfaces of an NTN system.

5.4.3.10 Security Mechanisms for Wireless Interfaces of the NTN

Security on the wireless interfaces of the NTN can be considered at two levels, node-centric security and interface-centric security. These two levels are discussed below.

Table 5.8 Candidate security mechanisms for wireline interfaces in an NTN system.

Wireline NTN Interface	Candidate Security Mechanisms
NTN Control Function-NTN Gateway	IKEv2 Certificate based procedure
NTN Control Function-Non-NTN gNB Functions	IKEv2 Certificate based procedure
NTN Gateway-Non-NTN gNB Functions	Encryption/ Ciphering based procedure
NTN Control Function-O&M	IKEv2 Certificate based procedure
Non-NTN Infrastructure gNB Functions-O&M	IKEv2 Certificate based procedure
Non NTN gNB-5G Core	Intra-Network based security, IKEv2

Table 5.9 Candidate node-centric security mechanisms for an NTN system.

Wireless NTN Interface	Candidate node-centric security mechanisms
UE- Non-Terrestrial gNB	Validation of the gNB and the UE - Watermarking
NTN Payload-NTN-Gateway	Mutual authentication and Intra-Network security
UE and NTN Payload	Watermarking
UE-5GC	NAS Security, Mutual Authentication

5.4.3.11 Node-centric Security

Several NTN nodes interact with each other. Internode communications in a typical TN occur within the internal wired network of a service provider. Hence, a certain level of security is inherently achieved by a TN. However, due to the wireless interfaces between various NTN nodes or Network Functions, the TN-like security level does not exist inherently. A rogue entity with certain knowledge about the wireless interface (e.g., placement of a 5G signaling message on a wireless interface) could detect a message and use it to send a malicious message by pretending to be a valid NTN node to disrupt operations of the NTN. To prevent such faking by a rogue entity, the 3GPP-defined intra-network security mechanisms can be used to ensure the validity of the nodes. A two-step process can be implemented. In the first step, the NTN operator can implement the node validation procedure (e.g., while setting up a connection between two nodes). In the second step, a secure connection between two NTN nodes can be established as discussed below in the interface-centric security. Table 5.9 summarizes the candidate security mechanisms for node-centric security.

The UE and the NTN Payload (bent pipe or non-terrestrial gNB) transmit and receive information through the service link, and it is essential to secure both the nodes to for data confidentiality, security and privacy. Before user traffic is transmitted, the UE must make sure it is sending information to a legitimate NTN Payload (Non-Terrestrial gNB). To enable this additional layer of protection, a procedure called watermarking can be performed in an implementation specific manner. In an example watermarking approach, as part of initial UE-gNB communications (e.g., network acquisition, random access, and RRC setup), the non-terrestrial gNB can send a unique signature to the UE, and thereby conveying to the UE that it is a legitimate base station and the UE can securely exchange information. In contrast, if the NTN Payload is a fake base station or an unregistered one, it will not send such signature to the UE, and the UE will understand that the gNB is not a legitimate one, and thus will not exchange information with such fake gNB.

Table 5.10 Candidate interface-centric security mechanisms for an NTN system.

Wireless NTN Interface	Candidate interface-centric security mechanisms
UE-NTN Payload Service/Access Link	5G Security keys based security, beamforming, robust modulation techniques, frequency agility capability
NTN Payload-NTN Gateway Feeder Link	IPsec and DTLS security mechanisms, spread spectrum modulation, use of robust waveforms, optimal antenna radiation pattern, lower order modulation schemes

A mutual authentication mechanism can be performed between the NTN Payload and the NTN-Gateway during the discovery phase. This procedure is similar to the mutual authentication mechanism between the UE and the 5GC or between the NRF and the NF to prevent any type of man in the middle attacks. The NTN-Gateway and the NTN Payload will mutually authenticate each other and make sure that both nodes are legitimate to exchange information.

5.4.3.12 Interface-centric Security

Interface-centric security aims to establish a secure connection between two NTN nodes for communication. The service link between the UE and the NTN Payload is secured using encryption and ciphering algorithms (SNOW 3G, AES and ZUC based encryption) [16]. The feeder link between the NTN Payload and NTN-Gateway is secured through F1 interface (F1-C and F1-U)-based security mechanisms like IPsec and DTLS-based security mechanisms [15].

Activation of AS/ RRC signaling security is done between the UE and the gNB to secure the uplink and downlink for communications. Activation of NAS signaling security is done between the UE and the 5GC to secure the uplink and downlink for secure NAS signaling. Once these security mechanisms are activated, the UE and the gNB/5GC can exchange signaling messages (AS and NAS). The UE can exchange signaling messages with the gNB using the transparent payload as a relay node. In the case of regenerative payload architecture, where NTN Payload is the gNB, the UE will exchange AS and NAS signaling messages through the service link. Furthermore, the interface in this scenario, the air interface between the UE and the gNB is secure through encryption and ciphering algorithms and the uplink and downlink for AS and NAS signaling messages are secure though the activation of AS and NAS signaling security. Similarly, for exchange of user traffic between the UE and gNB, user plane security is activated for securing the uplink and downlink channels.

A few security practices that can be used to make the service link and feeder link more robust are mentioned in Table 5.10.

Beamforming. Beamforming acts as a spatial filter between the transmitter and the receiver. The beamforming technique allows both the transmitter and receiver to concentrate most of the energy in each other's direction. The transmitter will transmit data in the direction of the receiver, and the receiver will listen for information only from the direction of the transmitter. This procedure suppress interference from all other directions, thereby protecting the receiver from jamming attacks.

Beamforming with an optimal antenna radiation pattern can be used, where most of the energy is concentrated at the center of the beam.

Robust Modulation Techniques. Modulation techniques make a signal more resilient to interference, while supporting long distance communications, thereby protecting the information from jamming attacks. NTN systems contain non-linear amplifiers that require modulation techniques which are tolerant to distortion. Modulation techniques like Frequency Shift Keying (FSK) and Phase Shift Keying (PSK) are ideal modulation techniques for NTN systems, as they reduce the effect of non-linear amplification in a high power amplifier [17].

Frequency Agility. Frequency Agility or Frequency Hopping Spread Spectrum technique allows the transmitter to transmit information in different frequencies of the bandwidth. In this process, the transmitter will transmit information at one frequency for a specific time and will hop onto a different frequency in a given bandwidth. This will make it harder for the attacker or eavesdropper to listen to the transmitted information, as there is no way for the attacker to know at which frequency the transmitter will transmit next and for how long. Thus, Frequency Agility mechanism can be used to protect data against eavesdropping attacks.

Direct Sequence Spread Spectrum. The DSSS procedure make the data larger in bandwidth through rapid phase transition. In the DSSS procedure, once the input signal is encoded, during the modulation process, a spreading generator is introduced. This spreading generator generates a spreading code of length L and an X NOR operation is performed between the modulated digital signal and the generated spreading code. This output signal is now transmitted over the air to the receiver. The received message can be decoded by the receiver only when it has the spreading code. This output signal behaves as noise, and the power spectral density of the DSSS output signal is similar to the noise floor. Thus, the signal is hidden under the noise, making the communication more secure and immune to jamming attacks.

A comprehensive overview of various node-centric and interface-centric mechanisms is described in Figure 5.25.

5.5 Evolution of the NTN in 5G-Advanced and 6G

NTN enhancements for Release 19 and beyond are discussed in this section. Moreover, Section 5.5.2 discusses the role of NTN in 6G.

5.5.1 5G NTN Enhancements in 3GPP Release 19 and Beyond

This section describes future NTN enhancements in Release 19 and beyond. Enhancements to key features like Quality of Service, support for regenerative payloads, and spectrum sharing are discussed.

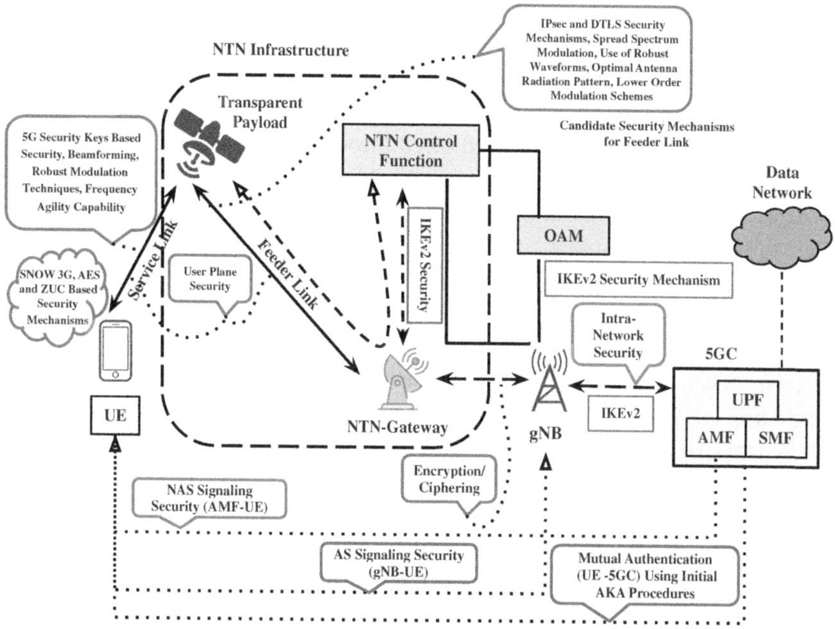

Figure 5.25 Security mechanisms for an NTN system.

5.5.1.1 Flexible QoS

5G Quality of Service Identifier (5QI) is a value that is associated to a set of QoS characteristics [12]. Examples of these QoS characteristics include Packet Delay Budget (PDB), Packet Error Rate (PER), and averaging window. The Packet Delay Budget provides an upper bound threshold for the packet delay between the UE and the UPF providing the N6 interface toward a Data Network. The Packet Error Rate (PER) provides an upper bound threshold of the rate of PDU that were processed by the link layer protocol of the sender but were not delivered to the upper level layer by the receiver [12]. Averaging window is the duration in which the guaranteed flow bit rate (GFBR) and the maximum flow bit rate are calculated at the UE, AN, and the 5GC, maximum data burst volume, priority level and the resource type. From an NTN perspective, especially for a GEO based NTN, the propagation delays are significantly greater than delays in terrestrial networks. Hence, the 3GPP introduced a new 5QI in Release 17, which has a PDB of 1100 ms.

Different NTNs operate at different altitudes depending on the NTN platforms they are deployed on. Considering the significant difference in altitudes, NTN platforms like HAPS would experience ultra low delays, LEO may experience short delays, MEO may experience a litter longer delays than LEOs and GEOs may experience long delays. Thus, NTNs may not be able to provide the same QoS as terrestrial networks. Relaxing and adding more flexibility to the QoS characteristics would allow more services to be supported by NTNs. For example, if a higher PER is allowed or a longer PDB is allowed through scaling of existing parameter settings of 5QIs, an NTN can serve more applications without violating the 3GPP QoS parameters and without introducing more 5QIs [18].

5.5.1.2 Regenerative Payloads

Regenerative NTN payloads and the associated architectural variants offer significant benefits for both end users and network operators. From an end user perspective, regenerative payloads offer reduced user and Control Plane latencies, single hop UE-gNB communications and ubiquitous service availability in areas where transparent payloads are not feasible (e.g., middle-of-the-ocean coverage, where no land mass is available to place an NTN-Gateway). The network operations aspect encompasses a range of deployment alternatives aimed at enhancing spectral efficiency, facilitating the operation of multiple core networks in collaboration with diverse service providers, optimizing the utilization of feeder and service link resources, enabling in-space edge compute capabilities, ensuring adherence to regulatory requirements across different regions, and extending the coverage of terrestrial networks to previously underserved regions [18].

There are a number of use cases that could be furthermore enhanced using regenerative payloads such as aeronautic and maritime broadband services, vehicular communications, operating modern farms in agriculture, providing communication and connectivity during the time of emergency, providing low latency medical care consultation services in geographic areas far away from hospitals.

5.5.1.3 Spectrum Sharing

Spectrum sharing between terrestrial and non-terrestrial networks would be an important feature in 5G advanced and 6G networks to achieve high data rates. For example, low-power indoor devices can use the licensed spectrum generally used by NTN systems. Similarly, in areas with no or low population like mountains and rural areas, the NTN systems can use the spectrum licensed by terrestrial networks. A centralized and intelligent spectrum coordination system can be set up to minimize interference between terrestrial networks and non-terrestrial networks. In one scenario, there can be geography-based or time-based use of the licensed spectrum of a spectrum owner, with one or more entities competing for using a part of the spectrum. The central spectrum coordination system would decide which entity gets to use that part of the spectrum [18].

Other spectrum related enhancements include (i) enabling efficient handover between terrestrial networks and NTNs, (ii) enabling frequencies above 10 GHz for NTNs with Q /V(40–50 GHz), and E (60–90 GHz) bands being future frequency candidates for NTNs, enabling low data rate voice over NTN capabilities, and using the store and forward behavior in IoT NTNs where a UE with sufficient power capabilities can store and forward data packets to an NTN platform when the NTN platform becomes addressable.

5.5.2 Role of the NTN in 6G

NTNs are expected to play a significant role in a number of industries like IoT, healthcare, transportation, and public safety [NGA White Paper]. They would also facilitate coverage and enhance the safety and emergency response in rural areas. 6G NTN will help fulfil the three primary use cases of NTN- service ubiquity, service continuity and service scalability. Another major area where the NTN would play a significant role is complementing the terrestrial networks in coverage and affordability of services with reasonable 6G performance. NTNs are also expected to contribute to the goal of digital world experience and the goal of security and privacy of customers through the design of secure NTN UEs and NTN platforms. Finally, NTNs are expected to support LTE and 5G NR-based IoT devices through ubiquitous network coverage and device connectivity. There would be tremendous scope of 6G based applications, services and devices using NTNs and will materialize in the 6G time frame [19].

Several NTN enhancements and innovations for 6G have been identified in [10]. They can be classified into two categories, architectural innovations and radio interface innovations. The architectural innovations include

- NTN backhaul to connect gNBs on UAVs to the terrestrial core network, new security mechanisms on top of the current security authentication and authorization mechanisms,
- NTN platforms functioning as multi-access edge computing (MEC) devices and providing compute power for functions such as scheduling, slice selection, and internode coordination, NTN network nodes with the capabilities to provide communication services without the support of terrestrial core network, and introduction of novel AI techniques to automate traffic routing and real time satellite related decisions.

For the radio interface innovations,

- NTNs could potentially use non RF-based free space optical (FSO) or laser radio technology, which would offer ultra-low latency communication services by effectively overcoming atmospheric distortion.
- NTNs could offer multiple access technologies for network slicing, where the user selects the best radio access technology for optimal quality of service, NTNs in 6G would need Random access channel enhancements to

prevent overloading one satellite, which could potentially be done by using non-orthogonal multiple access (NOMA) together with successive interference cancellation.
- NTNs could help increase spectral efficiency by using joint communication methods, which would make use of two more satellites for various purposes.

5.6 Key Takeaways

Limitations in terrestrial networks have led to the emergence of NTNs. The 3GPP introduced the NTN as a feature in Release 17 and has been enhancing the NTN specifications to add more capabilities and flexibility such as mobility related enhancements. While the 3GPP has defined an NTN architecture based on a transparent NTN payload in Release 17 and 18, regenerative NTN payloads may be supported in a future release. A transparent NTN payload acts as a relay and communicates with the UE on one link and an NTN-Gateway on the other link. The gNB is located on the Earth's surface when the NTN architecture utilizes a transparent NTN payload. In case of a regenerative NTN payload, the gNB or the gNB-DU is placed on the NTN payload.

The NTN faces several unique challenges compared to a TN. Examples of these NTN challenges include Long Propagation Delays, Variable Propagation Delays, High and Variable Doppler Shifts, Large Coverage Areas, Mobility of Cells, Different Types of Beams, Types of Payloads, Path Loss and Signal Strength, and Faraday Effect. The 3GPP has introduced solutions to address these challenges to design a 5G-based NTN. Example solutions include Synchronization enhancements, timing advance enhancements, protocol timer adjustments, HARQ enhancements, timing relationship enhancements, and QoS enhancement, among others.

The NTN possesses new interfaces compared to a TN, leading to the need for special security considerations. Work investigates the vulnerabilities in wireline and wireless interfaces in NTNs and provides various interfaces that are vulnerable to jamming attacks. This chapter presents preliminary findings of a quantitative analysis carried out on the impact of jamming attacks on the NTN's wireless interfaces.

5G has a comprehensive security framework, including security for service-based and non-service-based interfaces and integrity protection and ciphering for both the Access Stratum and Non-Access Stratum signaling. 5G also supports integrity protection and ciphering for the User Plane to protect user traffic. This chapter describes how 5G security framework can be applied to secure an NTN.

The NTN is expected to play an important role in 5G-Advanced and 6G systems with new architectures and capabilities beyond 5G-based NTNs.

References

1 38.811, G. T. (2021). Study on new radio (NR) to support non-terrestrial network. Technical report, 3GPP.
2 Americas, G. (2022). 5G & non-terrestrial networks. Technical report, 5G Americas.
3 3GPP, T. (2023). Study on narrow-band internet of things (NB-IOT)/enhanced machine type communication (EMTC) support for non-terrestrial networks (NTN). Technical report, 3GPP.
4 38.300, G. T. (2022). NR and NG-RAN overall description. Technical report, 3GPP.
5 38.821, G. T. (2022). Solutions for nr to support non-terrestrial networks (NTN). Technical report, 3GPP.
6 3GPP, T. (2022). Summary of rel-17 work items. Technical report, 3GPP.
7 3GPP, T. (2023). Physical channels and modulation. Technical report, 3GPP.
8 3GPP, T. (2023). Physical layer measurements. Technical report, 3GPP.
9 3GPP, T. (2023). Physical layer measurements. Technical report, 3GPP.
10 Stuhlfauth, R. (2023). 5G NTN takes flight: Technical overview of 5G non-terrestrial networks. Technical report, Rohde&Schwarz.
11 3GPP, T. (2022). Radio resource control (RRC) protocol specification. Technical report, 3GPP.
12 3GPP, T. (2022). System architecture for the 5G system (5GS). Technical report, 3GPP.
13 Tripathi, N. D. and Reed, J. H. (2014). Cellular communications- a comprehensive and practical guide. Technical report, Wiley.
14 Tripathi, N. D. and Reed, J. H. (2019). 5G cellular communications- journey and destination. Technical report, The Wireless University.
15 3GPP, T. (2023). Security architecture and procedures for 5G system. Technical report, 3GPP.
16 3GPP, T. (2023). 3GPP system architecture evolution (SAE); security architecture. Technical report, 3GPP.
17 Hsu, E. (2023). Modulation schemes for satellite communications. Technical report, Keysight Technologies.
18 3GPP (2023). NTN enhancements NSC NTN task group meeting. Technical report, 3GPP.
19 Alliance, N. G. (2023). NGA report, 6G technologies. Technical report, Next G Alliance.

6

Securing Next G Millimeter-wave and Terahertz Communications

Jacek Kibilda[1], Parth Pathak[2], Vijay Shah[3], Kai Zeng[4] and André Gomes[5]

[1]*Commonwealth Cyber Initiative, Virginia Tech, USA*
[2]*Department of Computer Science, George Mason University, USA*
[3]*Electrical and Computer Engineering, North Carolina State University, USA*
[4]*Department of Electrical and Computer Engineering, George Mason University, USA*
[5]*Department of Computer Science, Rowan University, USA*

6.1 Introduction

6th Generation wireless (6G) systems will have to unlock the potential of tens-of-gigahertz-wide contiguous spectrum channels lying in millimeter-wave (mmW) and Terahertz (THz) frequency bands to deliver on the promise of ultra-high-speed communication, high-precision ranging, and other emerging services. These bands, which following the Next G Alliance terminology we will refer to as *high-bands*, cover a range of frequencies between 30 and 300 GHz, and are distinctly characterized by significant pathloss and susceptibility to blockage related to atmospheric absorption, penetration, and diffraction losses [1, 2]. Therefore, to ensure coverage and service availability, systems operating in high-bands must utilize phased antenna arrays and reflective surfaces to create and steer highly directional beams.

Limited coverage and angular radiation patterns are a potential advantage from the perspective of security, e.g., [3–5], as they limit the range of locations from which a wireless attacker, such as an eavesdropper or jammer, may operate effectively. However, the high dependency on the channel, inefficiencies in beam design, and practical system-level solutions that mandate the need for initial access expose high-band communication systems to new security vulnerabilities.

6G Security: New Frontiers in Mobile Network Technologies and Verticals, First Edition.
Edited by Luiz A. DaSilva.

These vulnerabilities can be exploited by sophisticated malicious transceivers and by Reconfigurable Intelligent Surfaces (RISs), which can be deployed at scale with minimal cost and power consumption.

The broader goal of designing reliable, resilient, and trustworthy 6G networks, as stated in [6], will require understanding these vulnerabilities and addressing the possible attacks by designing solutions that account for the system architecture and properly utilize the innate characteristics of high-band communication. Beyond identifying and detailing new vulnerabilities and attacks affecting 6G mmW and THz systems, this chapter proposes a three-faceted security framework comprising physical layer authentication, localization, beam management, and Machine Learning (ML)-hardening that will enhance high-band 6G security.

6.2 Security Vulnerabilities in High-band Communication Systems

High-band communication, with its limited range and selective angular emission, innately reduces the system's exposure to more conventional attacks such as eavesdropping or jamming. Corroborating this intuition, ample fundamental studies have shown that link covertness scales with the directionality of transmission [7, 8]. Furthermore, an effective attack requires the adversary to be on the line-of-sight path between the legitimate transmitter and receiver [3, 9], limiting the opportunities for launching the attack and increasing the chances of its detection. However, not all that glitters is gold. One of the most severe vulnerabilities lies at the heart of the standalone directional communication system design: transmission in mmW and THz bands requires directional links and beam management to establish and manage those links. Attackers may seek to exploit the beam establishment protocols to disrupt or hijack the communication.

Figures 6.1 and 6.2 illustrate an attacker that utilizes an active transmitter, or a passive RIS, to produce confounding signals to perform either an uplink or downlink attack against the system's reference symbols. In Figure 6.1, the attacker targets the reception of the pilot signals transmitted by the user equipment (UEs) at the base station (BS), resulting in a wrong beamforming vector generated by the BS. In Figure 6.2, the attacker performs a beam-sweeping attack whereby it produces copies of the reference signals transmitted by the BS during beam sweeping meant to confuse the UE about the direction of the legitimate transmission. These signals can disrupt either the vanilla system or, at a potentially lower power budget, the ML algorithm implemented to control it. This section will review and assess these attacks and consider other prospective attacks that may affect the security of mmW systems.

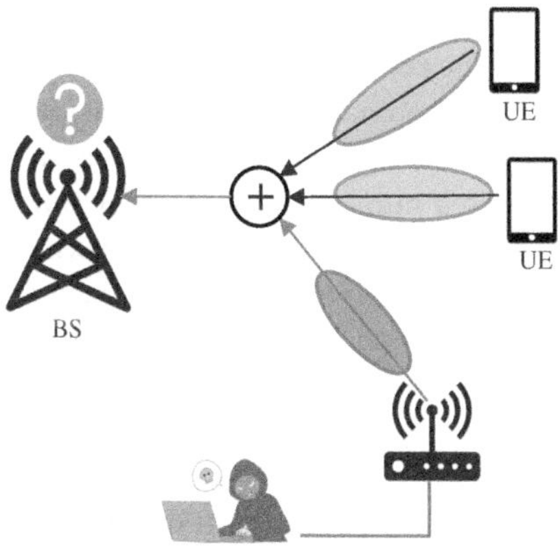

Figure 6.1 Pilot contamination attack.

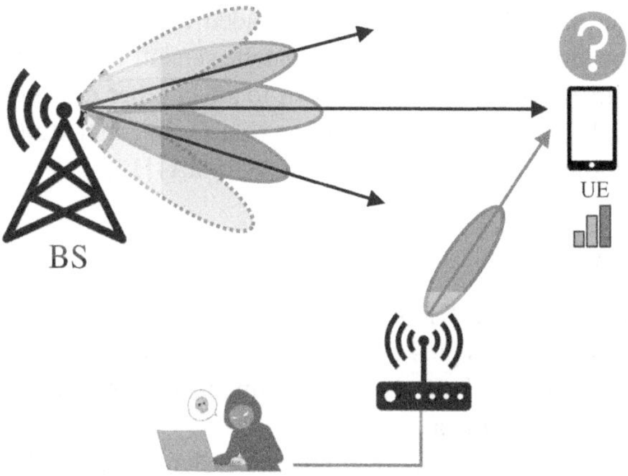

Figure 6.2 Beam sweeping attack.

6.2.1 Pilot Contamination Attack

The pilot contamination attack exploits the downlink beam (precoder) design for hybrid and digital massive Multiple-Input Multiple-Output (MIMO) systems. Designing a downlink beam requires high-quality Channel State Information (CSI) estimated based on the known training sequences transmitted by each UE in the uplink. Poor quality CSI will result in a precoder that does not accurately match the user channel and instead generates interference to other users. Since the pilot sequences are transmitted using unencrypted signals, an attacker may deliberately transmit known pilot sequences to disrupt the CSI acquisition, e.g., [10–14].

In [10], the feasibility of a simple attack where the attacker injects the same pilot sequence as the target user is considered. If the power introduced to the channel by the attacker is high enough, it dominates the channel estimation at the BS, leading to incorrect determination of the precoding vector. In [13], it is shown that a BS under attack can only serve a limited number of users whose number scales with the inverse of the square root of the number of attackers. Triggering the attack is relatively easy, as the training sequences are public information, and the attacker may also spend an arbitrary amount of time learning the beam training timing. Moreover, under the assumption of specific system-level knowledge, such as the BS location and/or channels of other users [11, 13], the attack can be made arbitrarily harmful.

In multi-user massive Multiple-Input Multiple-Output (mMIMO) systems, a passive RIS can be a particularly potent aid in triggering this attack [15–17]. When randomized phase shifts are altered between the training phase and data transmission, the performance degradation can be severe for multiple users at a time [15]. With just the knowledge of its own cascaded channel, the same malicious RIS can trigger an optimized and more potent version of the attack [16]. Similarly harmful attacks have also been shown to affect other directional transmission-based multiple access schemes, including Non-Orthogonal Multiple Access (NOMA) [12] and Rate-Splitting Multiple Access (RSMA) [17].

6.2.2 Beam Sweeping Attack

Standalone mmW systems, such as 5G NR or 802.11ad/ay, must implement beam sweeping to distribute the broadcast signal over the intended area and help end-users identify their spatial orientation [18], else will remain unaware of their directions and thereby not able to communicate at all.

Exhaustive probing of transmitter and receiver directions significantly reduces the covertness of communication, and the public nature of the beam-sweeping protocol and lack of integrity protection on broadcast signals allows the attacker to inject strong enough signals to potentially cause a beam misalignment, as illustrated in Figure 6.2. In [19], the attack as described above is triggered by an active jammer that performs its own beam sweeping with a jamming signal and with high enough power to change the Received Signal Strength (RSS) measured at the legitimate end-user device. The attack can be particularly potent even for modest power levels.

A more robust and potentially more difficult to detect version of the attack involves an attacker that produces training symbols identical to the ones produced by the BS to confuse not only the RSS-based beam decision but also the Reference Signal Received Power (RSRP)-based decision. In [20], the active jammer produces a noisy version of the training symbols. This attack requires an active jammer or amplify-forward relay that recycles the training sequences emitted by the legitimate BS, akin to the attack presented in [21]. The latter attack may also potentially utilize an RIS [22], in which case the attacker may have a smaller link budget, but a potentially cheaper device and easier to conceal, as it does not produce a signal of its own.

Figure 6.3 illustrates the potency of the two versions of the attack with an active jammer and a malicious RIS versus the baseline case, with both line-of-sight (LoS) and non-line-of-sight (NLoS) channels between the legitimate BS and UE. In the presence of the active jammer, the end-user selects beam 7 instead of beam 31. If the BS decides to proceed with this choice, the signal strength will be reduced by approximately 37 dB compared to what it would have been in the case of no attack, compromising the quality of the wireless link. Similarly, in the case of a malicious RIS attack, the end-user selects beam 45, which reduces the strength of the serving signal by about 20 dB compared to the baseline case.

6.2.3 Attack Against AI/ML-based Beam Management

Beam sweeping, as defined by the standard [23], is a highly resource-intensive procedure that introduces substantial redundancy to the system. This creates a strong incentive to reduce the sweeping overhead by considering ML-based approaches, e.g., [24–26]. Let us consider a receiver-based classifier $f(\mathbf{y}_M; \theta) : \mathcal{X}_M \rightarrow \mathbb{R}^N$, where $M \leq N$ is the size of the subset of all beams N, $\mathcal{X}_M$ is the space of input to the neural network and $\mathbf{y}_M$ is the vector of measurements for the M subset beams, and θ is the vector of the offline-trained model parameters. This classifier may be trained

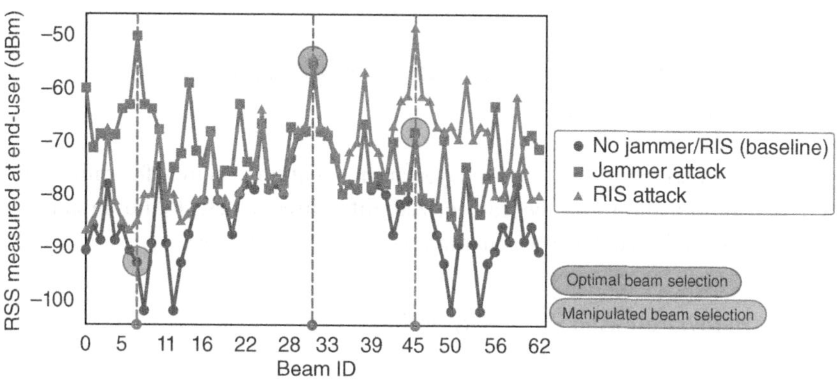

Figure 6.3 The effect of the beam sweeping attack on the link RSS for each of the 63 beams available at the BS. The jammer's transmit power is 5 dBm. The malicious RIS consists of 300 reflective elements. BS, UE, and the attacker are located at (0, 0), (500, 0), and (250, 100) meters.

to identify the best beam pair by considering a test sample containing received signal measurements after only probing M out of N beams.

The introduction of ML models exposes the beam management to attacks that rely on tiny (often imperceptible to humans) alterations to the input data [27]. These alterations can be introduced as unwanted signals or environment blockages during either the ML model training or deployment/testing. For instance, an attack on the training data, i.e., data poisoning, corrupts the training process of an ML model by introducing arbitrary and erroneous labels to the training dataset. In [28, 29], a poisoning attack against beam determination is proposed whereby the attacker, by smartly placing obstacles in the environment during the model training by a BS, creates a backdoor to the model. After the model deployment, the backdoor can be activated by reintroducing the obstacle(s) to the environment to arbitrarily change the beams used.

Another attack against ML-based beam management is applied solely during the model deployment. The vector of the attack is the input samples used for beam determination. The attacker's goal is to introduce a minimum perturbation to the environment, in the form of additional signals or obstacles, that will yield a wrong prediction, i.e.: $f(\mathbf{y}_M; \theta) \neq f(\mathbf{y}_M + \delta; \theta)$, where δ is a vector of perturbations. Here, state-of-the-art Adversarial Machine Learning (AML) techniques, such as Fast Gradient Sign Method (FGSM) or Projected Gradient Descent (PGD), play a prominent role in identifying the perturbations necessary to conduct the attack [30–33]. While potentially powerful in its impact, the attack is difficult to trigger as it requires the attacker to have access to the trained model (both the model classifier and its trained parameters θ) and at least partial information about the channel between the BS and UE to estimate the received signal strength.

6.3 PHY-layer Authentication for High-band Communication

Generally, existing physical (PHY)-layer authentication schemes for mmW and THz communication can be categorized into three types: Radio Frequency (RF)/hardware-based, location/channel-based, and hybrid ones.

6.3.1 Hardware-based RF Fingerprinting

RF/hardware-based schemes explore the imperfect features extracted from hardware or modulation to form an RF fingerprint that can identify devices. It is based on the observation in [34] that different beam patterns introduce different impairments to the waveform, which lead to impaired baseband I/Q samples. Then a convolutional neural network (CNN) can be trained based on baseband I/Q samples corresponding to different transmitter beam patterns/indices, and be used to differentiate transmitter beam patterns based on I/Q samples. This work motivates further study on differentiating different mmW transmitters based on I/Q samples collected on different transmission patterns.

Instead of using I/Q samples, authors in [35] propose to use the Signal-to-Noise Ratio (SNR) traces collected in the sector-level sweep (SLS) or beam sweeping process for mmW transmitter device identification. SLS trace-based fingerprinting can be directly applied to off-the-shelf devices without any extra hardware requirements. Usually, SLS traces depend on the transmitter beam pattern, orientation, and the wireless channel. In order to achieve an orientation and wireless channel oblivious device identification, the authors proposed a novel RF fingerprinting framework that extracts the adjacent beam correlation as features for identification. The fingerprinting scheme involves correlation analysis, surface fitting, curve pursuing, and binary classification. The authors conducted proof-of-concept experiments using off-the-shelf Talon AD7200 and MG360 WiGig devices and achieved high authentication accuracy at 99% with a small size of training samples.

6.3.2 Channel/Location-based Authentication

Location/channel-based schemes can be used to counter identity spoofing attacks based on location/channel features. Due to the sparsity of the channel in high-bands, virtual Angle of Arrival (AoA) can be used to detect and identify spoofing attacks [36–41]. In [36], authors proposed a channel-based identity spoofing attack detection scheme based on channel virtual (or called beamspace) representation for mmW communication. The principal components of channel virtual representation are used as a new channel feature. The proposed features are more sensitive to the transmitters' location than traditional channel features. The authors further proposed two detection strategies to tackle static and dynamic radio environments. For the static radio environment where the high temporal channel correlation is utilized, a Neyman-Pearson testing-based detection method is proposed to examine the correlation of consecutive channel measurements. The problem is transformed into a one-class classification for the dynamic radio environment. An online detection framework based on a feedforward neural network with a single hidden layer is presented. Simulation results showed that the beamspace-based methods outperform traditional CSI-based detection methods.

The virtual channel sparsity characteristics can also be utilized to detect pilot contamination attacks in NOMA and grant-free Internet of Things (IoT) mmW communication [12, 42]. By harnessing the sparsity and statistics of the MIMO virtual channel, the authors in [12] proposed two effective pilot contamination attack detection schemes to tackle static and dynamic environments, respectively. The basic idea is that a pilot contamination attack will change the channel sparsity characteristics, i.e., the sparsity degree will be reduced, or more signal paths/AoA will be detected at the receiver. The receiver can detect the existence of such an attack. In the static environment, the problem of pilot contamination attack detection is formulated as a binary hypothesis test of the virtual channel sparsity. For the dynamic environment, the statistic of the peaks in the virtual channel is leveraged to distinguish the contamination state from the normal state. A peak estimation algorithm and a machine learning-based detection framework were proposed to achieve high detection performance. A feature selection algorithm and an optimization model considering the detection accuracy and delay are presented to further optimize the proposed scheme. Simulation results showed that the proposed schemes can achieve 100% detection rate with 10^{-3} false alarm rate in the static environment and above 95% in the dynamic environment under various settings. Similar principles have been applied for pilot contamination attack detection in mmW grant-free access network [42], in which IoT nodes randomly access the network without waiting for any permission, and the non-orthogonal pilot sequences are used to identify active transmitter and estimate the corresponding channels. To tackle the uncertainty of random access of IoT nodes, the authors proposed a Single-Hidden-layer Multiple Measurement (SHMM) Siamese network based on the similarity of virtual channels.

6.3.3 Hybrid RF Fingerprinting

The hybrid schemes utilize both the hardware and location-induced features to identify a transmitter or detect an identity spoofing attack [43–45]. The authors in [44, 45] proposed to use the SNR traces collected during the SLS or beam sweeping process to identify a specific transmitter in a static environment. The observation is that SNR trace depends on transmitter location and intrinsic hardware impairment. It can be used for efficient detection even when the attacker is co-located with the legitimate transmitter. It is also readily available in off-the-shelf mmW devices. The detection problem is then formulated as a classification problem. The authors propose a novel neural network framework that utilizes generative adversarial networks (GANs) to enrich data samples to address the small sample learning and fast model construction challenges. The authors conducted experiments using off-the-shelf 802.11ad devices, Talon AD7200s, and MG360, and showed that the detection accuracy can reach 98% using small sample sizes under different scenarios.

6.4 Localization of Malicious High-band Devices

The motivation of the localization is that if a malicious actor can be localized, BSs and UEs can then leverage this information to proactively prevent any signal leakage towards such an actor. When BSs and UEs proactively avoid transmitting signals toward the malicious actor, many PHY-layer attacks can be mitigated effectively. While localization in mmW networks has been extensively studied [46–48], the prior approaches are limited to active transceivers that can measure and provide some form of channel feedback, e.g., measured SNR, CSI, etc. However, the problem is non-trivial when localizing passive devices, such as RIS, which can neither perform explicit channel measurements nor actively transmit.

A RIS can be localized in two ways: (i) multiple (at least three) cooperating BSs estimate their Angle of Departure (AoD) or Time of Flight (ToF) to a RIS, enabling location estimation through triangulation and (ii) one BS estimates both AoD and ToF to a RIS, making it possible to directly estimate a RIS location through angle and distance. Since a RIS cannot measure the channel and/or transmit actively, a UE-assisted cooperative RIS localization scheme can be designed. Such localization has two fundamental blocks: BS to RIS AoD estimation and BS to RIS ToF estimation.

6.4.1 BS to RIS AoD Estimation

The beam sweeping process can be exploited for the purpose of localization, making it not only protocol compatible but also making the localization transparent to the (possibly malicious) RIS. Given the passive nature of RIS, our UE-assisted BS to RIS AoD estimation is based on two AoA estimation Figure 6.4: (i) BS-UE AoA estimation and (ii) BS-RIS-UE AoA estimation.

The AoA estimation relies on channel impulse response (CIR) measurement at the UE. CIR has been used for LoS AoA estimation in prior works such as [49, 50], albeit these works do not involve passive entities such as the RIS. Here, the UE can measure the channel metric (e.g., CSI or SNR) when the BS performs

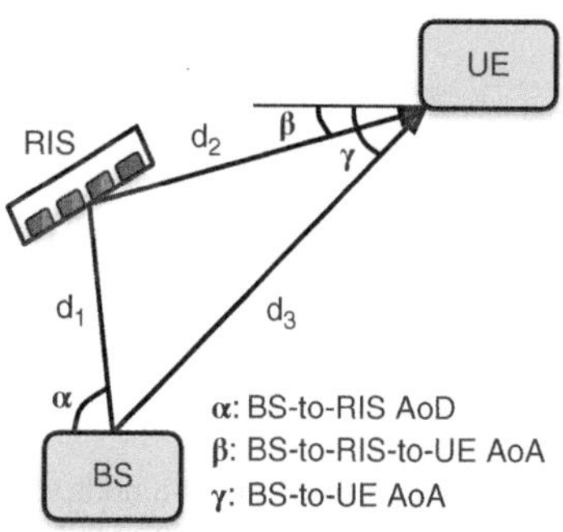

Figure 6.4 UE-assisted BS to RIS AoD estimation based on BS-UE AoA and BS-RIS-UE AoA estimation.

beam sweeping, collecting the channel quality for each transmit sector. Existing protocol-specific periodic beam sweeping processes can be leveraged for this purpose. To perform the localization, the UE can estimate the antenna gain pattern corresponding to BS's transmit sectors. The information can be represented as the azimuth and elevation gain for all angles for each sector. The UE can then use the SNR and sector gain pattern of all sectors to determine the angle where maximum signal power is received, i.e., AoA.

One key challenge in AoA estimation is that the received signal at the UE is a combination of the signal directly coming from BS and the signal coming from BS to RIS to the UE. However, we can exploit the power delay profile (PDP) derived from CIR to separate the two incoming signals in Figure 6.5. When we have the PDP of a received signal, each path can be represented as a peak at different arrival times because they have different traveling distances. The relative time difference between the peaks remains the same because the difference in physical path length remains the same even if beam patterns change or the measurements of PDP are incoherent. The peak amplitudes on the PDP can change according to different gains of beam patterns towards the path directions. When it has K multipath with different time delay τ, the received signal with the mth beam pattern can be expressed as: $y(t) = \sum_{i=1}^{K} \sum_{\theta} A_m(\theta_i) \cdot h_i(\theta_i) \cdot x(t - \tau_i)$, where $A_m(\theta_i)$ and $h_i(\theta_i)$ are the gain of the mth beam and the complex channel gain, respectively, when we have an incident angle of θ_i. It shows that when we have different gains due to different beam patterns, the received signal is changed by only the gain pattern of the beam pattern. If we measure the PDP with multiple different known BS patterns $m = 1, \ldots, M$ which should be $M \geq 4K$, each measurement provides a set of peaks $Amp_m = \{a_{m,1}, a_{m,2}, \ldots, a_{m,k}\}$. Hence, we can collect the amplitude set $Amp_{m,k} = \{a_{1,k}, a_{2,k}, \ldots, a_{M,k}\}$ with M beam patterns for k^{th} path. Due to the different gains of mth beam for different AoA, the peaks will have different amplitudes with

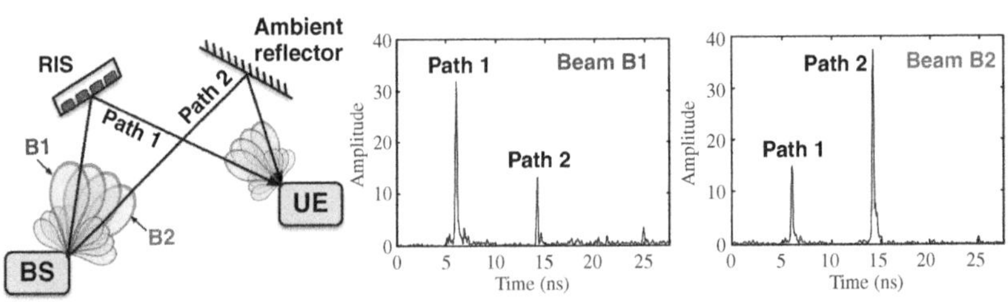

Figure 6.5 Different BS, UE and RIS sectors result in different received signal amplitude for RIS and ambient reflector paths, enabling us to not only separate them in time but also identify the RIS-associated path through amplitude variation.

the same physical paths. Hence, we can utilize this to estimate the AoA by comparing the normalized amplitude set $Amp'_{m,k} = \frac{Amp_{m,k}}{min(Amp_{m,k})}$ and beam pattern gain $G'_{m,\theta} = \frac{G_{m,\theta}}{min(G_{m,\theta})}$, where $G_{m,\theta}$ denotes the gain of mth beam pattern toward θ. Thus, the angle of arrival (θ_k) of kth path can be found from the following optimization: $\theta_k = \arg\min_\theta |Amp'_{m,k} - G'_{m,\theta}|^2$, which searches for discrete angle θ_k that provides a minimum error between the measured CIR and beam pattern. Because this is a non-linear error curve fitting problem, a Least Mean Square (LMS) algorithm can be used to find the approximate solution. As shown in Figure 6.4, the AoA for BS-UE and BS-RIS-UE paths can then be used by the BS to calculate the BS to UE AoD.

Figure 6.5 shows the PDP for two different transmit sectors from BS. Here, we use our mmW software radio (acting as BS) to transmit beam sweeping frames while a UE receives the frames over different sectors and measures the CIR and calculates the PDP. We carefully designed the experiments to include two reflections in the environment – a passive ambient reflector (i.e., wall) and a RIS. We observe that, depending on the antenna gain of different sectors, different amplitudes for two reflected paths are observed in the PDP. We find that using the proposed AoA finding algorithm, we can calculate the AoA of two paths at the UE with a mean error of 6.2°. The error can be further reduced with the use of optimized beam patterns compared to the non-uniform sectors used in commodity devices.

6.4.2 ToF Estimation and Different RIS Beam Sweeping Capabilities

A RIS can use a multi-lobe beam (multi-beam) or sweep through its own set of reflection beams for each incoming BS beam. The above-mentioned BS to RIS AoD estimation naturally extends to both RIS beam sweeping capabilities. In the case of multi-beam, the proposed method can still accurately estimate the AoA at the UE as long as the UE is covered through the multi-beam. In the case where the RIS sweeps through multiple beams, the AoA estimation is also readily applicable and, in fact, can achieve a high accuracy of AoA estimation if the RIS beams are narrower compared to the multi-beam.

It is worth noting that BS to RIS AoD estimation can be sufficient to mitigate or at least alleviate some of the attacks discussed previously by restricting BS's transmissions in directions other than the malicious RIS. However, there can be cases where precise localization of the RIS is necessary from an operational security perspective (e.g., manual site inspection). In such cases, the AoD can be combined with ToF information to localize a RIS using a single BS. Although direct ToF estimation between BS to RIS is not directly measurable, a UE-assisted process (similar to the AoD estimation) can be applied for ToF estimation. Here, the ToF from BS to UE and ToF from BS to RIS to UE can be measured by the UE using CIR.

This can be combined with the BS to RIS AoD estimate (or RIS to UE AoA estimation) to determine the BS to RIS ToF. Our proposed localization methods are protocol-compatible and can leverage the existing protocol messages for this purpose. The 5G NR protocol supports multi-Round Trip Time (RTT) measurements [51] where the time measurements can be performed with serving as well as the neighboring BSs. The AoA and AoD estimation can be carried out using the beam sweeping frames as described earlier and can be combined with ToF estimated through multi-RTT measurements.

6.5 Securing AI/ML-driven Beam Management

Addressing beam management security concerns requires new beam management protocol designs whereby the transmit power of the probing sequences are optimized [52], or the probing sequence is randomized [20]. This section focuses on AI/ML-specific challenges and adaptations. Particularly, two primary security measures are considered: i) the adaptive reduction of the probing opportunities in the presence of adversaries and ii) robust/adversarial training to harden the AI/ML models used for beam determination.

6.5.1 Adaptive Reduction of Beam Sweeping Opportunities

The vast literature on reducing the probing opportunities, e.g., [26, 34, 53–58], is focused solely on improving the efficiency of the beam sweeping process. Intuitively, reducing the sweeping space should increase the secrecy of a high-band link. To validate this intuition, a strategy for adaptive beam management to reduce legitimate link exposure to potential attacks is proposed [59, 60]. The proposed strategy leverages Deep Reinforcement Learning (DRL) for its ability to utilize the context information available during beam management, such as previous best beam information, and its adaptability to generalize even in the presence of a dynamic environment [61]. DRL has been shown to perform particularly well in adapting different beam management tasks [26, 57, 58, 62].

The proposed adaptive beam management can be formally described as follows using the Markov Decision Process (MDP) framework [63]. At a discrete time t, an agent selects an action $\mathbf{a}_t \in \mathcal{A}$ to be performed on the environment corresponding to selecting a beam sweeping sequence from a set of beam sweeping sequences available in the system $\mathcal{A}$. Each beam sweeping sequence is a subset of beams l taken from a fixed analog beamforming codebook $\mathcal{L}$ defined for the given antenna array, akin to practical implementations specified by the 3GPP for 5G NR [23]. The state of the environment at time t then consists of three features $s_t = (v_t, \mathbf{a}_t, l_t)$, where $v_t \in \mathbb{R}^+$ represents the average SNR from the legitimate communication link, $\mathbf{a}_t \in \mathcal{A}$ represents the selected sweeping sequence and $l_t \in \mathcal{L}$ denotes the ID of the best beam at timestep t. Each time the sweeping sequence $\mathbf{a}_t$ in state s_t is selected, the environment returns the reward r_t and updates the state to s_{t+1}. The proposed solution relies on Q-values to determine and store a matching between the state and action pairs.

In order to optimize the secrecy of beam management, the proposed reward function adapts the definition of secrecy pressure, proposed in [64], to scenarios where the legitimate link uses beam sweeping, and a discrete number of sensor nodes take signal-level measurements. Formally, this can be described as:

$$r_t(\mathbf{a}_t, s_t) = \frac{1}{|\mathbf{a}_t| \times |\mathcal{X}|} \sum_{l \in \mathbf{a}_t} \sum_{x \in \mathcal{X}} \max\{0, \hat{C}_B(v_t, l) - \hat{C}_E(x, l)\}, \tag{6.1}$$

where $\mathcal{X}$ is the set of SNR measurements captured at time t, $l \in \mathbf{a}_t$ represents the beam being swept by the legitimate transmitter from codebook $\mathbf{a}_t$, and $\hat{C}_B(v_t, l)$ denotes the rate of the legitimate link based on the measured SNR v_t for beam l, and $\hat{C}_E(x, l)$ the rate of the eavesdropper link based on the measured SNR $x \in \mathcal{X}$ when beam l is being used at the transmitter. $\hat{C}_B(v_t, l)$ and $\hat{C}_E(x, l)$ can be interpreted as rate estimates based on the measured SNR at the receiver and sensors, respectively. The proposed reward is expressed in bits per second, and higher values indicate increased secrecy regarding the legitimate link. The best action is selected following an ϵ-greedy strategy. The value of the $\epsilon \in (0, 1)$ parameter determines the preference between exploration and exploitation [63]. The neural network is updated following the Double Deep Q-Network (DDQN) approach to adapt to dynamic environments without overfitting [65].

The proposed beam management strategy is validated in a live experimental setting, leveraging STAMINA [66, 67], a Software-Defined Radio (SDR) platform for experimentation with mmW beam management. Figure 6.6 illustrates the beam sweep secrecy and beam determination accuracy for a scenario where three mmW radio units represent the legitimate transmitter, legitimate receiver, and an

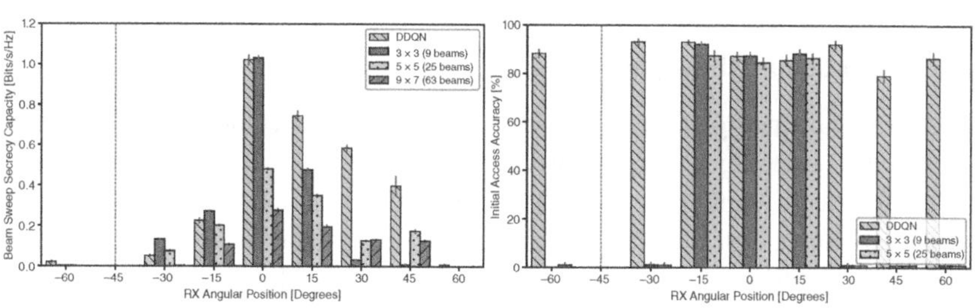

Figure 6.6 Performance of the sweep secrecy capacity with respect to the different angular positions of the receiver. The presence of the sensor node (red dashed vertical line) compromises the secrecy of the legitimate link. The right-hand-side figure shows the changes to the accuracy of the beam determination procedure with the changes to the receiver position.

eavesdropper that is located a distance away at a −45° angle to the left of the transmitter. The experimental radio units up/down convert signals from/to an Intermediate Frequency (IF) in the n46 band (5.3 GHz) to/from the n257 band (28 GHz), and are equipped with 8×8 phased array antennas that operate over a predefined codebook $\mathcal{L}$ comprising 63 calibrated beams arranged in a 9×7 grid, ranging from ±45° in the azimuth and ±35° in the elevation plane. To ensure that the proposed method can identify sweeping sequences that optimize the secrecy in realtime over such a large codebook, the set of beam sweeping sequences available in the system $\mathcal{A}$ comprises only sweeping sequences of size 3×3, 5×5 beams centered around each beam and an exhaustive search through the space of 63 available beams.

In the left-hand side of Figure 6.6, we can observe that the 3×3 sweeping sequence centered around the antenna boresight can achieve the highest sweep secrecy. As the receiver moves outside of the angular range of the transmitter, it starts to leverage sidelobes or reflections to communicate. The adaptive beam management strategy can achieve better performance until the angle where the mmW radios become misaligned, and communication is no longer feasible. We can also observe the influence of the eavesdropper sensor on the sweep secrecy capacity, as it drops dramatically in the vicinity of the sensor, indicating that in that direction, the communication has no secrecy and should be avoided. The right graph in Figure 6.6 shows the beam determination accuracy. For the region within the range of ±45° degrees for the relative orientation between mmW transmitter and receiver, the different sweeping sequences and the proposed strategy perform similarly to each other. However, as the receiver moves away from the boresight, accurate beam determination is considerably more challenging, especially when shorter sweeping sequences are used. The proposed DDQN strategy maintains accuracy throughout the considered region while optimizing secrecy. The experimental results validate the operation of the proposed strategy and demonstrate its improvement over link secrecy while avoiding eavesdroppers and maintaining link quality.

6.5.2 Hardening the AI/ML

The Artificial Intelligence (AI)/ML-driven adaptation is vulnerable to adversarial machine learning attacks, as was noted in Section 6.2. In general, an adversary interested in exploiting the AI/ML capabilities will seek to add perturbations to the legitimate inputs to the model or the training datasets, resulting in either incorrect classification or reduced confidence in the model. In the case of beam management, misclassification will result in selecting a sub-optimal beam for transmission or potentially a beam that aligns the legitimate node with a malicious one [30]. A couple of ways can be pursued to defend against and mitigate AML attacks against beam management, depending on whether the training or input data is believed to be vulnerable.

If the training data is the subject of the attack, the legitimate system must be able to detect the poisoned model and remove the impact of the poisoned samples. In the case of beam management, the model may be poisoned during the training process by smartly altering the environment in which the model is trained and will yield wrong beam predictions only if the trigger is present in the input data [28]. Detecting the potential presence of backdoor triggers in a model may be done prior to the model deployment and without the knowledge of the dataset or its labels. In [28], the model is analyzed for backdoors by tracking the anomalies in the model's confidence of predictions in response to random stimuli. Overly high confidence in predictions made for random stimuli may suggest the presence of a backdoor. The detoxification of the model can be done similarly to adversarial training by updating the model weights to optimize the minimax criterion, i.e., minimizing the worst-case loss. In other words, the updated model is re-trained to yield correct beam predictions even for backdoored samples. Unfortunately, the accuracy of such a detoxification process hinges on the knowledge of the adversarial samples.

The adversarial training is generally applied to increase the robustness of the AI/ML model to adversarial inputs. The principal idea behind adversarial training is for the legitimate user to perform the training on data that is enhanced by additional data corresponding to the adversarial samples [68]. The creation of adversarial training samples will generally depend on the adversarial model. For instance, in the mmW beam prediction model considered in [31], the additional data points are created using the FGSM attack against beam prediction. The adversarial training for the FGSM uses the gradient of the loss function associated with the legitimate model to generate the adversarial samples. However, training with adversarial samples will generally reduce model accuracy on the original task. Moreover, adversarial training may require a lot of samples and regular re-training depending on the complexity of the attack and the adversary's ability to adapt to the attack. This training overhead can be addressed using the defensive distillation [69]. The defensive distillation does not use adversarial training samples. Instead, it relies on transferring knowledge between two network architectures. One network trains on the dataset with hard labels, i.e., for each sample, a specific class is assigned and produces a soft output in the form of a probability distribution over the label set. This soft output of the first network is used to replace the original labels in the dataset, and the newly created dataset with soft labels is fed into the second neural network with the same neural architecture. The second network can extract extra information from probability vector labels to substantially improve the robustness of the model to adversarial perturbations. In other words, defensive distillation improves the generalizability of the trained model. In the context of beam management, the defensive distillation was applied to mitigate the FGSM attack against ML-controlled beam management [32]. Alternatives to adversarial training and defensive distillation involve collecting data and training on randomized beam sweeping sequences to decrease the attacker's capabilities to intercept information about the beam management [70].

6.6 Key Takeaways

In this section, we extract key takeaways from the security threats and our proposed security framework and discuss their implications for securing high-band 6G.

Protocols and system design vulnerabilities are the Achilles heel of high-band 6G systems. There is ample evidence corroborating the intuition that highly directional transmission yields greater security. However, establishing and maintaining connectivity in standalone high-band communication systems requires beam management and training, which have been shown to be vulnerable to tampering through over-the-air attacks triggered by an active jammer or even passive RIS. These attacks generally exploit the turn-based protocol for establishing communication beams, but smarter attacks have been proposed that directly target vulnerabilities in neural networks that make beam selection and design decisions.

Physical layer features will be key to authenticate legitimate transmissions. An important defense mechanism against over-the-air attacks against beam management involves the ability to authenticate signals from legitimate transceivers. Large beam codebooks and sparse channels in mmW and THz offer an increased feature space that can be used for transceiver authentication. Where specific features may be difficult to track with classical methods due to the size and noise in the data, such as I/Q, CSI, or SNR samples, neural network-based approaches can help sort through the data and classify the transceivers.

Location information about malicious nodes can inform the beamforming design. High-band systems may update their beamforming vectors not only to concentrate transmission/reception to obtain higher SNR but also to create nulls in the direction of malicious nodes. This requires localization capabilities. In mmW and THz communication, the channel sparsity can be used to help separate legitimate from non-legitimate paths, and the large beam patterns can be utilized to estimate the AoA. For passive attackers, such as RIS, an additional ToF estimation can be made.

AI/ML can be used to increase efficiency and mitigate attacks, albeit it must be appropriately trained to ensure robustness to targeted attacks. The beam management algorithms may be adapted to effectively increase secrecy. The AI/ML models used in these adaptations should also be subjected to adversarial training or robust optimization to improve their robustness to attacks exploiting small perturbations to the input data.

References

1 Xiao, M., Mumtaz, S., Huang, Y., Dai, L., Li, Y., Matthaiou, M., Karagiannidis, G. K., Björnson, E., Yang, K., I, C.-L., and Ghosh, A. (2017). Millimeter wave communications for future mobile networks. *IEEE Journal on Selected Areas in Communications*, 35(9): 1909–1935.

2 Rappaport, T. S., Xing, Y., Kanhere, O., Ju, S., Madanayake, A., Mandal, S., Alkhateeb, A., and Trichopoulos, G. C. (2019). Wireless communications and applications above 100 GHz: Opportunities and challenges for 6G and beyond. *IEEE Access*, 7: 78729–78757.

3 Ma, J., Shrestha, R., Adelberg, J., Yeh, C.-Y., Hossain, Z., Knightly, E., Jornet, J. M., and Mittleman, D. M. (2018). Security and eavesdropping in terahertz wireless links. *Nature*, 563(7729): 89–93.

4 He, S., Zhang, Y., Wang, J., Zhang, J., Ren, J., Zhang, Y., Zhuang, W., and Shen, X. (2021). A survey of millimeter-wave communication: Physical-layer technology specifications and enabling transmission technologies. *Proceedings of the IEEE*, 109(10): 1666–1705.

5 Jamali, M. V. and Mahdavifar, H. (2021). Covert millimeter-wave communication: Design strategies and performance analysis. *IEEE Transactions on Wireless Communications*, 21(6): 3691–3704.

6 ATIS Next G Alliance (2022). Next G Alliance report: Roadmap to 6G. Technical report, Next G Alliance.

7 Wang, L., Elkashlan, M., Duong, T. Q., and Heath, R. W. (2014). Secure communication in cellular networks: The benefits of millimeter wave mobile broadband. In *IEEE International Workshop on Signal Processing Advances in Wireless Communications (SPAWC)*, pages 115–119.

8 Wang, C. and Wang, H.-M. (2016). Physical layer security in millimeter wave cellular networks. *IEEE Transactions on Wireless Communications*, 15(8): 5569–5585.

9 Ju, Y., Wang, H.-M., Zheng, T.-X., and Yin, Q. (2017). Secure transmissions in millimeter wave systems. *IEEE Transactions on Communications*, 65(5): 2114–2127.

10 Kapetanović, D., Zheng, G., Wong, K.-K., and Ottersten, B. (2013). Detection of pilot contamination attack using random training and massive MIMO. In *IEEE Annual International Symposium on Personal, Indoor, and Mobile Radio Communications (PIMRC)*, pages 13–18.

11 Akgun, B., Krunz, M., and Koyluoglu, O. O. (2018). Vulnerabilities of massive MIMO systems to pilot contamination attacks. *IEEE Transactions on Information Forensics and Security*, 14(5): 1251–1263.

12 Wang, N., Jiao, L., Alipour-Fanid, A., Dabaghchian, M., and Zeng, K. (2019a). Pilot contamination attack detection for NOMA in 5G mm-wave massive mimo networks. *IEEE Transactions on Information Forensics and Security*, 15: 1363–1378.

13 Akbar, N., Yan, S., Khattak, A. M., and Yang, N. (2020). On the pilot contamination attack in multi-cell multiuser massive MIMO networks. *IEEE Transactions on Communications*, 68(4): 2264–2276.

14 Yang, J., Ji, X., Wang, F., Huang, K., and Guo, L. (2021). A novel pilot spoofing scheme via intelligent reflecting surface based on statistical CSI. *IEEE Transactions on Vehicular Technology*, 70(12): 12847–12857.

15 Huang, H., Zhang, Y., Zhang, H., Cai, Y., Swindlehurst, A. L., and Han, Z. (2024). Disco Intelligent Reflecting Surfaces: Active Channel Aging for Fully-Passive Jamming Attacks. *IEEE Transactions on Wireless Communications*, 23(1): 806–819.

16 de Sena, A. S., Kibilda, J., Mahmood, N. H., Gomes, A., and Latva-aho, M. (2024). Malicious RIS Versus Massive MIMO: Securing Multiple Access Against RIS-Based Jamming Attacks. *IEEE Wireless Communications Letters*, 13(4): 989–993.

17 De Sena, A. S., Gomes, A., Kibiłda, J., Mahmood, N. H., DaSilva, L. A., and Latva-Aho, M. (2024). Malicious RIS Meets RSMA: Unveiling the Robustness of Rate Splitting to RIS-Induced Attacks. In *IEEE Global Communications Conference (GLOBECOM)*, pages 3576–3581.

18 Giordani, M., Polese, M., Roy, A., Castor, D., and Zorzi, M. (2018). A tutorial on beam management for 3GPP NR at mmWave frequencies. *IEEE Communications Surveys & Tutorials*, 21(1): 173–196.

19 Dinh-Van, S., Hoang, T. M., Cebecioglu, B. B., Fowler, D. S., Mo, Y. K., and Higgins, M. D. (2023). A Defensive Strategy Against Beam Training Attack in 5G mmWave Networks for Manufacturing. *IEEE Transactions on Information Forensics and Security*, 18: 2204–2217.

20 Darsena, D. and Verde, F. (2022). Anti-jamming beam alignment in millimeter-wave MIMO systems. *IEEE Transactions on Communications*, 70(8): 5417–5433.

21 Li, J., Lazos, L., and Li, M. (2024). SecBeam: Securing mmWave Beam Alignment Against Beam-Stealing Attacks. In *IEEE Conference on Communications and Network Security (CNS)*, pages 1–9.

22 Gomes, A., de Sena, A. S., Mahmood, N. H., Latva-aho, M., DaSilva, L., and Kibiłda, J. (2024). Beam Management Manipulation with Adversarial

Reconfigurable Intelligent Surfaces. In *IEEE Global Communications Conference (GLOBECOM)*, pages 3231–3236.

23 3GPP (2021). Physical layer procedures for control (Release 17). Technical Report 38.213 V17.0.0, 3GPP.

24 Klautau, A., Batista, P., González-Prelcic, N., Wang, Y., and Heath, R. W. (2018). 5G MIMO data for machine learning: Application to beam-selection using deep learning. In *Information Theory and Applications Workshop (ITA)*, pages 1–9.

25 Zhang, J., Huang, Y., Wang, J., You, X., and Masouros, C. (2020). Intelligent interactive beam training for millimeter wave communications. *IEEE Transactions on Wireless Communications*, 20(3): 2034–2048.

26 Cousik, T. S., Shah, V. K., Erpek, T., Sagduyu, Y. E., and Reed, J. H. (2024). Deep Learning for Fast and Reliable Initial Access in AI-Driven 6G mmWave Networks. *IEEE Transactions on Network Science and Engineering*, 11(6): 5668–5680.

27 Heaven, D. (2019). Why deep-learning AIs are so easy to fool. *Nature*, 574: 163–166.

28 Zhang, Z., Tian, M., Li, C., Huang, Y., and Yang, L. (2022a). Poison neural network-based mmWave beam selection and detoxification with machine unlearning. *IEEE Transactions on Communications*, 71(2): 877–892.

29 Zhang, Z., Yang, R., Zhang, X., Li, C., Huang, Y., and Yang, L. (2022b). Backdoor federated learning-based mmWave beam selection. *IEEE Transactions on Communications*, 70(10): 6563–6578.

30 Kim, B., Sagduyu, Y., Erpek, T., and Ulukus, S. (2021). Adversarial attacks on deep learning based mmWave beam prediction in 5G and beyond. In *IEEE Statistical Signal Processing Workshop (SSP)*, pages 590–594.

31 Catak, F. O., Kuzlu, M., Catak, E., Cali, U., and Unal, D. (2022). Security concerns on machine learning solutions for 6G networks in mmWave beam prediction. *Physical Communication*, 52: 101626.

32 Kuzlu, M., Catak, F. O., Cali, U., Catak, E., and Guler, O. (2023). Adversarial security mitigations of mmWave beamforming prediction models using defensive distillation and adversarial retraining. *International Journal of Information Security*, 22(2): 319–332.

33 Krishna, G. S., Supriya, K., Singh, S., and Baidya, S. (2023). Adversarial Security and Differential Privacy in mmWave Beam Prediction in 6G networks. In *Cyber Security in Networking Conference (CSNet)*, pages 5–11.

34 Polese, M., Restuccia, F., and Melodia, T. (2021). DeepBeam: Deep waveform learning for coordination-free beam management in mmWave networks. In *Proceedings of the International Symposium on Theory, Algorithmic Foundations, and Protocol Design for Mobile Networks and Mobile Computing*, pages 61–70.

35 Wang, N., Li, W., Jiao, L., Alipour-Fanid, A., Xiang, T., and Zeng, K. (2022). Orientation and channel-independent RF fingerprinting for 5G IEEE 802.11ad devices. *IEEE Internet of Things Journal*, 9(11): 9036–9048.

36 Li, W., Wang, N., Jiao, L., and Zeng, K. (2021). Physical layer spoofing attack detection in mmWave massive mimo 5G networks. *IEEE Access*, 9: 60419–60432.

37 Wang, N., Li, W., Wang, P., Alipour-Fanid, A., Jiao, L., and Zeng, K. (2020c). Physical layer authentication for 5G communications: Opportunities and road ahead. *IEEE Network*, 34(6): 198–204.

38 Wang, N., Li, W., Alipour-Fanid, A., Jiao, L., Dabaghchian, M., and Zeng, K. (2021b). Pilot contamination attack detection for 5G mmWave grant-free iot networks. *IEEE Transactions on Information Forensics and Security*, 16: 658–670.

39 Wang, N., Li, W., Alipour-Fanid, A., Dabaghchian, M., and Zeng, K. (2020b). Compressed-sensing-based pilot contamination attack detection for NOMA-IoT Communications. *IEEE Internet of Things Journal*, 7(8): 7764–7772.

40 Wang, N., Wang, P., Alipour-Fanid, A., Jiao, L., and Zeng, K. (2019b). Physical-layer security of 5G wireless networks for IOT: Challenges and opportunities. *IEEE Internet of Things Journal*, 6(5): 8169–8181.

41 Wang, N., Jiao, L., Wang, P., Dabaghchian, M., and Zeng, K. (2018). Efficient identity spoofing attack detection for IoT in mm-Wave and massive MIMO 5G communication. In *IEEE Global Communications Conference (GLOBECOM)*, pages 1–6.

42 Wang, N., Li, W., Alipour-Fanid, A., Jiao, L., Dabaghchian, M., and Zeng, K. (2021c). Pilot contamination attack detection for 5G mmWave grant-free IoT networks. *IEEE Transactions on Information Forensics and Security*, 16: 658–670.

43 Wang, N., Li, W., Wang, P., Alipour-Fanid, A., Jiao, L., and Zeng, K. (2020d). Physical layer authentication for 5G communications: Opportunities and road ahead. *IEEE Network*, 34(6): 198–204.

44 Wang, N., Jiao, L., Wang, P., Li, W., and Zeng, K. (2021a). Exploiting beam features for spoofing attack detection in mmWave 60-GHz IEEE 802.11ad networks. *IEEE Transactions on Wireless Communications*, 20(5): 3321–3335.

45 Wang, N., Jiao, L., Wang, P., Li, W., and Zeng, K. (2020a). Machine learning-based spoofing attack detection in mmWave 60 GHz IEEE 802.11ad networks. In *IEEE Conference on Computer Communications (INFOCOM)*, pages 2579–2588.

46 Garcia, D., Lacruz, J. O., Jiménez Mateo, P., and Widmer, J. (2020). POLAR: Passive object localization with IEEE 802.11ad using phased antenna arrays. In *IEEE Conference on Computer Communications (INFOCOM)*, pages 1838–1847.

47 Ruan, Y., Chen, L., Zhou, X., Guo, G., and Chen, R. (2022). Hi-Loc: Hybrid Indoor Localization via Enhanced 5G NR CSI. *IEEE Transactions on Instrumentation and Measurement*, 71: 1–15.

48 Menta, E. Y., Malm, N., Jäntti, R., Ruttik, K., Costa, M., and Leppänen, K. (2019). On the performance of AoA–based localization in 5G ultra–dense networks. *IEEE Access*, 7: 33870–33880.

49 Pefkianakis, I. and Kim, K.-H. (2018). Accurate 3D Localization for 60 GHz Networks. In *Proceedings of the 16th ACM Conference on Embedded Networked Sensor Systems*, SenSys '18, page 120–131.

50 Sur, S., Pefkianakis, I., Zhang, X., and Kim, K.-H. (2018). Towards Scalable and Ubiquitous Millimeter-Wave Wireless Networks. In *Proceedings of the 24th Annual International Conference on Mobile Computing and Networking*, MobiCom '18, page 257–271.

51 Dwivedi, S., Shreevastav, R., Munier, F., Nygren, J., Siomina, I., Lyazidi, Y., Shrestha, D., Lindmark, G., Ernstrom, P., Stare, E., Razavi, S. M., Muruganathan, S., Masini, G., Busin, A., and Gunnarsson, F. (2021). Positioning in 5G Networks. *IEEE Communications Magazine*, 59(11): 38–44.

52 Zhang, J., Li, M., Yan, S., Liu, C., Chen, X., Zhao, M.-J., and Whiting, P. (2021). Joint beam training and data transmission design for covert millimeter-wave communication. *IEEE Transactions on Information Forensics and Security*, 16: 2232–2245.

53 Aykin, I. and Krunz, M. (2020). Efficient beam sweeping algorithms and initial access protocols for millimeter-wave networks. *IEEE Transactions on Wireless Communications*, 19(4): 2504–2514.

54 Ganji, V. S. S., Lin, T.-H., Espinal, F. A., and Kumar, P. (2022). BeamSurfer: Minimalist Beam Management of Mobile mm-Wave Devices. *IEEE Transactions on Wireless Communications*, 21(11): 8935–8949.

55 Gan, C., Yang, J., and Shen, C. (2020). Thresholded Wirtinger Flow for Fast Millimeter Wave Beam Alignment. In *Conference on Signals, Systems, and Computers (Asilomar)*, pages 32–36.

56 Yan, H. and Cabric, D. (2019). Compressive initial access and beamforming training for millimeter-wave cellular systems. *IEEE Journal of Selected Topics in Signal Processing*, 13(5): 1151–1166.

57 Narengerile, N., Thompson, J., Patras, P., and Ratnarajah, T. (2021). Deep Reinforcement Learning-Based Beam Training for Spatially Consistent Millimeter Wave Channels. In *IEEE International Symposium on Personal, Indoor and Mobile Radio Communications (PIMRC)*, pages 579–584.

58 Zhang, Z., Zhang, J., Zhang, Y., Yu, L., Gao, F., Shi, Q., Liu, G., Yuan, Z., and Fan, W. (2024). Deep reinforcement learning based dynamic beam selection in dual-band communication systems. *IEEE Transactions on Wireless Communications*, 23(4): 2591–2606.

59 Baron-Hyppolite, A., Abreu, J. V. F., Santos, J. F., DaSilva, L. A., and Kibilda, J. (2023). Adaptive beam management for secure mmWave communications using software-defined radios. In *IEEE Military Communications Conference (MILCOM)*, pages 243–244.

60 Baron-Hyppolite, A., Santos, J. F., DaSilva, L., and Kibiłda, J. (2024). Eavesdropper avoidance through adaptive beam management in SDR-based MmWave communications. In *International Symposium on Wireless Communication Systems (ISWCS)*, pages 1–6.

61 Sutton, R. S. and Barto, A. G. (2018). *Reinforcement Learning: An Introduction*. The MIT Press, Cambridge, Massachusetts, 2 edition.

62 Lei, W., Lu, C., Huang, Y., Rao, J., Xiao, M., and Skoglund, M. (2022). Adaptive Beam Sweeping With Supervised Learning. *IEEE Wireless Communications Letters*, 11(12): 2650–2654.

63 Winder, P. (2020). *Reinforcement Learning*. O'Reilly Media. Accessed 17 Jan. 2024.

64 Mucchi, L., Ronga, L., Zhou, X., Huang, K., Chen, Y., and Wang, R. (2017). A New Metric for Measuring the Security of an Environment: The Secrecy Pressure. *IEEE Transactions on Wireless Communications*, 16(5): 3416–3430.

65 van Hasselt, H., Guez, A., and Silver, D. (2016). Deep Reinforcement Learning with Double Q-Learning. *Proceedings of the AAAI Conference on Artificial Intelligence*, 30(1).

66 Santos, J. F., Fathalla, E., Da Silva, A. P., Da Silva, L. A., and Kibilda, J. (2023). STAMINA: Implementation and Evaluation of Software-Defined Millimeter Wave Initial Access. In *IEEE International Conference on Communications (ICC)*, pages 346–352.

67 Fathalla, E. S., Zargarzadeh, S., Xin, C., Wu, H., Jiang, P., Santos, J. F., Kibilda, J., and Da Silva, A. P. (2023). Beam Profiling and Beamforming Modeling for mmWave NextG Networks. In *International Conference on Computer Communications and Networks (ICCCN)*, pages 1–10.

68 Adesina, D., Hsieh, C.-C., Sagduyu, Y. E., and Qian, L. (2022). Adversarial machine learning in wireless communications using RF data: A review. *IEEE Communications Surveys & Tutorials*, 25(1): 77–100.

69 Papernot, N., McDaniel, P., Wu, X., Jha, S., and Swami, A. (2016). Distillation as a defense to adversarial perturbations against deep neural networks. In *Symposium on Security and Privacy (SP)*, pages 582–597.

70 Tuna, O. F. and Kadan, F. E. (2024). Security of AI-driven beam selection for distributed MIMO in an adversarial setting. *IEEE Access*, 12: 42028–42041.

7

Securing Cyber-physical Systems Using Homomorphic Encryption

Atul Khatri[1], Ruizhi Cheng[2], Songqing Chen[2] and Bo Han[2]

[1]*College of Engineering and Computing, George Mason University, USA*
[2]*Department of Computer Science, George Mason University, USA*

7.1 Introduction

Cyber-physical systems (CPS) represent an emerging paradigm in computing, characterized by the intricate integration of digital and physical domains. These systems have found diverse applications, such as medical systems [1, 2], intelligent transportation [3], extended reality (XR) [4, 5], Internet of Things (IoT) [6, 7], *etc.* The transformative potential of CPS, evident in its profound implications for daily life, has captivated stakeholders in academia, industry, and governance [8].

While several technological strides have facilitated the rise of CPS, the onset of NextG (6G), the generation succeeding 5G, is set to be a pivotal enabler. The present 5G technology, despite its advancements, has revealed certain technical limitations during its global deployment. This revelation has ignited concerted research endeavors exploring the horizons of NextG [9]. NextG Internet, with its envisioned capabilities, is poised to offer more than just enhanced data transmission rates. It aspires to leverage the expansive wireless spectrum, such as millimeter wave (mmWave), terahertz (THz), and optical wireless bands [10]. Such comprehensive spectrum coverage, combined with anticipated advancements in latency and connection density, promises to furnish the robust connectivity backbone that CPS demands.

The proliferation of NextG-enabled CPS has ushered in an era where society is deeply interwoven with ubiquitous computing devices. For instance, the biometric

6G Security: New Frontiers in Mobile Network Technologies and Verticals, First Edition.
Edited by Luiz A. DaSilva.

information embedded in modern electronics, such as fingerprint and facial recognition, serves crucial purposes such as identity verification, facilitating financial transactions, and supporting law enforcement activities. However, the vast data exchanges in the CPS raise significant security and privacy concerns. When sensitive data is being transmitted over the Internet network, a security flaw or an attack at any network node can compromise the entire system. Critical infrastructures stand out as enticing targets for cyber adversaries, amplifying the stakes of potential breaches.

Data encryption serves as the primary defense against these threats. Public-key cryptography protocols, such as Diffie–Hellman [11] and RSA [12], have been widely adopted. However, they also have fatal weaknesses, the most prominent of which is the need to store private keys securely. Once the private key is cracked, the protected system may completely collapse. Fully Homomorphic Encryption (FHE) [13] offers a transformative approach, addressing the inherent pitfalls of conventional encryption methodologies. FHE allows for computations on ciphertexts without necessitating decryption, thus obviating the challenges tied to key storage. In the era of NextG-enabled CPS, characterized by an explosion of data generation and exchange, FHE emerges as a potent tool for establishing robust, privacy-centric systems.

In this chapter, our primary objective is to elucidate the potential and challenges of leveraging FHE as a robust protective mechanism for NextG-enabled CPS. We embark on our exploration with a meticulous chronological assessment of FHE, mapping its inception through to the cutting-edge advancements in the field. To facilitate a grounded comprehension, we subsequently delve into the most recent efforts surrounding the implementation and standardization of FHE. Following this, our discourse shifts to the tangible applications of FHE in safeguarding diverse facets of CPS. Our examination is spearheaded with a case study, focusing on the deployment of FHE to safeguard mobile spatial computing [14], an emergent paradigm allowing users to engage with their perceived 3D space through physical actions. Our preliminary results demonstrate that even though FHE is a potential solution for providing end-to-end protection on mobile spatial computing, several challenges remain. For example, it can lead to a decrease in the accuracy of spatial mapping and localization, the two main components of spatial computation, while significantly increasing computational latency and data volume. Expanding our scope, we present an exhaustive review of recent advancements in applying FHE to diverse CPS applications, spanning biometric-based authentication, critical infrastructures, and IoT. Our survey reveals that, despite the discernible progress in integrating FHE across various CPS applications, its inherent computational and data overheads continue to pose substantial barriers to its widespread adoption, warranting further investigation and optimization.

7.2 Background

In this section, we list some widely accepted definitions of CPS followed by its application domains. We also provide a historical background of FHE, describe the state of contemporary research, and offer a comparison of various open-source implementations.

7.2.1 Cyber-physical Systems

Baheti *et al.* describe CPS as a newer generation of systems integrating both physical and computational capabilities that interact with its users through novel modalities [15]. Lee states that CPS consists of networked computing devices that monitor and control various physical processes which, in turn, affect computation through a tight feedback loop [16]. The Industrial Cyber-Physical Systems (ICyPhy) group defines CPS as a new engineering discipline that lies at the intersection of the physical and the cyber; combining engineering models and methods from various fields of engineering with those of computer science and engineering. Rajkumar *et al.* describe CPS as "physical and engineered systems whose operations are monitored, coordinated, controlled and integrated by a computing and communication core" [17]. Thus, in general, CPS can be understood to be the coupling of physical phenomena with computational capabilities resulting in dynamic feedback loops that enhance system performance and efficiency across diverse applications.

CPS has applications across various industries encompassing many facets of everyday life. For instance, Dey *et al.* define medical cyber-physical systems (MCPS) as a critical application domain of CPS where networked medical devices are increasingly used in hospitals to deliver continuous high-quality healthcare [1]. Milenković *et al.* describe the usage of wireless sensor networks (WSN) for personal health monitoring, backed by advancements in sensors, low-powered integrated circuits, and wireless networks [18]. Wang *et al.* introduce a secure health care application architecture CPeSC3 (cyber physical enhanced secured WSN-integrated cloud computing for health care) that is applicable to health care monitoring and decision support systems [19].

CPS is also an integral part of various Critical Infrastructures (CI). Modern electrical power distribution networks incorporate Smart Grids, a progressive advancement from traditional electrical grids, to deliver power efficiently. According to Yu *et al.*, smart grids employ complex monitoring, control, and communication mechanisms to reliably deliver energy and enhance the operational efficiency of electric power generators and distributors [20]. According to the latest report by the National Institute of Standards and Technology (NIST), smart grids exhibit characteristics, such as increased reliability and resiliency to service disruptions that distinguish them from traditional grids [21]. The share of renewable energy in global power generation is projected to rise from 29% in 2022 to 35% in 2025 [22]. But renewable energy sources such as wind power and solar power exhibit variability that is subject to time of day and weather patterns [23]. Smart grids address the problem of distributed power generation which will be necessary for the continued rise in global renewable energy usage.

The transportation sector suffers from issues such as high accident rates and severe traffic congestion. The consequences include delayed delivery of consumer products, effects on modern supply chains, and economic losses due to damage to vehicles and products. Crash-related injuries and loss of human life are the most severe outcomes. Intelligent Transportation Systems (ITS) aim to address these issues and develop more efficient and safer mobility solutions. CPS is an important component of ITS. ITS depends upon vehicle-to-vehicle (V2V) communication through vehicular ad hoc network (VANET) to provide safety and comfort services to passengers and drivers [24]. The IEEE 802.11p Wireless Access in Vehicular Environments (WAVE) standard has been used to provide Dedicated Short Range Communication (DSRC) for V2V communication [25]. The multitude of sensors in modern vehicles coupled with V2V will result in the continuous production of a large amount of data. That data can be used to make informed decisions on the design and application of ITS, which, in turn, makes it safer and more efficient [26].

Since the physical world is inherently unpredictable and uncertain, there is a need to consolidate CPS to be more reliable and predictable [16]. This is a stringent requirement for CPS deployed to various mission-critical services. Applications such ITS and SG require increased system autonomy and will benefit from self-organizing/reorganizing cellular networks [27]. Attacks in Medical CPS could lead to disruptions in providing adequate patient care and the deterioration of patients' health. Service disruptions in SG will be an impediment to industrial operations and can affect the daily lives of regular people. ITS can suffer from fluctuations in the Quality of Service (QoS) of networks which can result in traffic congestion, increased travel time, and loss of life or property. Security is a major concern for integrity in CPS and is facilitated by mechanisms such as cryptography, access control, and intrusion detection [28]. Malicious actors could exploit security vulnerabilities in various CPS and extract sensitive user data such as Social Security Number (SSN), financial institution credentials, and other miscellaneous customer information. These issues require a reassessment of conventional security approaches and the investigation of innovative security methods to ensure both security and privacy in CPS.

7.2.2 Fully Homomorphic Encryption

Rivest, Adleman, and Dertouzos introduced "privacy transformation" in 1978 and it forms the foundation of Fully Homomorphic Encryption [29]. Fully Homomorphic encryption allows arbitrary computation over encrypted data. In an additively homomorphic encryption scheme, encryption followed by homomorphic addition is equivalent to addition followed by encryption. Similarly, in a multiplicatively homomorphic scheme, encryption followed by homomorphic multiplication is equivalent to multiplication followed by encryption

$$[x] \oplus [y] = [x + y] \text{ and } [x] \otimes [y] = [x \cdot y],$$

where $[m]$ represents a ciphertext for some plaintext m. In contrast, a partially homomorphic encryption (PHE) scheme allows us to perform a single operation, either addition or multiplication, over encrypted data for an arbitrary number of times. Additionally, a somewhat homomorphic encryption scheme (SHE) facilitates restricted computation over a predetermined set of operations. For example, the SHE Boneh-Goh-Nissim (BGN) encryption scheme facilitates unbounded homomorphic addition and a single homomorphic multiplication operation [30].

FHE allows arbitrary addition and multiplication to be performed over encrypted data. Since arbitrary functions can be described as Boolean circuits, FHE can theoretically evaluate any function over encrypted data. Craig Gentry published the first fully homomorphic scheme in 2009 in his thesis [13] that led to a wealth of work over the past decade. According to Peikert, FHE schemes can be classified according to their theoretical underpinnings [31]. He categorizes FHE schemes into 3 generations:

1. First-generation FHE schemes based on Gentry's with security based on ad-hoc average-case assumptions about ideal lattices or the *approximate greatest common divisor* (GCD) problem [32].
2. Second-generation FHE schemes are based on *learning with errors* (LWE) and *ring learning with errors* (RLWE) [33].
3. Third-generation FHE schemes are marked by practical improvements and further refinement of security assumptions [34].

The present stage of FHE research is characterized by its emphasis on open-source implementations, practical applications in real-world scenarios, and the establishment of standardized methodologies and security prerequisites. The following section offers a comprehensive summary of the evolution of fully homomorphic encryption.

Vanilla Fully Homomorphic Encryption Scheme. Gentry's innovative approach is based on a mathematical structure called a lattice. He initially constructs a partially homomorphic encryption scheme capable of performing homomorphic computations on low-degree polynomials. During each homomorphic evaluation, the ciphertext accumulates noise, and when the noise reaches a certain threshold, the ciphertext becomes impossible to decrypt. Gentry addresses this issue by introducing bootstrapping, which reduces the rate of noise accumulation in ciphertexts. A bootstrappable scheme possesses the capability to perform its own decryption. In this scheme, a noisy ciphertext is encrypted twice and then decrypted "homomorphically" using an encryption of the private key. Gentry uses the analogy of a locked glovebox to illustrate bootstrapping, where the glovebox represents encryption, the lock represents the decryption key, and the manipulation of items inside the box using gloves represents homomorphic evaluation [35]. The accumulation of noise is represented by the gloves becoming defective over time. When the gloves become defective, the original box and its key are placed inside a second locked glovebox, which can be unlocked. Object manipulation continues inside the second glovebox. However, fully implementing the homomorphic properties of Gentry's original scheme is impractical, mainly due to the high computational cost associated with bootstrapping. Gentry's bootstrapping method pioneered the initial phase of rapid innovation in FHE [36]. Smart and Vercauteren introduced an FHE scheme with reduced key and ciphertext sizes, and they also implemented a version known as Somewhat Homomorphic Encryption [37]. However, the security assumptions of this system were compromised by Cramer et al.'s key-recovery attack in 2016 [38]. Gentry and Halevi achieved the first complete implementation of Gentry's FHE scheme [39], but the execution time for a single bootstrapping operation was approximately 30 minutes. Although running the implementation on a GPU resulted in faster execution times, it remains impractical for real-world deployment.

In another study, Smart and Vercauteren present a SHE scheme that incorporates Single-Instruction Multiple-Data (SIMD) operations, facilitating parallelization in FHE through ciphertext packing [40]. SIMD instructions enable simultaneous computation by encoding multiple plaintext values within a single ciphertext vector. Ciphertext packing allows for batch computation in various FHE schemes. Furthermore, the permutation of plaintext slots within a packed ciphertext can be employed to achieve evaluation speedups [41]. These concepts are visually depicted in Figure 7.1, where the left-to-right representation demonstrates the processing of multiple values within a vector using a single instruction, along with the possibility of slot permutation.

Recent Developments. Gentry's work a subsequent wave of innovation led to the development of FHE schemes with diverse theoretical foundations [31]. These schemes can be categorized into different groups based on their underlying security assumptions. The first category relies on assumptions such as Learning with

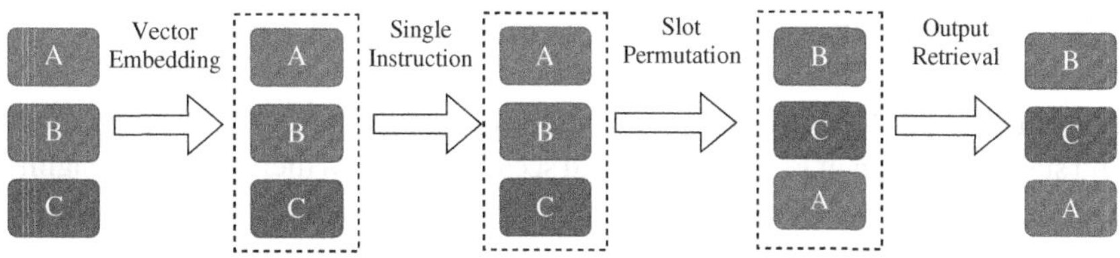

Figure 7.1 From left to right. A vector embeds multiple values. A singular operation concurrently manipulates all the values. Slot permutation rearranges the values within the vector. Individual output values are retrieved.

Errors (LWE) [42] and Ring Learning with Errors (RLWE) [43]. The second category is based on the approximate Greatest Common Divisor (GCD) problem [32]. Finally, another category of schemes builds upon NTRU-like assumptions [44].

Brakerski *et al.* introduced a leveled FHE scheme based on LWE assumptions [45]. The security of their scheme is based on the worst-case hardness of "short vector problems" on arbitrary lattices. The authors show that a SHE scheme can be based on LWE using a novel *relinearization* technique. Their work demonstrates the usage of a novel *dimension-modulus reduction* technique that results in ciphertext shortening and reduced decryption complexity, without introducing any additional assumptions. The Brakerski-Gentry-Vaikuntanathan (BGV) scheme is more performant than previous FHE schemes and is based on weaker security assumptions [46]. Other works include the extension of BGV to incorporate batching techniques of Smart and Vercauteren [41–47], and using simpler bootstrapping techniques [48]. Gentry *et al.* also demonstrated a working implementation of holomorphic encryption that can evaluate the AES-128 circuit [49].

Brakerski-Fan-Vercauteren (BFV) scheme is a modification of the "scale-invariant" scheme introduced by Brakerski which does not require modulus switching [50]. This scheme ports Brakerski's LWE scheme to the ring-LWE setting and implements SIMD for batch computing. The authors introduce two optimized versions of relinearization that results in smaller relinearization key and faster computations.

Dijk *et al.* describe a SHE scheme using only elementary modular arithmetic, and use Gentry's approach to convert it into an FHE scheme [51]. While Gentry's implementation involves working with ideal lattices over a polynomial ring, the DGHV scheme uses addition and multiplication over the integers. This scheme's security depends upon the hardness of the approximate integer GCD problem. An improvement to the DGHV scheme's impractical large public key size was suggested by Coron *et al.* [52]. The authors reduced the size of the public key by encrypting the public key elements with a quadratic form instead of a linear form. This preserves the semantic security of the system while being based on a stronger variant of the approximate GCD problem.

7.3 Standardization and Implementation of FHE

Current research focuses on practical application, open-source implementation, and standardization of FHE methods and security. Notably, there have been significant technical advancements resulting in various open-source software libraries.

7.3.1 Standardization

The Homomorphic Encryption Standardization Consortium consisting of government, industry, and academia experts was established in 2017. They initiated regular workshops with the goal of creating a community standard for Homomorphic Encryption (HE) and as a result, the Homomorphic Encryption Standard was developed in 2018 [53]. This standard specifies security assumptions and potential attacks and suggests security parameters for implementations in prominent FHE applications.

The Consortium has categorized contemporary FHE into three models of homomorphic computation: Boolean circuits, modular (exact) arithmetic, and approximate number arithmetic [54]. The modular (exact) arithmetic approach builds upon the progress made in second-generation FHE through R/LWE-based schemes, particularly BGV [46] and BFV [50, 102]. This approach involves performing arithmetic operations on circuits using values that are encrypted within a specific range defined by a modulus t. Recent advancements include the introduction of the residue number system (RNS) variants of BFV, which offer faster modular arithmetic operations [55–57]. Enhancements to the plaintext space in the BFV scheme enable homomorphic evaluation of deep circuits with rational number inputs [58] and improved bootstrapping techniques [59].

The Boolean circuit-based approach in FHE involves evaluating Boolean circuits using encrypted bits. Two prominent schemes based on this approach are FHEW and TFHE. FHEW, proposed by Ducas and Micciancio in their Fasted Homomorphic Encryption in the West (FHEW) scheme [60], introduces a ring variant of Alperin-Sheriff and Peikert's bootstrapping for the GSW scheme and a novel homomorphic NAND gate. By evaluating this NAND gate on a single bit, the bootstrapping process in FHEW runs in less than a second. TFHE, another Boolean circuit-based scheme called Fast Fully Homomorphic Encryption Over the Torus [61], extends the GSW scheme. It encodes polynomials over the torus and evaluates them over Boolean circuits [62]. The techniques developed for TFHE also apply to the FHEW cryptosystem. Recent advancements include improved bootstrapping methods for TFHE [61, 63], expanding the bootstrapping operation in FHEW to multiple bits [64, 65], multi-value bootstrapping for both TFHE and FHEW [66], and refreshing multiple ciphertexts using a single refresh procedure [67].

The leading approach in approximate number arithmetic is represented by the CKKS scheme, also known as Homomorphic Encryption for Arithmetic of Approximate Numbers (HEAAN) [68]. In this article, the scheme itself is referred to as CKKS, while HEAAN is used to denote the corresponding software. When evaluating an approximate circuit using ciphertext, the result obtained is an approximation rather than an exact value. These approaches are well-suited for efficient polynomial approximation and floating-point computations. Within the CKKS scheme, fast polynomial approximation enables the evaluation of functions such as multiplicative inverse, exponential, logistic, and discrete Fourier transform [68]. Furthermore, there have been additional works focusing on improving bootstrapping techniques [69, 70] and introducing variants based on Residue Number Systems (RNS) [71–73].

Genise et al. introduced a somewhat homomorphic variation of the GSW scheme [74]. This scheme utilizes an inhomogeneous variant of the NTRU assumption (iNTRU) for enhanced security, surpassing the LWE assumption. It demonstrates a more efficient evaluation of encrypted, nondeterministic finite automata compared to other existing schemes. Boura et al. proposed a unified framework named Chimera [75]. This framework combines the strengths of the BFV, CKKS, and TFHE schemes, enabling seamless switching between their plaintext spaces. It also presents a set of bridges to connect these schemes together.

7.3.2 Implementation

The advancements in the current generation of research on fully holomorphic encryption schemes have yielded efficient and publicly accessible libraries and software. These resources are continually evolving to align with cutting-edge FHE research and the recently formulated community standards for homomorphic encryption [86]. Despite the rapid advancements in the efficiency of FHE, the implementation of FHE schemes remains a persistent challenge for individuals without expertise in the field. Implementations necessitate custom development, demanding a high level of proficiency in programming and cryptography. Notably, the conversion of computations between plaintext and ciphertext entails non-trivial procedures, careful management of noise is required to ensure correctness during decryption, and security parameters must be manually chosen [87]. To address these complexities and make FHE more accessible and widely applicable, early efforts have been made to introduce compilers and tools [81, 88, 89]. A comprehensive examination of the primary implementations, compilers, and tools related to FHE is presented below. The details of these implementations are summarized in Table 7.1, which includes pertinent information such as the software's name, creator, programming language, and a brief description of the implemented schemes.

Table 7.1 Libraries and software for FHE.

Product	Creator	Language	Summary
SEAL [76]	Microsoft	C++	A popular FHE library widely used for modular arithmetic (BFV) and approximate arithmetic (CKKS).
HEib [77]	IBM	C++	A widely-utilized FHE library that incorporates BGV for modular arithmetic and CKKS for approximate arithmetic.
TFHE [78]	Gama et al.	C++	Incorporates an optimized variant of the GSW scheme based on ring operations.
HEAAN [79]	CryptoLab, Inc.	C++	Utilizes the CKKS scheme for performing approximate number arithmetic.
PALISADE [80]	New Jersey Institute of Technology	C++	A library for lattice cryptography that provides support for multiple protocols in FHE, including BGV, BFV, and StSt.
Cingulata [81]	CEA LIST	C++	A compiler and RTE designed for C++ programs utilizing FHE. It incorporates the BFV scheme and offers support for TFHE as well.
FV-NFLlib [82]	CryptoExperts	C++	Implements the FV scheme and is constructed using the NFLLib lattice cryptography library.
OpenFHE [83]	openfheorg	C++	A new open-source FHE software library that incorporates selected design ideas from prior FHE projects, such as PALISADE, HElib, and HEAAN, and includes several new design concepts and ideas.
Lattigo [84]	Laboratory for Data Security	Go	Implementation of BFV and HEAAN.
$\Lambda \circ \lambda$ [85]	E.Crockett and C.Peikert	Haskell	Implements a BGV-type FHE scheme.

Microsoft's Simple Encrypted Arithmetic Library (SEAL). The Simple Encrypted Arithmetic Library (SEAL), developed and maintained by the Cryptography Research Group at Microsoft Research since 2015, is a prominent homomorphic encryption library. SEAL is implemented in C++ and includes a .NET wrapper [76]. Over time, SEAL has undergone revisions in terms of the underlying encryption scheme. Initial versions of SEAL utilized a variant of the YASHE scheme [90]. However, in SEAL version 3.3 [76], the BFV and CKKS schemes are implemented. The release of SEAL 3.3

occurred in June 2019 and can be accessed on Microsoft's official website [76]. Notably, Intel AI has incorporated SEAL's implementation of the CKKS scheme into the Intel HE transformer for nGraph (nGraph-HE) [91]. The nGraph-HE serves as a homomorphic backend to Intel's graph compiler for artificial neural networks.

IBM's Homomorphic Encryption Library (HElib). The Homomorphic Encryption Library (HElib), developed by Shai Halevi and Victor Shoup, is an extensively used library for FHE [92]. HElib, implemented in C++, supports versions of the BGV and CKKS schemes. To function, HElib relies on the Library for Doing Number Theory (NTL) mathematical library [93]. Initially, HElib solely implemented the BGV scheme. However, re-implementation efforts in March 2018 resulted in significant speed enhancements, and by January 2019, HElib incorporated support for the CKKS scheme. The HElib library is publicly accessible on Github [94].

Fast Fully Homomorphic Encryption over the Torus (TFHE). Fast Fully Homomorphic Encryption over the Torus (TFHE) is open-source software available on Github, designed for FHE in C/C++ [78]. TFHE is built upon its predecessor library, TFHEW, although the latter is no longer actively maintained [60]. The implementation of TFHE allows for the homomorphic evaluation of Boolean circuits on single-bit messages. Notably, TFHE relies on the Fastest Fourier Transform in the World (FFTW) library [95]. TFHE adopts a ring variant of the GSW scheme [34] with incorporated optimizations [60, 62, 96]. Moreover, there exists a CUDA-accelerated Fully Homomorphic Encryption Library called cuFHE [97]. cuFHE implements TFHE on CUDA-enabled GPUs using C++. It also provides an optional Python wrapper for convenience. Additionally, another library known as NuCypher fully homomorphic encryption (NuFHE) is written in Python and implements TFHE utilizing CUDA and OpenCL [98].

HEAAN. The Homomorphic Encryption for Approximate Numbers (HEAAN) library, which implements the CKKS scheme, is a software package utilized for homomorphic encryption [79]. HEAAN is implemented in C++ and relies on the NTL library [93]. Since its initial release in 2016, HEAAN has undergone various optimizations to enhance its performance, including the incorporation of bootstrapping [99]. The HEAAN library is publicly accessible on Github [79].

PALISADE. PALISADE is a C++ library developed by the New Jersey Institute of Technology (NJIT) that leverages lattice cryptography primitives to support a range of cryptographic functionalities. These include public-key encryption, various Single-Input Homomorphic Encryption schemes, Leveled Homomorphic Encryption (LHE) schemes, Proxy Re-Encryption (PRE), and other multiparty capabilities [80]. PALISADE version 1.5.0 was released in March 2019 and can be accessed on GitLab. PALISADE implements different versions of several Homomorphic Encryption (HE) schemes, namely LTV [100], Stehle-Steinfeld (StSt)

[101], BGV [46], and BFV [102], along with PRE schemes [80]. It is worth noting that the inclusion of LTV is mainly for historical purposes, as the scheme is no longer considered secure [80].

Cingulata. Cingulata (pronounced "tchingulata," also known as Armadillo) is a compiler toolchain and runtime environment specifically designed for C++ programs utilizing FHE [81]. It offers a high-level C++ interface to facilitate the development of FHE applications using a "low-level" implementation of the scheme. Initially, Cingulata exclusively supported the BFV scheme. However, in June 2019, support for TFHE in gate bootstrapping mode was added. Cingulata is publicly accessible on Github [81].

(FV-)NFLlib. FV-NFLlib is an implementation of the FV scheme in C++ [82]. FV-NFLlib is available on GitHub and was last updated in 2020.

OpenFHE. OpenFHE is a newly developed open-source software library for FHE that integrates selected design concepts from previous FHE projects, including PALISADE, HELib, and HEAAN [83]. The library incorporates several design features and ideas to enhance its functionality. Noteworthy design features introduced by OpenFHE include the following:

1. From the outset, OpenFHE assumes that all implemented FHE schemes will support bootstrapping and scheme switching.
2. OpenFHE offers support for multiple hardware acceleration backends through the utilization of a standard Hardware Abstraction Layer (HAL).
3. OpenFHE provides both user-friendly and compiler-friendly modes of operation. In the user-friendly mode, the library automatically handles maintenance operations such as modulus switching, key switching, and bootstrapping. In the compiler-friendly mode, these decisions are left to an external compiler.

These design features enhance the usability and flexibility of OpenFHE, making it suitable for a variety of use cases in FHE research and development.

Lattigo. Lattigo is a lattice-based cryptographic library implemented in Go [84]. Lattigo implements schemes such as BFV [55, 57] and HEAAN [68, 99].

$\Lambda\circ\lambda$ (LOL). The $\Lambda\circ\lambda$: Functional Lattice Cryptography library (also known as "LOL") is a comprehensive general-purpose software package for lattice cryptography. It incorporates an implementation of a BGV-type FHE scheme [103]. Implemented in Haskell, $\Lambda\circ\lambda$ offers high-level methods for effectively implementing lattice cryptography primitives, including the BGV-type FHE scheme.

A Language and Compiler for Homomorphic Encryption Made easY (ALCHEMY) works in conjunction with $\Lambda\circ\lambda$. It enables the compilation of plaintexts written in a modular domain-specific language, utilizing $\Lambda\circ\lambda$ as the backend implementation [87, 89]. The ALCHEMY compiler simplifies the implementation process for users by automatically generating many of the required parameters, thus reducing the burden on users in configuring the system.

7.3.3 FHE Implementation in NextG Use Case: Healthcare

In this subsection, we examine how the aforementioned FHE libraries and implementations are being adapted to support critical use cases in NextG environments, with a particular focus on healthcare, one of the most transformative and security-sensitive domains in 6G systems [104]. Key features of 6G, such as ultra-low latency, massive device connectivity, and high throughput, significantly enhance the potential for real-time medical data exchange and intelligent care delivery [105]. While these advancements enable unprecedented capabilities, they also raise significant concerns regarding data privacy, regulatory compliance, and system security [106]. FHE has thus emerged as a promising cryptographic solution, allowing computations to be performed directly on encrypted medical data without compromising confidentiality [107, 108].

To achieve this, recent efforts have leveraged a wide range of FHE libraries to enable privacy-preserving healthcare applications aligned with the demands of NextG systems. One notable application is the secure analysis of physiological signals collected from wearable or implantable medical sensors. These devices continuously stream biometric data such as electrocardiograms (ECG) and blood pressure, which can be transmitted over high-speed networks (*e.g.*, 6G) to edge servers or federated AI agents for encrypted real-time inference. Studies have demonstrated that the *Microsoft SEAL* library [73] enables accurate encrypted inference on ECG signals, achieving over 85% heart disease prediction accuracy while preserving end-to-end data confidentiality [109]. Meanwhile, FHE is gaining momentum in secure federated learning frameworks [110], where multiple healthcare institutions collaboratively train diagnostic models without exchanging raw patient data. Recent achievement using *HElib* and *PALISADE* [80] has demonstrated privacy-preserving aggregation of encrypted features extracted from chest X-ray and CT scans [111]. Furthermore, FHE libraries have been adopted to support encrypted medical imaging workflows, enabling deep learning models to operate directly on homomorphically encrypted images. For instance, the PPPML-HMI framework [112] performs encrypted diagnostic inference on chest CT scans using *OpenFHE* [83]. Finally, in AI-assisted telesurgical systems, FHE has been applied to secure auxiliary data streams such as consent logs, drug dosage records, and real-time patient vitals. While latency-critical control loops still rely on conventional encryption, recent work [113] has demonstrated that *OpenFHE* [83] and *TFHE* [78] can be used to execute encrypted lookups and consent verification tasks with fast responsiveness. Collectively, these developments illustrate the technical viability and practical utility of incorporating FHE libraries into real-world, privacy-preserving healthcare systems in NextG networks.

7.4 Recent Advancements in Securing CPS Using FHE

In this section, we delve into recent efforts in harnessing FHE to safeguard CPS applications. We initiate our exploration with a case study on mobile spatial computing, followed by a comprehensive survey in various CPS domains, including facial recognition, control systems, and IoT.

7.4.1 Critical Infrastructures

Critical infrastructures, such as power plants, power distribution networks, and transportation systems, hinge on the intricate workings of CPS for routine operations. These entities are acutely susceptible to CPS attacks [114]. Ensuring the security of signals, both from sensors to controllers and from controllers to actuators, is paramount in safeguarding against malicious intrusions.

Research efforts have delved into the development of mechanisms to monitor and counteract CPS attacks. Pasqualetti *et al.* presented a comprehensive mathematical model that encapsulates CPS, associated threats, and monitoring systems [115]. Their innovative approach encompassed both centralized and decentralized detection and identification monitors, emphasizing fault detection and isolation. Another noteworthy study by Lee *et al.* articulated a robust estimation framework tailored for continuous-time linear dynamical systems [116]. This framework is adept at estimating system states amidst sensor attacks, ensuring bounded estimation errors. Nevertheless, some attacks, especially stealthy ones, remain elusive to detection methodologies [115]. Intricacies such as zero-dynamics attacks can be executed even when the attacker possesses imperfect model knowledge [117]. Moreover, adversaries who monitor both the input and output channels of a control system are equipped to construct near-perfect models, thereby amplifying threats to system integrity [118].

One potential solution lies in ensuring encrypted communication among controllers, sensors, and actuators. Implementing a wholly encrypted feedback loop can effectively shield signal exchanges from malicious entities. However, this approach warrants that the controller houses a secret decryption key—a potential vulnerability if compromised. To address this, FHE presents an attractive proposition, enabling the controller to operate directly on encrypted signals without necessitating stored keys.

Kogiso *et al.* charted a pioneering path by encrypting linear controllers via a hybrid homomorphic encryption framework, synergizing public-key RSA and ElGamal encryption techniques [119]. Their approach succeeded in encrypting both communication signals and several system parameters without compromising inherent system functionality. A salient limitation emerged in the controller's inability to perform additions on encrypted signals—a crucial operation for state updates. The researchers skillfully navigated this by adding decrypted signals externally and reintroducing the encrypted sum into the controller, albeit with amplified system intricacy and network bandwidth demands. Simulations corroborated the effective masking of system parameters and in-controller signal processing. Moreover, models constructed by observing encrypted channels were discerned to diverge significantly from genuine system dynamics.

Building upon this, Junsoo *et al.* advocated the application of FHE for controller encryption, aiming to enhance the cybersecurity posture of CPS [120]. Their methodology hinged on a tiered HE mechanism, facilitating indefinite homomorphic operations, though at the cost of the bootstrapping overhead. The authors' insight is built upon an LWE-based homomorphic encryption scheme. While most existing HE schemes have fixed plaintext spaces and offer only modular operations, Cheon *et al.* [121] proposed a HE scheme allowing floating point operations between ciphertexts. The ciphertext output differs from the original message, but it can be viewed as an approximate value in floating-point computations, especially for sufficiently small tolerances. The homomorphic multiplication of a real matrix with a ciphertext decreases its level. Thus, the ability to perform homomorphic computations on encrypted data is limited to a specific depth, beyond which the accumulated noise in the ciphertext makes it impossible to decrypt.

The authors take inspiration from Ducas and Micciancio's bootstrapping scheme [60] and adapt their idea to a leveled HE scheme. Their key insight is that refreshing a ciphertext and transforming it to a new ciphertext of a larger level allows for an infinite number of homomorphic function evaluations. While this approach suffers from a considerable bootstrapping overhead, it was ameliorated by deploying multiple controllers running in parallel. This ensured the continuous availability of at least one controller for input generation, while others underwent bootstrapping. The authors also introduced a novel *multiplication tree* structure to prolong the lifespan of the encrypted variables, thereby increasing the time between two successive bootstrapping operations. Experimental validations underscored the efficacy of their approach in securely governing a quadruple water tank system [122]. The generated outputs adeptly shadowed reference signals, and controller transitions were observed to be transient-free.

7.4.2 Internet of Things (IoT)

IoT encompasses a diverse array of objects equipped with networking capabilities, united by a set of communication protocols that ensure the efficient exchange of information, underpinning system interoperability [123]. A prominent offshoot of

IoT is the Industrial IoT (IIoT), often synonymous with "Industry 4.0" [124]. It aims to harness advanced data analytics tools and methodologies, sifting through real-time data from a network of devices, aiming to unearth insights that can amplify the efficiency, safety, and reliability of the industrial domain. Given IIoT's profound implications for contemporary manufacturing and supply chain logistics, it invariably emerges as a lucrative target for cyber threats. This vulnerability has been starkly demonstrated by notorious cyber-attacks such as Mirai [125] and Stuxnet [126]. Consequently, bolstering security and privacy within IIoT has gained paramount importance.

A myriad of strategies have been proposed to augment privacy and security within the IoT realm. Given the inherent resource constraints of contemporary IoT devices, encapsulating limited processing capabilities and modest memory, recent scholarly endeavors have gravitated towards lightweight cryptographic paradigms [127]. However, they have been demonstrated to have security issues [128–131].

In response to these challenges, there has been a growing interest in harnessing Fully Homomorphic Encryption (FHE) to bolster IoT security. Lu *et al.* elucidated the utility of the BGN homomorphic encryption model [132] to protect the confidentiality of range queries and device-specific data, particularly in fog-enhanced IoT scenarios [133]. Parallelly, Fiore *et al.* advocated for the integration of multikey homomorphic authenticators, enabling the verification of data that is outsourced to potentially mistrusted cloud entities [134]. Nonetheless, the inherent computational intensity associated with FHE, especially during bootstrapping phases, can be prohibitive for devices constrained by limited resources.

In a bid to enhance the computational efficiency of FHE within the IoT domain, Li *et al.* introduced a streamlined schema tailored for IIoT privacy preservation [135]. This approach can be succinctly described through its triad of participants: data originators, cloud service intermediaries, and end data users. By prioritizing data encryption at its genesis, establishing secure communication pathways, and mandating cloud entities to process only encrypted data, their methodology achieves a judicious equilibrium between stringent privacy measures and computational agility.

Let D_i represent an individual node in a set of n data owners, x_i be a data value generated by node D_i, and c_i be the corresponding ciphertext. Encryption ensures data owner privacy and each node is oblivious to the data generated by all other nodes. Each data item also has a unique label (*e.g.*, timestamp) which doesn't need to be secure. A program label $\mathcal{P}$ is a tuple $(f, l_1, l_2, \dots, l_n)$ where f is the function to be evaluated homomorphically and l_i is the label for the i-th data point. During the preprocessing phase, secret key s_k, public key p_k, and verification key v_k are generated. The encryption algorithm E generates ciphertext

$$\mathbf{c} = Enc(s_k, l, \mathbf{x}), \tag{7.1}$$

where *Enc* represents the encryption algorithm used, $\mathbf{x}$ is the dataset, and l is the label for each data value. Similarly, the cloud server generates a new ciphertext by homomorphically evaluating f over the ciphertext $\mathbf{c}$ based on request $\mathcal{P}$

$$\mathbf{c}' = Eval(p_k, f, \mathbf{c}). \tag{7.2}$$

A data user can use s_k to decrypt the received ciphertext into the constituent result and labels, and can then decide whether to accept or reject it. The decryption operation is

$$Dec(s_k, \mathcal{P}, \mathbf{c}'). \tag{7.3}$$

Since IIoT devices operate in a resource-constrained environment, it is imperative to reduce computational complexity to realize real-time guarantees. The computational overhead in data owners stems from the need to perform *Enc* on each data value. $Enc(s_k, l, \mathbf{x})$ can be decomposed into

$$c_l = Enc(s_k, l), \;\; c_i = Enc(c_l, \mathbf{x}). \tag{7.4}$$

c_l can be reused for encrypting data values generated by the same node. Similarly, data owners need to perform *Dec* to retrieve each result and the associated label. The $Dec(s_k, \mathcal{P}, \mathbf{c})$ operation can be decomposed into

$$sk_{\mathcal{P}} = Dec(s_k, \mathcal{P}), \;\; x_i' = Dec(sk_{\mathcal{P}}, \mathbf{c}). \tag{7.5}$$

This method of splitting the computations lies at the heart of the authors' lightweight privacy-preserving scheme.

The arrangement of the authors' simulation assessment consists of air quality monitoring sensors distributed at 15 sites in Bristol, U.K. A mobile application (data user) queried the server, located at the University of West England (UWE), for information from a sensor located at AURN *St Pauls* (data owner). The query $\mathcal{P} = (f_1, 05082019)$ consisted of a *const-multiplication* algorithm f_1 and a timestamp $l = 05082019$. Against a reference value of (55.8, 5.9, 32.6, 20.3, 6.0), the obtained values were found to be (59.9, 10.3, 32.3, 15.5, 7.0). Additionally, they also demonstrated the differences in the computational time of *Enc*, *Dec*, and *Eval* for differing key sizes.

7.4.3 Biometric-based Authentication

Biometric-based authentication system, such as face recognition, is a critical subset of CPS applications. As CPS platforms become more ingrained in everyday life, the imperative for robust biometric authentication mechanisms, such as face recognition, escalates. Biometric-based user authentication comprises two phases: enrollment and testing [136, 137]. During enrollment, user-specific biometric data is collected and pre-processed, and the high-dimensional features are extracted. Finally, these feature vectors are stored in a database along with their identity labels. In the testing phase, user data is compared with the stored profile to verify their identity.

Biometric information is also encoded as feature vectors by deep-learning models, such as convolutional neural network (CNN). Attacks targeting the database of biometric templates can significantly compromise the privacy of enrolled users. Mai *et al.* demonstrated that access to unencrypted facial image representations allows a malicious actor to reconstruct the original facial image [138]. The reconstructed facial images were successfully matched by state-of-the-art face recognition systems. Cryptographic methods have been utilized to perform secure matching over encrypted biometric templates. Soutar *et al.* proposed biometric encryption of fingerprints [139]. Visual cryptography [140] has been used to verify fingerprint [141], iris templates [142], and facial images [143]. Fuzzy vault [144] based schemes have been proposed to verify fingerprint [145] and iris templates [146]. Alternative protection schemes based on pattern recognition have also been proposed. These include non-invertible transformation functions [147], and cancelable biometrics [148, 149]. These methods trade-off matching performance for the security of the template database.

Matchings between feature vectors necessitate the computation of distances between enrollment and test vectors, often using metrics such as Euclidean distance and Cosine similarity. Since these metrics can be decomposed into a series of additions and multiplications, FHE can be utilized to compute them entirely in the encrypted domain. Most existing works utilizing homomorphic encryption are limited to binary templates and are based on PHE schemes [150–159]. Additionally, due to the high computational complexity of HE schemes, these works are computationally impractical. For instance, a naive implementation of FHE requires 48.7 MB of memory and 12.8 seconds to match a single pair of 512-dimensional encrypted face feature vectors [160].

Cheon *et al.* utilized a SHE-based scheme with SIMD operations to match encrypted iris templates with user attempts [161]. They successfully demonstrated Hamming distance matching with 2400-bit iris templates, achieving completion times of 0.47 seconds for the server and 0.1 seconds for the user. Boddeti utilized batching and dimensionality reduction techniques to compare two 512-dimensional face representations with 16KB of memory in 2.5 milliseconds [162]. Engelsma *et al.* adopted an identical scheme for 1:1 matching of 192-dimensional encrypted fingerprint representations in 1.25 milliseconds [163]. These approaches are a significant improvement over earlier FHE-based matching schemes. However, the space and time complexity remains infeasible for 1: m matching applications such as image search. For example, an encrypted matching against a gallery of size $m = 100$ million, each representation having 512-dimensionality, would take over 38 hours and 9 TB memory [162, 164].

Addressing this, Engelsma *et al.* proposed Homomorphically Encrypted Representation Search (HERS) [164] which is designed for efficient 1:m encrypted matching. HERS operates by compressing the image representation towards its estimated intrinsic dimensionality with minimal loss of accuracy. This is followed by encrypting the compressed representation using a FHE scheme. Finally, searches are performed entirely in the encrypted domain.

There is a wide body of work regarding low dimensional estimation of data manifolds that reside in a higher dimensional space. These include classical methods such as Principal Component Analysis (PCA), Multidimensional Scaling (MDS), and Isomap. In addition, there are methods utilizing deep networks to learn about more complex non-linear mappings. Gong *et al.* estimate the intrinsic dimensionality of image representations and develop a deep neural network based non-linear mapping, dubbed *DeepMDS* [165]. HERS is meant to be an extendable wrapper on top of an existing image search system and hence models its dimensionality reduction component after DeepMDS.

In a typical matching operation, a query representation $\boldsymbol{q} \in \mathbb{R}^d$ is matched against a database of m template feature vectors $\boldsymbol{P} = [\boldsymbol{p}_1, \boldsymbol{p}_2, \dots, \boldsymbol{p}_m] \in \mathbb{R}^{d\times m}$. Common distance metrics such as Euclidean distance and cosine similarity are used to compute a score that decides the result of the matching task. Both of these metrics involve matrix-vector products of the form $\boldsymbol{P}^T\boldsymbol{q}$. Thus, the matching process consists of md scalar multiplications and md scalar additions. The key idea behind HERS is utilizing encryption to secure the template database and the query while maintaining the ability to match efficiently at scale, without any loss of accuracy. Even if an attacker gains access to the template database, without a decryption key, they cannot reconstruct the original image or glean other characteristics inherent to the representation.

HERS utilizes the Fan-Vercauteren (FV) scheme for FHE. FV scheme requires a private decryption key, a public encryption key, and evaluation keys. Since the FV scheme works solely on integers, the real-valued representations $\boldsymbol{q} \in \mathbb{R}^d$ need to be encoded to integer representations $\boldsymbol{q} \in \mathbb{Z}^d$. Each representation's real-valued features were quantized into integers with a precision of 0.004. The authors utilized a custom encoding scheme for efficient 1:m matching by leveraging SIMD primitives [166] of the FV scheme.

During the enrollment phase, the client generates a public-private key pair. For a template $\boldsymbol{p}$, the client quantizes the representation features, encodes the quantized template into a plaintext polynomial, and encrypts the plaintext into ciphertext using the public key. The client transmits the ciphertext, along with template metadata, to the server, which then adds it to the existing template database. In the search phase, for a query $\boldsymbol{p}$, the client quantizes the representation features, encodes the quantized query into a plaintext polynomial, and encrypts the plaintext into ciphertext using the private key. The client transmits the ciphertext to the server, which then computes the encrypted similarity scores and sends the result back to the client. The client decrypts the response using the private key and obtains the index of the nearest match. During enrollment, the server lacks access to raw database representations, and in the search phase, it lacks access to search query representation or index scores.

The addition/multiplication of two plaintext integers translates to the addition/multiplication of two polynomials of degree n in the encrypted domain. This leads to a n-fold and $\mathcal{O}(n^2)$-fold increase in computational complexity for addition and multiplication respectively. A naive FV scheme without SIMD has a high computational overhead rendering it impractical for real-world applications. In comparison to Bodetti's scheme [162], which scales linearly with the size of the database m, HERS scales better for larger databases due to a slower rate of increase in computational complexity by a factor of $\sim \mathcal{O}\left(\frac{d}{n}\right)$.

Since a smaller feature dimension d offers significant computational improvement in 1:m matching applications, the authors of HERS have reduced the dimensionality of feature vectors while retaining accuracy. Gong *et al.* [165] and Ansuini *et al.* [167] demonstrated that representations learned by deep convolutional neural networks are highly redundant with the actual intrinsic dimensionality being 20× to 30× smaller than the ambient space the representation is embedded in. Gong *et al.* introduced a non-linear mapping, dubbed DeepMDS [165], to compress the representation from the ambient space to its intrinsic space. DeepMDS has two primary limitations that constrain its direct applicability in HERS. First, DeepMDS has to be trained separately for each dataset. Because it was trained and evaluated on the same dataset, it isn't suitable for HERS, which demands a generalized compression approach applicable across various datasets in real-world scenarios. Second, the feature representations learned by DeepMDS do not preserve the local representation structure sufficiently. These limitations result in reduced matching accuracy and do not serve the goal of HERS to compress the feature representation while retaining matching accuracy.

HERS introduces DeepMDS++, a deep neural network-based dimensionality reduction method. It comprises multiple nonlinear layers and has been trained to preserve pairwise distances when mapping an embedding from the ambient space to its intrinsic space. This enables HERS to perform efficient matching in the encrypted domain. Experimental results demonstrate that for a gallery of size $m = 100$ million, HERS requires 5.9 MB for the probe template and 280 GB for the gallery. This is in contrast with Boddetti's work [162], which requires 19K for the probe and 9 TB for the same gallery size. HERS was able to search against 100 million 32-dimensional representations within 740 seconds, which is 186× faster than [162] using 64× less memory. For a gallery of size 100 million, the server needs to transmit 100 MB of data back to the client. On the client side, the decryption of 100 million integers required 1 second. The accuracy of HERS was shown to be within ≈ 2% of unencrypted image search.

Beyond traditional user-to-system interactions, biometric-based authentication is becoming equally essential in other emerging device-based CPS use cases. These scenarios involve device-to-device, infrastructure-to-device, or ambient user recognition scenarios, where continuous and privacy-preserving identity assurance is crucial. In the following, we discuss three representative use cases: machine-type communication (MTC) [168], vehicular networks [169], and autonomous systems [170].

Machine-type Communication (MTC). Massive MTC in 6G envisions swarms of devices (industrial sensors, robots, smart appliances) operating autonomously with only sporadic human intervention [168]. In such contexts, biometric authentication plays a critical role when a human operator must assume control or authorize a sensitive action on an autonomous device [171]. For example, a factory robot control panel may require a technician's fingerprint or face scan to override automated routines, or an industrial robot may verify an operator's identity via biometrics before enabling collaborative mode [172]. These scenarios demand reliable human authentication without compromising privacy or system performance. Naively sending biometric samples to cloud servers introduces privacy vulnerabilities and relies on secure network connectivity. Credential-based schemes are also impractical in industrial CPS due to scalability and reliability concerns [173]. While 6G networks alleviate connectivity challenges through ultra-reliable links, strong cryptographic protection is still necessary to ensure biometric security during transit and computation. FHE provides a viable solution by enabling secure offloading of biometric verification. A local device can encrypt the captured biometric feature vector and transmit it to an edge or cloud server, which performs the matching against an encrypted template database without decrypting the data [174]. The encrypted result is returned and decrypted locally, ensuring that sensitive biometric identifiers remain concealed throughout.

Vehicular Networks. In the automotive sector, 6G-enabled Vehicle-to-Everything (V2X) networks are expected to incorporate biometric authentication to enhance both security and user experience [175]. Connected vehicles are increasingly equipped with biometric sensors, such as fingerprint scanners embedded in door handles and steering wheels, as well as in-cabin cameras capable of facial or iris recognition [176]. These sensors enable features such as keyless entry, engine start authorization, and personalized in-vehicle settings for recognized drivers [176]. A particularly compelling application is continuous driver verification, where in-cabin camera systems monitor facial or ocular features in real time to ensure the driver is both authorized and alert [177]. Such mechanisms can prevent theft or misuse by transitioning the vehicle to a safe mode if an unrecognized or inattentive driver is detected. Additionally, infrastructure-based biometric checkpoints are emerging as a novel use case. Pilot deployments have explored the use of facial or fingerprint recognition at toll booths and access gates to automate authentication, replacing the need for cards or RFID tags [178]. These applications necessitate frequent transmission of sensitive biometric data, such as facial images, fingerprints, and iris patterns, between vehicles and roadside or cloud infrastructure, introducing significant privacy and security challenges.

FHE offers a promising solution by ensuring that biometric data remains encrypted throughout the entire processing pipeline [179, 180]. In an FHE-enabled V2X scenario, a vehicle's onboard unit captures the biometric sample and immediately encrypts the derived feature vector using the owner's public key [181]. This encrypted data is then transmitted over the network to a roadside unit or cloud server, which performs the necessary authentication directly on the ciphertext. For instance, a toll system may homomorphically compare the encrypted facial descriptor against an encrypted reference template associated with the vehicle's account, generating an encrypted match score. Only the vehicle, which holds the corresponding private key, can decrypt the result and determine whether the authentication was successful, after which it may proceed to permit gate entry or complete a payment transaction. Throughout this process, the biometric data remains protected because neither the server nor any intermediate network node gains access to raw biometric content.

Autonomous Authentication. Beyond specific applications, NextG CPS environments demand a fundamentally new authentication paradigm: one that is continuous, context-aware, and frictionless [182, 183]. In conventional systems, authentication is typically a one-time event (*e.g.*, login), after which a user or device is implicitly trusted. By contrast, 6G-enabled CPS systems are highly dynamic and operate under zero-trust principles, where each access request or contextual change requires re-verification due to user mobility and the pervasiveness of security threats. Biometric traits are well-suited to support this continuous and autonomous authentication model. Such traits can be passively or periodically captured—for instance, a wearable ECG (Electrocardiogram) monitor that constantly verifies a user's heartbeat signature [184], or a smart vehicle that persistently checks the driver's identity through face or voice recognition [185], thereby enabling ongoing identity assurance without requiring user intervention. Implementing continuous biometric authentication in a zero-trust 6G environment necessitates the real-time analysis of highly sensitive data, and thus demands robust privacy-preserving mechanisms [183].

FHE plays a crucial role by allowing identity verification to occur directly on encrypted data streams. Using FHE, the underlying system can securely ingest encrypted biometric readings and conduct verification without exposing any plaintext data to the server or intermediate devices. The authentication process thus becomes both autonomous and privacy-preserving: the CPS continually requests encrypted proofs of identity and evaluates them entirely within the encrypted domain. However, one major challenge lies in handling frequent re-authentication under the computational burden imposed by encrypted operations [186]. Biometric verification with FHE is inherently resource-intensive [187], and excessive frequency may introduce latency or drain device batteries. To address this, we can leverage edge computing to synergistically offload and balance computational demands for FHE [188]. For instance, critical authentication tasks can be assigned to secure edge servers (*e.g.*, equipped with FHE accelerators [189]), while less urgent checks may be processed locally or retrieved from cached, previously verified results. Moreover, performance can be further enhanced through batching of biometric samples [164] or by employing more efficient cryptographic variants, such as somewhat homomorphic encryption for constrained computations [102].

7.4.4 Mobile Spatial Computing

Mobile spatial computing is an emerging computing paradigm in which a user can interact with their perceived 3D space by means of physical actions [190]. It contains two key components, mapping and localization, both of which involve two parties, the client and the server, and the communication between them. Any part of this service could be subject to attacks and/or information leakage. We assume a strong threat model, where the server could be compromised and actively probing the information supplied by the user. The information communicated between the client and server could be subject to eavesdropping or other attacks (*e.g.*,, a man-in-the-middle attack).

To address this problem, we explore leveraging FHE to shield spatial mapping and localization from privacy leakage and security attacks. There are three major challenges to adopting FHE for this purpose. First, it is non-trivial to integrate FHE into existing spatial mapping and localization systems, given their complex operations and different types of data to operate on. Second, the current generation of FHE algorithms cannot handle the non-polynomial operations required for spatial mapping and localization. Moreover, some of the schemes in FHE, such as CKKS [191], create only approximate results, which inevitably leads to a decrease in localization accuracy. Third, FHE has poor runtime performance and substantially inflates the size of encrypted data. Optimizing its efficacy, especially for resource-restricted devices such as smartphones, and efficiently transmitting encrypted data over wireless channels remain pressing challenges.

Therefore, we propose a more practical solution for protecting spatial mapping and localization using FHE. Both 2D/3D coordinates and the feature descriptors extracted from localization images and spatial maps can be encrypted using FHE. However, the process of finding matching features in the spatial map for localization features requires a comparison of distances between different pairs of feature descriptors. Even though Pegasus [192] can support comparison operations, it yields a high computational overhead. Moreover, encrypting both 2D/3D positions and feature descriptors will lead to high storage and transmission overhead. Given this backdrop, we propose to encrypt only the more compact 2D/3D coordinates, allowing the voluminous feature descriptors to remain unencrypted. Such a strategy not only renders feature matching operationally feasible but also markedly attenuates the visual fidelity of images reconstructed from the delineated features [193].

Another intricate puzzle is the estimation of the 6DoF pose utilizing FHE-encrypted data. Given that this conundrum essentially embodies a P*n*P issue [194], the prevalent resolution strategy hinges on the P3P methodology. P3P, however, demands the evaluation of *cosine* values and multiple non-polynomial calculations. Since FHE does not inherently support these operations, a viable detour is the adoption of the direct linear transformation (DLT) technique [195], which offers a linearized resolution to the PnP dilemma. DLT mandates a minimum of six 3D-2D correspondences to calculate the camera pose. A brief description of the DLT algorithm is as follows.

Given homogeneous coordinates of a 3D point in the world coordinate system $\mathbf{P} = (x, y, z, 1)^{\mathrm{T}}$, homogeneous coordinates of its 2D projected point in the image coordinate system $\mathbf{u} = (u, v, 1)^{\mathrm{T}}$, and camera parameters

$$\mathbf{K} = \begin{pmatrix} f_x & & c_x \\ & f_y & c_y \\ & & 1 \end{pmatrix}$$

the projection of the 3D point to the 2D point can be expressed as follows:

$$\lambda \begin{pmatrix} u \\ v \\ 1 \end{pmatrix} = \begin{pmatrix} f_x & & c_x \\ & f_y & c_y \\ & & 1 \end{pmatrix} (\mathbf{R} \mid \mathbf{t}) \begin{pmatrix} x \\ y \\ z \\ 1 \end{pmatrix} \tag{7.6}$$

Here, $\mathbf{R}$ is the rotation matrix of the camera pose, and $\mathbf{t}$ is the translation vector of the camera pose. In the kDLT algorithm, we first ignore the constraint that the rotation matrix must be orthogonal, so that $\left(\mathbf{R} \mid \mathbf{t}\right)$has12 unknown parameters $\mathbf{x} = (a_1, \cdots, a_{12})^{\mathrm{T}}$. From Eq. (7.6), we can get

$$\begin{pmatrix} \mathbf{P}^{\mathrm{T}} f_x & 0 & \mathbf{P}^{\mathrm{T}}(c_x - u) \\ 0 & \mathbf{P}^{\mathrm{T}} f_y & \mathbf{P}^{\mathrm{T}}(c_y - v) \end{pmatrix} \begin{pmatrix} a_1 \\ a_2 \\ \vdots \\ a_{12} \end{pmatrix} = 0. \tag{7.7}$$

From the above derivation, a pair of 3D-2D correspondence provides two equations, i.e., Eq. (7.7). When the number of corresponding points ≥ 6, we get the equation,

$$\mathbf{Ax} = \mathbf{0} \tag{7.8}$$

where the size of $\mathbf{A}$ is $2n \times 12$. We can perform singular value decomposition(SVD) on matrix $\mathbf{A}$ to solve Eq. (7.8), and get the camera pose.

To better understand the issues of integrating FHE into spatial mapping and localization, we conducted a pilot study using the SEAL [73] homomorphic encryption library released by Microsoft. It includes two different schemes BFV and CKKS [191]. CKKS scheme was chosen since it supports computation on encrypted real numbers [191]. We evaluated the computational overhead and inflation of encrypted data caused by FHE. For CKKS, the parameters *poly_modulus_degrees* and *coeff_modulus* were chosen to be 8192 and {60,40,40,60}, respectively. It was found that calculating the pair-wise distance between localization features and those in the spatial map took 7.3 ms on average with FHE-encrypted data in contrast with plain text, which took only 0.0076 ms. Similarly, the size of a feature descriptor extracted by SIFT [196] was 128 bytes while the size of an encrypted feature using the CKKS scheme was 128 Kbytes. This illustrates that it is impractical to encrypt feature descriptors with FHE.

Next, we evaluated the localization error introduced by estimating the 6DoF using DLT, instead of P3P. We measured the localization time, errors, and quality of recreated scenes after adding noisy features to the spatial map. We calculated the position error P_e as the Euclidean distance between an image's estimated position and the ground truth. We also calculated the orientation error O_e as the minimum rotation angle required to align the estimated rotation and ground. RANSAC (random sampling consensus) was used to remove outliers for both approaches. Figure 7.2 shows the normalized localization errors of DLT using the results of P3P as the baseline. It was found that compared to P3P, for 50% of the cases, the normalized localization error of DLT was higher than 1.3 for position and orientation,

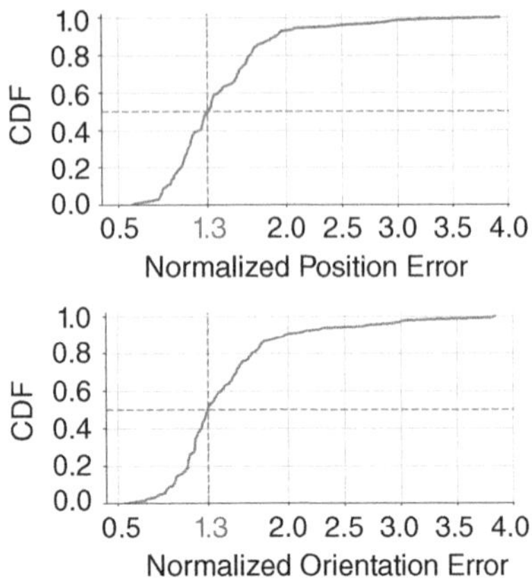

Figure 7.2 CDF of normalized localization errors when solving P*n*P using DLT, compared to P3P.

respectively. We conclude that FHE cannot support accurate 6DoF pose estimation methods such as P*n*P with RANSAC and that alternative solutions lead to high localization errors. Additionally, the computation latency of operations on FHE-encrypted data was found to be 1000 times higher than that of plain text. Finally, FHE resulted in size inflation of encrypted data resulting in increased network bandwidth and storage overhead.

7.5 Key Takeaways

In this book chapter, we have examined the potentialities of FHE, an emerging encryption paradigm that facilitates computations on ciphertexts without decryption, to fortify NextG-enabled CPS. Our discourse began with an exposition on the evolutionary trajectory and advancements of various FHE schemes. To delve deeper into the challenges associated with deploying FHE within CPS, we applied this cryptographic technique to mobile spatial computing, an important application within the CPS framework. Subsequently, we offered an exhaustive review of recent advancements pertaining to the utilization of FHE for securing a multitude of CPS applications. The insights gleaned from our case study, combined with our encompassing survey, underscore that while FHE emerges as a promising solution for securing CPS, its significant computational demands and data volume warrant further refinements. We hope that our study will inspire further research in this domain, ultimately achieving the fully secured NextG-enabled CPS landscape.

References

1 Dey, N., Ashour, A. S., Shi, F., Fong, S. J., and Tavares, J. M. R. (2018). Medical cyber-physical systems: A survey. *Journal of Medical Systems*, 42: 1–13.

2 Baker, S. B., Xiang, W., and Atkinson, I. (2017). Internet of Things for Smart Healthcare: Technologies, Challenges, and Opportunities. *IEEE Access*, 5: 26521–26544.

3 Xiong, G., Zhu, F., Liu, X., Dong, X., Huang, W., Chen, S., and Zhao, K. (2015). Cyber-physical-social System in Intelligent Transportation. *IEEE/CAA Journal of Automatica Sinica*, 2(3): 320–333.

4 Tsang, Y., Yang, T., Chen, Z., Wu, C., and Tan, K. (2022). How is extended reality bridging human and cyber-physical systems in the IoT-empowered logistics and supply chain management? *Internet of Things*, page 100623.

5 Jagatheesaperumal, S. K. and Rahouti, M. (2022). Building digital twins of cyber physical systems with metaverse for Industry 5.0 and beyond. *IT Professional*, 24(6): 34–40.

6 Pivoto, D. G., de Almeida, L. F., da Rosa Righi, R., Rodrigues, J. J., Lugli, A. B., and Alberti, A. M. (2021). Cyber-physical systems architectures for industrial internet of things applications in Industry 4.0: A literature review. *Journal of Manufacturing Systems*, 58: 176–192.

7 Greer, C., Burns, M., Wollman, D., and Griffor, E. (2019). Cyber-physical systems and internet of things. https://tsapps.nist.gov/publication/get_pdf.cfm?pub_id=927259 [accessed August 18, 2025].

8 CPS-VO | Cyber-Physical Systems Virtual Organization. https://cps-vo.org/. [accessed 18 August 2025].

9 Saad, W., Bennis, M., and Chen, M. (2019). A vision of 6G wireless systems: Applications, trends, technologies, and open research problems. *IEEE Network*, 34(3): 134–142.

10 Wang, C.-X., You, X., Gao, X., Zhu, X., Li, Z., Zhang, C., Wang, H., Huang, Y., Chen, Y., Haas, H., et al. (2023). On the road to 6G: Visions, requirements, key technologies and testbeds. *IEEE Communications Surveys & Tutorials.*

11 Maurer, U. M. and Wolf, S. (2000). The Diffie–Hellman Protocol. *Designs, Codes and Cryptography*, 19(2-3): 147–171.

12 Boneh, D. et al. (1999). Twenty years of attacks on the RSA cryptosystem. *Notices of the AMS*, 46(2): 203–213.

13 Gentry, C. (2009). *A fully homomorphic encryption scheme*. PhD thesis, Stanford University. crypto.stanford.edu/craig.

14 Han, B. (2019). Mobile immersive computing: Research challenges and the road ahead. *IEEE Communications Magazine*, 57(10): 112–118.

15 Baheti, R. and Gill, H. (2011). Cyber-physical systems. *The Impact of Control Technology*, 12(1): 161–166.

16 Lee, E. A. (2008). Cyber physical systems: Design challenges. In *2008 11th IEEE international symposium on object and component-oriented real-time distributed computing (ISORC)*, pages 363–369. IEEE.

17 Rajkumar, R., Lee, I., Sha, L., and Stankovic, J. (2010). Cyber-physical systems: the next computing revolution. In *Proceedings of the 47th Design Automation Conference*, pages 731–736.

18 Milenković, A., Otto, C., and Jovanov, E. (2006). Wireless sensor networks for personal health monitoring: Issues and an implementation. *Computer communications*, 29(13-14): 2521–2533.

19 Wang, J., Abid, H., Lee, S., Shu, L., and Xia, F. (2011). A secured health care application architecture for cyber-physical systems. *arXiv preprint arXiv:1201.0213.*

20 Yu, X. and Xue, Y. (2016). Smart grids: A cyber–physical systems perspective. *Proceedings of the IEEE*, 104(5): 1058–1070.

21 Gopstein, A., Nguyen, C., O'Fallon, C., Hastings, N., Wollman, D., et al. (2021). *NIST framework and roadmap for smart grid interoperability standards, release 4.0.* Department of Commerce. National Institute of Standards and Technology.

22 IEA. Executive summary – electricity market report 2023 – analysis - IEA. https://www.iea.org/reports/electricity-market-report-2023/executive-summary [accessed 18 August 2025].

23 OSTI (2023). Integrating variable renewable energy: Challenges and solutions (technical report) | osti.gov. https://www.osti.gov/biblio/1097911 [accessed 18 August 2025].

24 Rawat, D. B., Song, M., and Shetty, S. (2015). *Dynamic spectrum access for wireless networks.* Springer.

25 Arena, F., Pau, G., and Severino, A. (2020). A review on IEEE 802.11 p for intelligent transportation systems. *Journal of Sensor and Actuator Networks*, 9(2): 22.

26 Zhu, L., Yu, F. R., Wang, Y., Ning, B., and Tang, T. (2018). Big data analytics in intelligent transportation systems: A survey. *IEEE Transactions on Intelligent Transportation Systems*, 20(1): 383–398.

27 Shi, J., Wan, J., Yan, H., and Suo, H. (2011). A survey of cyber-physical systems. In *2011 International Conference on Wireless Communications and Signal Processing (WCSP)*, pages 1–6. IEEE.

28 Humayed, A., Lin, J., Li, F., and Luo, B. (2017). Cyber-physical systems security – A survey. *IEEE Internet of Things Journal*, 4(6): 1802–1831.

29 Rivest, R. L., Adleman, L., Dertouzos, M. L., et al. (1978). On data banks and privacy homomorphisms. *Foundations of Secure Computation*, 4(11): 169–180.

30 Boneh, D., Goh, E., and Nissim, K. (2005a). Evaluating 2-dnf formulas on ciphertexts. In Kilian, J., editor, *Theory of Cryptography, Second Theory of Cryptography Conference, TCC 2005,* Cambridge, MA, USA, February 10-12,

2005, *Proceedings*, volume 3378 of *Lecture Notes in Computer Science*, pages 325–341. Springer.

31 Peikert, C. (2016). A decade of lattice cryptography. *Found. Trends Theor. Comput. Sci.*, 10(4): 283–424.

32 Howgrave-Graham, N. (2001). Approximate integer common divisors. In *Revised Papers from the International Conference on Cryptography and Lattices*, CaLC '01, page 51–66, Berlin, Heidelberg. Springer-Verlag.

33 Brakerski, Z. and Vaikuntanathan, V. (2011). Fully homomorphic encryption from ring-LWE and security for key dependent messages. In *Annual Cryptology Conference*, pages 505–524. Springer.

34 Canetti, R. and Garay, J. A., editors, (2013). *Advances in cryptology–CRYPTO 2013: 33rd annual cryptology conference,* Santa Barbara, CA, USA, August 18–22, 2013. Proceedings, *Part I*, volume 8042 Springer.

35 Gentry, C. (2010). Computing arbitrary functions of encrypted data. *Communications of the ACM*, 53(3): 97–105.

36 Naehrig, M., Lauter, K., and Vaikuntanathan, V. (2011). Can homomorphic encryption be practical? In *Proceedings of the 3rd ACM Workshop on Cloud Computing Security Workshop*, CCSW '11, page 113–124, New York, NY, USA. Association for Computing Machinery.

37 Smart, N. P. and Vercauteren, F. (2010). Fully homomorphic encryption with relatively small key and ciphertext sizes. In Nguyen, P. Q. and Pointcheval, D., editors, *Public Key Cryptography – PKC 2010*, pages 420–443, Berlin, Heidelberg. Springer Berlin Heidelberg.

38 Cramer, R., Ducas, L., Peikert, C., and Regev, O. (2016). Recovering Short Generators of Principal Ideals in Cyclotomic Rings. In Fischlin, M. and Coron, J.-S., editors, *Advances in Cryptology – EUROCRYPT 2016*, pages 559–585, Berlin, Heidelberg. Springer Berlin Heidelberg.

39 Gentry, C. and Halevi, S. (2011). Implementing gentry's fully-homomorphic encryption scheme. In Paterson, K. G., editor, *Advances in Cryptology – EUROCRYPT 2011*, pages 129–148, Berlin, Heidelberg. Springer Berlin Heidelberg.

40 Smart, N. P. and Vercauteren, F. (2014). Fully Homomorphic SIMD Operations. *Designs, Codes and Cryptography*, 71: 57–81.

41 Gentry, C., Halevi, S., and Smart, N. P. (2012). Fully homomorphic encryption with polylog overhead. In *Proceedings of the 31st Annual International Conference on Theory and Applications of Cryptographic Techniques*, EUROCRYPT'12, page 465–482, Berlin, Heidelberg. Springer-Verlag.

42 Regev, O. (2009). On lattices, learning with errors, random linear codes, and cryptography. *Journal of the ACM*, 56(6).

43 Lyubashevsky, V., Peikert, C., and Regev, O. (2013). On ideal lattices and learning with errors over rings. *Journal of the ACM*, 60(6).

44 Hoffstein, J., Pipher, J., and Silverman, J. H. (1998). Ntru: A ring-based public key cryptosystem. In *Proceedings of the Third International Symposium on Algorithmic Number Theory*, ANTS-III, page 267–288, Berlin, Heidelberg. Springer-Verlag.

45 Brakerski, Z. and Vaikuntanathan, V. (2014). Efficient fully homomorphic encryption from (standard) LWE. *SIAM Journal on Computing*, 43(2): 831–871.

46 Brakerski, Z., Gentry, C., and Vaikuntanathan, V. (2012). (leveled) fully homomorphic encryption without bootstrapping. In Goldwasser, S., editor, *Innovations in Theoretical Computer Science 2012*, Cambridge, MA, USA, *January 8-10, 2012*, pages 309–325. ACM.

47 Gentry, C., Halevi, S., Peikert, C., and Smart, N. P. (2012a). Ring switching in BGV-style homomorphic encryption. In *International Conference on Security and Cryptography for Networks*, pages 19–37. Springer.

48 Gentry, C., Halevi, S., and Smart, N. P. (2012b). Better bootstrapping in fully homomorphic encryption. In *International Workshop on Public Key Cryptography*, pages 1–16. Springer.

49 Gentry, C., Halevi, S., and Smart, N. P. (2012d). Homomorphic evaluation of the AES circuit. In *Annual Cryptology Conference*, pages 850–867. Springer.

50 Brakerski, Z. (2012). Fully homomorphic encryption without modulus switching from classical GAPSVP. In Safavi-Naini, R. and Canetti, R., editors, *Advances in Cryptology – CRYPTO 2012*, pages 868–886, Berlin, Heidelberg. Springer Berlin Heidelberg.

51 Van Dijk, M., Gentry, C., Halevi, S., and Vaikuntanathan, V. (2010). Fully homomorphic encryption over the integers. In *Advances in Cryptology–EUROCRYPT 2010: 29th Annual International Conference on the Theory and Applications of Cryptographic Techniques, French Riviera,* May 30–June 3, 2010. *Proceedings 29*, pages 24–43. Springer.

52 Coron, J.-S., Mandal, A., Naccache, D., and Tibouchi, M. (2011). Fully homomorphic encryption over the integers with shorter public keys. In *Annual Cryptology Conference*, pages 487–504. Springer.

53 Albrecht, M., Chase, M., Chen, H., Ding, J., Goldwasser, S., Gorbunov, S., Halevi, S., Hoffstein, J., Laine, K., Lauter, K., Lokam, S., Micciancio, D., Moody, D., Morrison, T., Sahai, A., and Vaikuntanathan, V. (2019). Homomorphic encryption standard. Cryptology ePrint Archive, Paper 2019/939.https://eprint.iacr.org/2019/939.

54 Hallman, R. A., Laine, K., Dai, W., Gama, N., Malozemoff, A. J., Polyakov, Y., and Carpov, S. (2018). Building applications with homomorphic encryption. In *Proceedings of the 2018 ACM SIGSAC Conference on Computer and Communications Security*, CCS '18, page 2160–2162, New York, NY, USA. Association for Computing Machinery.

55 Bajard, J., Eynard, J., Hasan, M. A., and Zucca, V. (2016). A full RNS variant of FV like somewhat homomorphic encryption schemes. *IACR Cryptol. ePrint Arch.*, page 510.

56 Bajard, J., Eynard, J., Martins, P., Sousa, L., and Zucca, V. (2019). An HPR variant of the FV scheme: Computationally cheaper, asymptotically faster. *IACR Cryptology ePrint Archive*, page 500.

57 Halevi, S., Polyakov, Y., and Shoup, V. (2018). An improved RNS variant of the BFV homomorphic encryption scheme. *IACR Cryptology ePrint Archive*, page 117.

58 Chen, H., Laine, K., Player, R., and Xia, Y. (2018b). High-precision arithmetic in homomorphic encryption. In Smart, N. P., editor, *Topics in Cryptology – CT-RSA 2018*, pages 116–136, Cham. Springer International Publishing.

59 Chen, H. and Han, K. (2018). Homomorphic lower digits removal and improved FHE bootstrapping. *Cryptology ePrint Archive*, Paper 2018/067. https://eprint.iacr.org/2018/067.

60 Ducas, L. and Micciancio, D. (2015). FHEW: bootstrapping homomorphic encryption in less than a second. In Oswald, E. and Fischlin, M., editors, *Advances in Cryptology - EUROCRYPT 2015 - 34th Annual International Conference on the Theory and Applications of Cryptographic Techniques*, Sofia, Bulgaria, April 26-30, 2015, *Proceedings, Part I*, volume 9056 of *Lecture Notes in Computer Science*, pages 617–640. Springer.

61 Chillotti, I., Gama, N., Georgieva, M., and Izabachène, M. (2020). TFHE: Fast fully homomorphic encryption over the torus. *Cryptology ePrint Archive*, 33(1): 34–91.

62 Chillotti, I., Gama, N., Georgieva, M., and Izabachène, M. (2016). Faster fully homomorphic encryption: Bootstrapping in less than 0.1 seconds. In Cheon, J. H. and Takagi, T., editors, *Advances in Cryptology - ASIACRYPT 2016 - 22nd International Conference on the Theory and Application of Cryptology and Information Security, Hanoi, Vietnam, December 4-8, 2016, Proceedings, Part I*, volume 10031 of *Lecture Notes in Computer Science*, pages 3–33.

63 Chillotti, I., Gama, N., Georgieva, M., and Izabachène, M. (2017b). Faster Packed Homomorphic Operations and Efficient Circuit Bootstrapping for TFHE. In Takagi, T. and Peyrin, T., editors, *Advances in Cryptology – ASIACRYPT 2017*, pages 377–408, Cham. Springer International Publishing.

64 Biasse, J.-F. and Ruiz, L. (2015). FHEW with Efficient Multibit Bootstrapping. In Lauter, K. and Rodríguez-Henríquez, F., editors, *Progress in Cryptology – LATINCRYPT 2015*, pages 119–135, Cham. Springer International Publishing.

65 Bonnoron, G., Ducas, L., and Fillinger, M. (2017). Large fhe gates from tensored homomorphic accumulator. *Cryptology ePrint Archive*, Paper 2017/996. https://eprint.iacr.org/2017/996.

66 Carpov, S., Izabachène, M., and Mollimard, V. (2019). New techniques for multi-value input homomorphic evaluation and applications. In *The Cryptographer's Track at RSA Conference.*

67 Micciancio, D. and Sorrell, J. (2018). Ring packing and amortized FHEW bootstrapping. *IACR Cryptology ePrint Archive*, 2018: 532.

68 Takagi, T. and Peyrin, T., editors (2017). *Advances in Cryptology - ASIACRYPT 2017 - 23rd International Conference on the Theory and Applications of Cryptology and Information Security, Hong Kong, China, December 3-7, 2017, Proceedings, Part I*, volume 10624 of *Lecture Notes in Computer Science*. Springer.

69 Chen, H., Chillotti, I., and Song, Y. (2018a). Improved bootstrapping for approximate homomorphic encryption. *IACR Cryptology ePrint Archive*, page 1043.

70 Cheon, J. H., Han, K., Kim, A., Kim, M., and Song, Y. (2018a). Bootstrapping for approximate homomorphic encryption. *IACR Cryptology ePrint Archive*, page 153.

71 Blatt, M., Gusev, A., Polyakov, Y., Rohloff, K., and Vaikuntanathan, V. (2019). Optimized homomorphic encryption solution for secure genome-wide association studies. *IACR Cryptology ePrint Archive*, page 223.

72 Cheon, J. H., Han, K., Kim, A., Kim, M., and Song, Y. (2018b). A full RNS variant of approximate homomorphic encryption. *IACR Cryptology ePrint Archive*, page 931.

73 microsoft-seal (2018). Microsoft SEAL(release 3.0). http://sealcrypto.org/.

74 Genise, N., Gentry, C., Halevi, S., Li, B., and Micciancio, D. (2019). Homomorphic encryption for finite automata. *IACR Cryptology ePrint Archive*, page 176.

75 Boura, C., Gama, N., and Georgieva, M. (2018). Chimera: a unified framework for b/fv, TFHE and HEAAN fully homomorphic encryption and predictions for deep learning. *IACR Cryptol. ePrint Arch.*, page 758.

76 Microsoft SEAL(release 3.3). https://github.com/Microsoft/SEAL [accessed 18 August 2025].

77 HElib: An Implementation of Homomorphic Encryption. https://github.com/HomEnc/HElib [accessed 18 August 2025].

78 TFHE: Fast Fully Homomorphic Encryption Library. https://tfhe.github.io/tfhe/ [accessed 18 August 2025].

79 HEAAN Software Library. https://github.com/snucrypto/HEAAN [accessed 18 August 2025].

80 The PALISADE Lattice Cryptography Library. https://palisade-crypto.org/software-library/ [accessed 18 August 2025].

81 CEA-LIST-cingulata. Cingulata. https://github.com/CEA-LIST/Cingulata [accessed 18 August 2025].

82 FV-NFLlib. https://github.com/CryptoExperts/FV-NFLlib [accessed 18 August 2025].

83 Badawi, A. A., Bates, J., Bergamaschi, F., Cousins, D. B., Erabelli, S., Genise, N., Halevi, S., Hunt, H., Kim, A., Lee, Y., Liu, Z., Micciancio, D., Quah, I., Polyakov, Y., R.V., S., Rohloff, K., Saylor, J., Suponitsky, D., Triplett, M., Vaikuntanathan, V., and Zucca, V. (2022). OpenFHE: Open-Source Fully Homomorphic Encryption Library. Cryptology ePrint Archive, Paper 2022/915.https://eprint.iacr.org/2022/915

84 Lattigo 1.3.0. https://github.com/tuneinsight/lattigo [accessed 18 August 2025].

85 lol-apps. Λ∘λ: Functional Lattice Cryptography. https://hackage.haskell.org/package/lol-apps [accessed 18 August 2025].

86 Homomorphic Encryption Standardization. http://homomorphicencryption [accessed 10 December 2025].

87 Crockett, E., Peikert, C., and Sharp, C. (2018). ALCHEMY: A language and compiler for homomorphic encryption made easy. In Lie, D., Mannan, M., Backes, M., and Wang, X., editors, *Proceedings of the 2018 ACM SIGSAC Conference on Computer and Communications Security, CCS 2018,* Toronto, ON, Canada, *October 15-19, 2018*, pages 1020–1037. ACM.

88 RAMPARTS: RApid Machine-learning Processing Applications and Reconfigurable Targeting of Security. https://galois.com/project/ramparts/ [accessed 18 August 2025].

89 A Language and Compiler for Homomorphic Encryption Made easY (ALCHEMY). https://github.com/cpeikert/ALCHEMY [accessed 18 August 2025].

90 Bos, J. W., Lauter, K. E., Loftus, J., and Naehrig, M. (2013). Improved security for a ring-based fully homomorphic encryption scheme. In Stam, M., editor, *Cryptography and Coding - 14th IMA International Conference, IMACC 2013,* Oxford, UK, December 17-19, 2013. *Proceedings*, volume 8308 of *Lecture Notes in Computer Science*, pages 45–64. Springer.

91 Boemer, F., Lao, Y., Cammarota, R., and Wierzynski, C. (2019). nGraph-HE: a graph compiler for deep learning on homomorphically encrypted data. In Palumbo, F., Becchi, M., Schulz, M., and Sato, K., editors, *Proceedings of the 16th ACM International Conference on Computing Frontiers, CF 2019,* Alghero, Italy, *April 30 - May 2, 2019*, pages 3–13. ACM.

92 Halevi, S. and Shoup, V. (2018). Faster homomorphic linear transformations in HELIB. In Shacham, H. and Boldyreva, A., editors, *Advances in Cryptology - CRYPTO 2018 - 38th Annual International Cryptology Conference,* Santa Barbara, CA, USA, August 19-23, 2018, *Proceedings, Part I*, volume 10991 of *Lecture Notes in Computer Science*, pages 93–120. Springer.

93 NTL: A library for doing number theory. https://www.shoup.net/ntl [accessed 18 August 2025].

94 HElib: An Implementation of Homomorphic Encryption. https://github.com/HomEnc/HElib [accessed 18 August 2025].

95 Frigo, M. (1999). A fast Fourier transform compiler. In Ryder, B. G. and Zorn, B. G., editors, *Proceedings of the 1999 ACM SIGPLAN Conference on Programming Language Design and Implementation (PLDI),* Atlanta, Georgia, USA, *May 1-4, 1999*, pages 169–180. ACM.

96 Chillotti, I., Gama, N., Georgieva, M., and Izabachène, M. (2017a). Faster Packed Homomorphic Operations and Efficient Circuit Bootstrapping for TFHE. In *Proceeding of International Conference on the Theory and Applications of Cryptology and Information Security.*

97 CUDA-accelerated Fully Homomorphic Encryption (cuFHE). https:/github.com //vernamlab/cuFHE [accessed 18 August 2025].

98 NuCypher Fully Homomorphic Encryption (NuFHE). https://github.com/nucypher/nufhe [accessed 18 August 2025].

99 Cheon, J. H., Han, K., Kim, A., Kim, M., and Song, Y. (2018c). A full RNS variant of approximate homomorphic encryption. In Cid, C. and Jr., M. J. J., editors, *Selected Areas in Cryptography - SAC 2018 - 25th International Conference,* Calgary, AB, Canada, August 15-17, 2018, *Revised Selected Papers*, volume 11349 of *Lecture Notes in Computer Science*, pages 347–368. Springer.

100 López-Alt, A., Tromer, E., and Vaikuntanathan, V. (2012). On-the-fly multiparty computation on the cloud via multikey fully homomorphic encryption. In Karloff, H. J. and Pitassi, T., editors, *Proceedings of the 44th Symposium on Theory of Computing Conference, STOC 2012,* New York, NY, USA, *May 19-22, 2012*, pages 1219–1234. ACM.

101 Stehlé, D. and Steinfeld, R. (2011). Making NTRU as secure as worst-case problems over ideal lattices. In Paterson, K. G., editor, *Advances in Cryptology - EUROCRYPT 2011 - 30th Annual International Conference on the Theory and Applications of Cryptographic Techniques,* Tallinn, Estonia, May 15-19, 2011. *Proceedings*, volume 6632 of *Lecture Notes in Computer Science*, pages 27–47. Springer.

102 Fan, J. and Vercauteren, F. (2012). Somewhat practical fully homomorphic encryption. *Cryptology ePrint Archive.*

103 Crockett, E. and Peikert, C. (2015). $\Lambda\circ\lambda$: A Functional Library for Lattice Cryptography. *IACR Cryptol. ePrint Arch.*, 2015: 1134.

104 Ullah, S., Li, J., Chen, J., Ali, I., Khan, S., Ahad, A., Ullah, F., and Leung, V. C. (2024b). A survey on emerging trends and applications of 5G and 6G to healthcare environments. *ACM Computing Surveys*, 57(4): 1–36.

105 DFKI. EBerlin 6G Conference 2024: How 6G and AI transform medicine of tomorrow.https://www.dfki.de/en/web/news/how-6g-and-ai-transform-medicine-of-tomorrow. [accessed on 10 December 2025].

106 Rani, P. (2025). From Data to Diagnosis: Unleashing AI and 6G in Modern Medicine. *EKSPLORIUM-BULETIN PUSAT TEKNOLOGI BAHAN GALIAN NUKLIR*, 46(1): 69–103.

107 Lee, C. H., Lim, K. H., and Eswaran, S. (2025). A comprehensive survey on secure healthcare data processing with homomorphic encryption: Attacks and defenses. *Discover Public Health*, 22(1): 1–29.

108 Sheu, R.-K., Lin, Y.-C., Pardeshi, M. S., Huang, C.-Y., Pai, K.-C., Chen, L.-C., and Huang, C.-C. (2023). Adaptive Autonomous Protocol for Secured Remote Healthcare Using Fully Homomorphic Encryption (AutoPro-RHC). *Sensors*, 23(20): 8504.

109 Frimpong, E., Nguyen, K., Budzys, M., Khan, T., and Michalas, A. (2024). GuardML: Efficient Privacy Preserving Machine Learning Services Through Hybrid Homomorphic Encryption. In *Proceedings of ACM/SIGAPP Symposium on Applied Computing.*

110 Yang, Q., Liu, Y., Chen, T., and Tong, Y. (2019). Federated machine learning: Concept and applications. *ACM Transactions on Intelligent Systems and Technology (TIST)*, 10(2): 1–19.

111 Wibawa, F., Catak, F. O., Kuzlu, M., Sarp, S., and Cali, U. (2022). Homomorphic encryption and federated learning based privacy preserving CNN training: COVID-19 detection use-case. In *Proceedings of European Interdisciplinary Cybersecurity Conference (EICC).*

112 Zhou, J., Zhou, L., Wang, D., Xu, X., Li, H., Chu, Y., Han, W., and Gao, X. (2024). Personalized and privacy-preserving federated heterogeneous medical image analysis with PPPML-HMI. *Computers in Biology and Medicine*, 169: 107861.

113 Malik, A., Ratha, N., Yalavarthi, B., Sharma, T., Kaushik, A., and Jutla, C. (2024). Confidential and protected disease classifier using fully homomorphic encryption. In *Proceedings of IEEE Conference on Artificial Intelligence (CAI).*

114 Slay, J. and Miller, M. (2007). Lessons learned from the Maroochy water breach. In *International Conference on Critical Infrastructure Protection*, pages 73–82. Springer.

115 Pasqualetti, F., Dörfler, F., and Bullo, F. (2013). Attack detection and identification in cyber-physical systems. *IEEE Transactions on Automatic Control*, 58(11): 2715–2729.

116 Lee, C., Shim, H., and Eun, Y. (2015). Secure and robust state estimation under sensor attacks, measurement noises, and process disturbances: Observer-based combinatorial approach. In *2015 European Control Conference (ECC)*, pages 1872–1877. IEEE.

117 Teixeira, A., Shames, I., Sandberg, H., and Johansson, K. H. (2015). A secure control framework for resource-limited adversaries. *Automatica*, 51: 135–148.

118 Park, G., Shim, H., Lee, C., Eun, Y., and Johansson, K. H. (2016). When adversary encounters uncertain cyber-physical systems: Robust zero-dynamics attack with disclosure resources. In *2016 IEEE 55th Conference on Decision and Control (CDC)*, pages 5085–5090. IEEE.

119 Kogiso, K. and Fujita, T. (2015). Cyber-security enhancement of networked control systems using homomorphic encryption. In *2015 54th IEEE Conference on Decision and Control (CDC)*, pages 6836–6843. IEEE.

120 Kim, J., Lee, C., Shim, H., Cheon, J. H., Kim, A., Kim, M., and Song, Y. (2016). Encrypting controller using fully homomorphic encryption for security of cyber-physical systems. *IFAC-PapersOnLine*, 49(22): 175–180.

121 Cheon, J. H., Kim, A., Kim, M., and Song, Y. S. (2016b). Floating-point homomorphic encryption. *IACR Cryptology ePrint Archive*, 2016: 421.

122 Johansson, K. H. (2000). The quadruple-tank process: A multivariable laboratory process with an adjustable zero. *IEEE Transactions on Control Systems Technology*, 8(3): 456–465.

123 Ashton, K. et al. (2009). That 'Internet of Things' Thing. *RFID Journal*, 22(7): 97–114.

124 Lasi, H., Fettke, P., Kemper, H.-G., Feld, T., and Hoffmann, M. (2014). Industry 4.0. *Business & information systems engineering*, 6: 239–242.

125 Antonakakis, M., April, T., Bailey, M., Bernhard, M., Bursztein, E., Cochran, J., Durumeric, Z., Halderman, J. A., Invernizzi, L., Kallitsis, M., et al. (2017). Understanding the MIRAI botnet. In *26th USENIX Security Symposium (USENIX Security 17)*, pages 1093–1110.

126 Langner, R. (2011). Stuxnet: Dissecting a cyberwarfare weapon. *IEEE Security & Privacy*, 9(3): 49–51.

127 Zhou, L., Yeh, K.-H., Hancke, G., Liu, Z., and Su, C. (2018). Security and privacy for the industrial internet of things: An overview of approaches to safeguarding endpoints.

128 Biham, E. and Dunkelman, O. (2000). Cryptanalysis of the A5/1 GSM stream cipher. In *Progress in Cryptology—INDOCRYPT 2000: First International Conference in Cryptology in India Calcutta, India, December 10–13, 2000 Proceedings 1*, pages 43–51. Springer.

129 Petrovic, S. and Fuster-Sabater, A. (2000). Cryptanalysis of the A5/2 algorithm. *Cryptology EPrint Archive.*

130 Wagner, D., Simpson, L., Dawson, E., Kelsey, J., Millan, W., and Schneier, B. (1999). Cryptanalysis of oryx. In *Selected Areas in Cryptography: 5th Annual International Workshop, SAC'98* Kingston, Ontario, Canada, August 17–18, *1998 Proceedings 5*, pages 296–305. Springer.

131 Verdult, R., Garcia, F. D., and Balasch, J. (2012). Gone in 360 seconds: Hijacking with hitag2. In *21st USENIX Security Symposium (USENIX Security 12)*, pages 237–252.

132 Boneh, D., Goh, E.-J., and Nissim, K. (2005b). Evaluating 2-DNF Formulas on Ciphertexts. In *Theory of Cryptography: Second Theory of Cryptography Conference, TCC 2005,* Cambridge, MA, USA, February 10-12, 2005. *Proceedings 2*, pages 325–341.

133 Lu, R. (2018). A new communication-efficient privacy-preserving range query scheme in Fog-enhanced IoT. *IEEE Internet of Things Journal*, 6(2): 2497–2505.

134 Fiore, D., Mitrokotsa, A., Nizzardo, L., and Pagnin, E. (2016). Multi-key homomorphic authenticators. In *International conference on the theory and application of cryptology and information security*, pages 499–530. Springer.

135 Li, S., Zhao, S., Min, G., Qi, L., and Liu, G. (2021). Lightweight privacy-preserving scheme using homomorphic encryption in industrial internet of things. *IEEE Internet of Things Journal*, 9(16): 14542–14550.

136 Jain, A., Bolle, R., and Pankanti, S. (1996). *Introduction to Biometrics*. Springer.

137 Jain, A. K. and Nandakumar, K. (2012). Biometric authentication: System security and user privacy. *Computer*, 45(11): 87–92.

138 Mai, G., Cao, K., Yuen, P. C., and Jain, A. K. (2018). On the reconstruction of face images from deep face templates. *IEEE Transactions on Pattern Analysis and Machine Intelligence*, 41(5): 1188–1202.

139 Soutar, C., Roberge, D., Stoianov, A., Gilroy, R., and Kumar, B. V. (1998). Biometric encryption using image processing. In *Optical Security and Counterfeit Deterrence Techniques II*, volume 3314, pages 178–188. SPIE.

140 Naor, M. and Shamir, A. (1995). Visual cryptography. In *Advances in Cryptology—EUROCRYPT'94: Workshop on the Theory and Application of Cryptographic Techniques Perugia, Italy, May 9–12, 1994 Proceedings 13*, pages 1–12. Springer.

141 Rao, Y. S., Sukonkina, Y., Bhagwati, C., and Singh, U. K. (2008). Fingerprint based authentication application using visual cryptography methods (improved id card). In *TENCON 2008-2008 IEEE Region 10 Conference*, pages 1–5. IEEE.

142 Revenkar, P., Anjum, A., and Gandhare, W. (2010). Secure iris authentication using visual cryptography. *arXiv preprint arXiv:1004.1748*.

143 Ross, A. and Othman, A. (2010). Visual cryptography for biometric privacy. *IEEE Transactions on Information Forensics and Security*, 6(1): 70–81.

144 Juels, A. and Sudan, M. (2006). A fuzzy vault scheme. *Designs, Codes and Cryptography*, 38: 237–257.

145 Uludag, U., Pankanti, S., and Jain, A. K. (2005). Fuzzy vault for fingerprints. In *International Conference on Audio-and Video-Based Biometric Person Authentication*, pages 310–319. Springer.

146 Lee, Y. J., Park, K. R., Lee, S. J., Bae, K., and Kim, J. (2008). A new method for generating an invariant iris private key based on the fuzzy vault system. *IEEE Transactions on Systems, Man, and Cybernetics, Part B (Cybernetics)*, 38(5): 1302–1313.

147 Ratha, N. K., Connell, J. H., and Bolle, R. M. (2001). Enhancing security and privacy in biometrics-based authentication systems. *IBM Systems Journal*, 40(3): 614–634.

148 Patel, V. M., Ratha, N. K., and Chellappa, R. (2015). Cancelable biometrics: A review. *IEEE Signal Processing Magazine*, 32(5): 54–65.

149 Jin, Z., Hwang, J. Y., Lai, Y.-L., Kim, S., and Teoh, A. B. J. (2017). Ranking-based locality sensitive hashing-enabled cancelable biometrics: Index-of-max hashing. *IEEE Transactions on Information Forensics and Security*, 13(2): 393–407.

150 Sadeghi, A.-R., Schneider, T., and Wehrenberg, I. (2009). Efficient privacy-preserving face recognition. In *International Conference on Information Security and Cryptology*, pages 229–244. Springer.

151 Troncoso-Pastoriza, J. R. and Pérez-González, F. (2012). Fully homomorphic faces. In *2012 19th IEEE International Conference on Image Processing*, pages 2657–2660. IEEE.

152 Gomez-Barrero, M., Maiorana, E., Galbally, J., Campisi, P., and Fierrez, J. (2017). Multi-biometric template protection based on homomorphic encryption. *Pattern Recognition*, 67: 149–163.

153 Drozdowski, P., Buchmann, N., Rathgeb, C., Margraf, M., and Busch, C. (2019). On the application of homomorphic encryption to face identification. In *2019 International Conference of the Biometrics Special Interest Group (BIOSIG)*, pages 1–5. IEEE.

154 Kim, T., Oh, Y., and Kim, H. (2020). Efficient privacy-preserving fingerprint-based authentication system using fully homomorphic encryption. *Security and Communication Networks*, 2020: 1–11.

155 Yang, T., Zhang, Y., Sun, J., and Wang, X. (2021). Privacy enhanced cloud-based facial recognition. *Neural Processing Letters*, pages 1–9.

156 Barni, M., Droandi, G., and Lazzeretti, R. (2015). Privacy protection in biometric-based recognition systems: A marriage between cryptography and signal processing. *IEEE Signal Processing Magazine*, 32(5): 66–76.

157 Lagendijk, R. L., Erkin, Z., and Barni, M. (2012). Encrypted signal processing for privacy protection: Conveying the utility of homomorphic encryption and multiparty computation. *IEEE Signal Processing Magazine*, 30(1): 82–105.

158 Xiang, C., Tang, C., Cai, Y., and Xu, Q. (2016). Privacy-preserving face recognition with outsourced computation. *Soft Computing*, 20: 3735–3744.

159 Asuncion, A. E. C., Guadalupe, B. C. T., and Yu, W. E. S. (2019). Implementation and analysis of homomorphic facial image encryption and manipulation. In *Proceedings of the 2019 4th International Conference on Multimedia Systems and Signal Processing*, pages 158–166.

160 Troncoso-Pastoriza, J. R., González-Jiménez, D., and Pérez-González, F. (2013). Fully private noninteractive face verification. *IEEE Transactions on Information Forensics and Security*, 8(7): 1101–1114.

161 Cheon, J. H., Chung, H., Kim, M., and Lee, K.-W. (2016a). Ghostshell: Secure biometric authentication using integrity-based homomorphic evaluations. *Cryptology ePrint Archive*. https://eprint.iacr.org/2016/484.pdf.

162 Boddeti, V. N. (2018). Secure face matching using fully homomorphic encryption. In *2018 IEEE 9th International Conference on Biometrics Theory, Applications and Systems (BTAS)*, pages 1–10. IEEE.

163 Engelsma, J. J., Cao, K., and Jain, A. K. (2019). Learning a fixed-length fingerprint representation. *IEEE transactions on pattern analysis and machine intelligence*, 43(6): 1981–1997.

164 Engelsma, J. J., Jain, A. K., and Boddeti, V. N. (2022). Hers: Homomorphically encrypted representation search. *IEEE Transactions on Biometrics, Behavior, and Identity Science*, 4(3): 349–360.

165 Gong, S., Boddeti, V. N., and Jain, A. K. (2019). On the intrinsic dimensionality of image representations. In 2019 IEEE. In *CVF Conference on Computer Vision and Pattern Recognition (CVPR)*, pages 3982–3991.

166 Juvekar, C., Vaikuntanathan, V., and Chandrakasan, A. (2018). {GAZELLE}: A low latency framework for secure neural network inference. In *27th USENIX Security Symposium (USENIX Security 18)*, pages 1651–1669.

167 Ansuini, A., Laio, A., Macke, J. H., and Zoccolan, D. (2019). Intrinsic dimension of data representations in deep neural networks. *Advances in Neural Information Processing Systems*, 32.

168 Mahmood, N. H., Böcker, S., Moerman, I., López, O. A., Munari, A., Mikhaylov, K., Clazzer, F., Bartz, H., Park, O.-S., Mercier, E., et al. (2021). Machine type communications: Key drivers and enablers towards the 6G era. *EURASIP Journal on Wireless Communications and Networking*, 2021(1): 134.

169 Nguyen, V.-L., Hwang, R.-H., Lin, P.-C., Vyas, A., and Nguyen, V.-T. (2022). Toward the age of intelligent vehicular networks for connected and autonomous vehicles in 6G. *IEEE Network*, 37(3): 44–51.

170 Wu, Y. (2022). Ethically responsible and trustworthy autonomous systems for 6G. *IEEE Network*, 36(4): 126–133.

171 Ullah, S., Bazai, S. U., Imran, M., Ilyas, Q. M., Mehmood, A., Saleem, M. A., Rafique, M. A., Haider, A., Khan, I., Iqbal, S., et al. (2024a). Recent developments in authentication schemes used in machine-type communication devices in machine-to-machine communication: Issues and challenges. *Computers, Materials & Continua*, 79(1).

172 Almohamade, S. S., Clark, J. A., and Law, J. (2020).Behaviour-based biometrics for continuous user authentication to industrial collaborative robots. 185–197. Springer.

173 Haas, S., Ulz, T., and Steger, C. (2017). Secured offline authentication on industrial mobile robots using biometric data. In *Robot World Cup*, pages 143–155. Springer.

174 Pradel, G. and Mitchell, C. (2021). Privacy-preserving biometric matching using homomorphic encryption. https://arxiv.org/pdf/2111.12372. [accessed on 18 August 2025].

175 Hasan, M., Mohan, S., Shimizu, T., and Lu, H. (2020). Securing Vehicle-to-Everything (V2X) Communication Platforms. *IEEE Transactions on Intelligent Vehicles*, 5(4): 693–713.

176 Abdelkader, G., Elgazzar, K., and Khamis, A. (2021). Connected vehicles: Technology review, state of the art, challenges and opportunities. *Sensors*, 21(22): 7712.

177 Derman, E. and Salah, A. A. (2018). Continuous real-time vehicle driver authentication using convolutional neural network based face recognition. In *Proceedings of IEEE International Conference on Automatic Face & Gesture Recognition (FG)*.

178 gaotek (2025). Comprehensive Guide for Biometrics Enabled Connected Vehicles (V2X). https://gaotek.com/comprehensive-guide-for-biometrics-connected-vehicles-v2x/. [accessed on August 18, 2025].

179 Sun, X., Yu, F. R., Zhang, P., Xie, W., and Peng, X. (2020). A Survey on Secure Computation Based on Homomorphic Encryption in Vehicular Ad Hoc Networks. *Sensors*, 20(15): 4253.

180 Chen, J., Li, K., and Philip, S. Y. (2021). Privacy-Preserving Deep Learning Model for Decentralized VANETs Using Fully Homomorphic Encryption and Blockchain. *IEEE Transactions on Intelligent Transportation Systems*, 23(8): 11633–11642.

181 Nguyen, C.-H., Manh, B. D., Hoang, D. T., Nguyen, D. N., and Dutkiewicz, E. (2024). Towards secure ai-empowered vehicular networks: A federated learning approach using homomorphic encryption. In *Proceedings of IEEE Vehicular Technology Conference*.

182 Sudarsan, S. V., Schelén, O., and Bodin, U. (2021). Survey on delegated and self-contained authorization techniques in CPS and IoT. *IEEE Access*, 9: 98169–98184.

183 Cheng, R., Chen, S., and Han, B. (2023). Toward zero-trust security for the metaverse. *IEEE Communications Magazine*, 62(2): 156–162.

184 Lobodzinski, S. S. and Laks, M. M. (2012). New devices for very long-term ECG monitoring. *Cardiology Journal*, 19(2): 210–214.

185 Patel, V. M., Chellappa, R., Chandra, D., and Barbello, B. (2016). Continuous user authentication on mobile devices: Recent progress and remaining challenges. *IEEE Signal Processing Magazine*, 33(4): 49–61.

186 Alnaim, A. K. and Alwakeel, A. M. (2025). Zero-trust mechanisms for securing distributed edge and fog computing in 6G networks. *Mathematics*, 13(8).

187 Alberto Torres, W. A., Bhattacharjee, N., and Srinivasan, B. (2015). Privacy-preserving biometrics authentication systems using fully homomorphic encryption. *International Journal of Pervasive Computing and Communications*, 11(2): 151–168.

188 Su, Y., Li, J., Li, Y., and Su, Z. (2022). Edge-enabled: A scalable and decentralized data aggregation scheme for iot. *IEEE Transactions on Industrial Informatics*, 19(2): 1854–1862.

189 Samardzic, N., Feldmann, A., Krastev, A., Manohar, N., Genise, N., Devadas, S., Eldefrawy, K., Peikert, C., and Sanchez, D. (2022). CraterLake: A hardware accelerator for efficient unbounded computation on encrypted data. In *Proceedings of Annual International Symposium on Computer Architecture (ISCA).*

190 Wu, N., Cheng, R., Chen, S., and Han, B. (2022). Preserving privacy in mobile spatial computing. In *Proceedings of the 32nd Workshop on Network and Operating Systems Support for Digital Audio and Video*, pages 1–7.

191 Cheon, J. H., Kim, A., Kim, M., and Song, Y. (2017). Homomorphic encryption for arithmetic of approximate numbers. In *Advances in Cryptology–ASIACRYPT 2017: 23rd International Conference on the Theory and Applications of Cryptology and Information Security, Hong Kong,* China, December 3-7, 2017, *Proceedings, Part I 23*, pages 409–437. Springer.

192 Lu, W.-j., Huang, Z., Hong, C., Ma, Y., and Qu, H. (2021). Pegasus: bridging polynomial and non-polynomial evaluations in homomorphic encryption. In *2021 IEEE Symposium on Security and Privacy (SP)*, pages 1057–1073. IEEE.

193 Wu, H. and Zhou, J. (2021). Privacy leakage of sift features via deep generative model based image reconstruction. *IEEE Transactions on Information Forensics and Security*, 16: 2973–2985.

194 Fischler, M. A. and Bolles, R. C. (1981). Random sample consensus: A paradigm for model fitting with applications to image analysis and automated cartography. *Communications of the ACM*, 24(6): 381–395.

195 Hartley, R. and Zisserman, A. (2003). *Multiple view geometry in computer vision.* Cambridge University Press.

196 Lowe, D. G. (2004). Distinctive image features from scale-invariant keypoints. *International Journal of Computer Vision*, 60: 91–110.

8

Securing Power-constrained Devices with Energy-preserving Cryptography

Khaled Khasawneh, Sai Dinakarrao, Sunanda Roy, Anhton Nguyen and Pavlos Zouridakis

Department of Electrical and Computer Engineering, George Mason University, USA

8.1 Introduction

The rapid proliferation of low-power devices, including Internet of Things (IoT) and embedded systems, is driving a paradigm shift in the way we interact with technology. These devices are becoming increasingly integral to various sectors such as healthcare, transportation, and smart cities [1]. Therefore, the security of these devices is paramount as they often handle sensitive data and perform critical functions.

However, one of the primary challenges hindering their wide-scale deployment is ensuring secure operations within constrained environments. Traditional cryptographic algorithms, which are effective in well-provisioned desktop and server settings, are often computationally intensive for resource-poor, battery-operated devices such as IoT and 5G/NextG devices. Thus, the challenge is further compounded by the need to maintain a balance between security and energy efficiency, as these devices rely heavily on limited power sources.

In response to this challenge, this chapter has pioneered novel techniques aimed at enhancing the security of these devices. Our approach focuses on overcoming the limitations of lightweight cryptography by integrating energy harvesting to boost the device power budget, leverage undervolting to reduce the power consumption of the on-device computations, and utilize re-configurable cryptography to reduce the computational overhead of traditional cryptography.

6G Security: New Frontiers in Mobile Network Technologies and Verticals, First Edition.
Edited by Luiz A. DaSilva.

These innovations are crucial for ensuring secure and efficient operation of NextG devices in constrained environments. By addressing the unique challenges of low-power devices, this chapter contributes to the advancement of secure IoT and embedded systems, paving the way for their widespread adoption in various industries. The techniques developed in this chapter provide a foundation for future research and development in the field of lightweight cryptography and energy-efficient security solutions, ensuring that the growing ecosystem of low-power devices can operate safely and reliably.

8.2 Task 1: Kinetic Energy Harvesting and Gait-based Node Authentication

IoT has grown significantly due to wireless tech advancements. Diverse applications bring security challenges, needing high-level network design and device-specific considerations. Limited resources in IoT devices make them vulnerable. Authentication is crucial for data integrity and user experience. Existing methods like cryptography, biometrics, and cloud offloading have limitations due to computation, interface, or security issues.

To address these challenges, we propose to use a biometric input, that is unique to the user, easy and reliable to collect, unobtrusive to the user [2], and robust i.e., difficult to spoof [3]. Considering these factors, we introduce a user authentication method that utilizes a user's walking gait (or form) [4]. Gait-based authentication can be performed indirectly (using external images of an individual) or directly (using the user's metrics) [5]. A user's movement at a low level, the way that they contact the floor surface and apply force to it, are unique identifiers [6] and that information can be obtained sufficiently with a wrist-worn device [7] or handheld devices [2]. Our preliminary analysis of gait signals between different users in our training dataset that has over 65,000 samples has shown an average correlation of no more than 0.1447. Such a low correlation suggests only a weak linear relationship between the two users. This indicates that the gait signatures are unique and can effectively be used for user authentication. A user's gait can be acquired by inertial motion sensors, which are already available on every handheld or wrist-worn device. Existing gait-based user authentication techniques require the use of Deep Neural Networks (DNN). A combination of convolutional and recursive neural networks is used in [2] with good accuracy, but it is very expensive in terms of computations and memory usage. In general, DNNs have gotten better at classification and produce high accuracy in pattern recognition problems, and for that reason they are used for biometric authentication methodologies. In pursuit of high accuracy, DNNs have gotten bigger, more complex and more computationally

Table 8.1 Comparison of gait-based authentication and RSA.

Algorithm	Accuracy	Immune to hacking	Continuous authentication	Passive input collection
RSA	100%	✗	✗	✗
eenet-lite+dqn	85.66%	✓	✓	✓

expensive. To address that, recent research has looked to reduce the amount of operations and memory required through pruning [8], quantization [9, 10] and other techniques [11, 12]. Also, in many cases, DNNs are able to extract the necessary features for classification from a small subset of the layers available. For that reason, early termination [13–19] techniques have been introduced. Considering this, we deploy early-exit neural networks (EENets) for gait-based user authentication. However, traditional EENets are limited by static exit selection which is not effective when different signals requires different exits for efficiency. Table 8.1 summarizes some of the differences between RSA and gait-based authentication.

On the other hand, IoT devices are deployed outside of the power grid and rely on batteries or energy-harvesting. As a result, it is possible that the execution of code that is running on devices in such settings gets interrupted, when the power supply is exhausted. Power depletion attacks describe cases when malevolent users force a device to run code that will exhaust the device's power source, rendering it incapable of performing any other tasks. In such cases, on restarting, the device has to redo the application execution (authentication), which can lead to redundant computations. Recent work enables intermittent computation, through task based computing [20], checkpointing [21, 22], or aiming to reduce the energy cost of checkpointing [23, 24]. It is therefore important to be able to save partial work, so that the energy spent up to the point of interruption is not wasted; also in order to avoid non-termination scenarios as well as power depletion attacks. Non-termination refers to a situation where a device keeps running code but it can never get to complete the computations required to run the code in its entirety.

To address these challenges, we make the following cardinal contributions.

- We introduce Early-exit Neural Networks (EENets) [25], in conjunction with performance-awareness intermittent computing for user authentication. EENets are a class of neural networks that are trained on both precision and cost. When a certain confidence threshold is met, computations terminate early. That leads to considerable savings in computational cost.
- We complement the EENet with a Deep Quality-Learning Network (DQN) for dynamic exit selection of EENets. The DQN is a reinforcement learning

algorithm that can be trained to choose a specific exit from the EENet, in case it is deemed beneficial to do so.

- Our approach was tested on the whuGAIT dataset [2] and it achieves >85% authentication accuracy with nearly 5.9× lower computations compared to traditional DNNs.

8.2.1 Proposed Approach

8.2.1.1 System Overview

In this section, we describe our proposed methodology on user authentication for lightweight IoT devices, using gait-based authentication with early exits and intermittent computations. The overview of our technique can be seen in Figure 8.1. The input gait signal is captured by the embedded inertial sensors on the user's device and is sent to the EENet for inference. The EENet is configured to work in conjunction with the DQN. We can set the DQN to prioritize accuracy, inference execution time or energy consumption and set the parameters of its reward function accordingly. Based on the input from the EENet and the reward function it directs the EENet to take the most appropriate exit. The result of that process is the prediction of the EENet with the authorization decision. One major intention of this work is to ensure the compatibility and efficiency on devices that are disconnected from the electrical power grid. To avoid termination due to exhaustion of the IoT device's power source, we introduce an intermittent computations

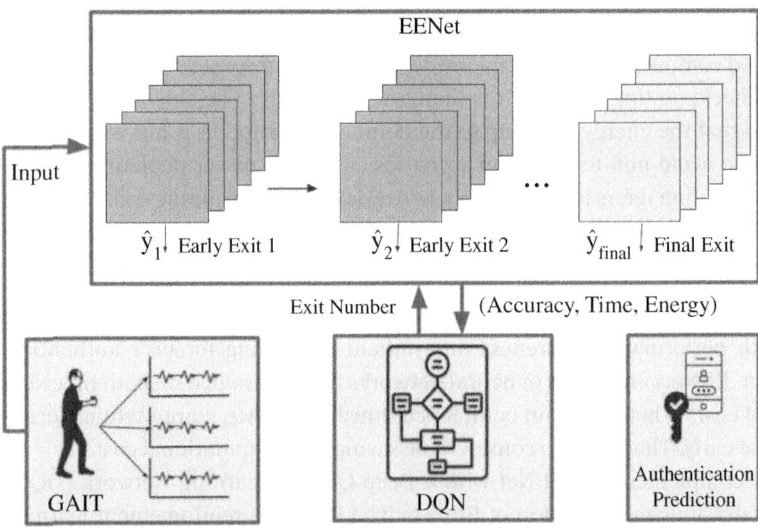

Figure 8.1 Overview of our gait-based authentication method utilizing EENet and DQN.

technique that can be deployed on any IoT device and perform gait-based authentication efficiently.

The rest of this section includes more details on the input signal with the gait data in Section 8.2.2, how the EENet framework is trained in Section 8.2.2.1, the structure of our lightweight gait authentication CNN called EENet-lite in Section 8.2.2.2, the configuration of our DQN reinforcement learning method in Section 8.2.2.3, and our intermittent computation methodology in Section 8.2.2.4.

8.2.2 Gait Signal

In order to use gait as an input signal, one needs to determine how to measure a user's walking form. This is very important because it is required to extract the gait signals efficiently with low overheads and low complexity.

In this work, we consider six measurements: the linear acceleration on the X, Y and Z axes and the gyroscopic acceleration, which is the rotation rate around the X, Y and Z axes to represent the gait signal. We use linear acceleration instead of absolute acceleration, as it allows to disregard the effect of gravity on the measurements. Gravitational force adds unwanted noise to the measurements, since it is always present, with only small differences depending on altitude and latitude and doesn't contribute to identifying the user by their gait. Therefore, one can ignore absolute acceleration in favor of linear acceleration.

These gait measurements can be extracted from lightweight embedded sensors such as MPU9250 (Invensense MPU-9250 Product Specification Revision 1.1) that can be found on smartphones, tablets, smartwatches, and fitness trackers. For instance, a single MPU9250 sensor can collect all the necessary data with minimal consumption (3.4 mA). We adopt the pre-processing methodology presented in [2] for gait collection. The result is a tuple of six elements: three for linear acceleration and three for gyroscopic acceleration as mentioned. Each element comprises 256 samples. More specifically, each sample is collected by identifying two continuous steps. Each step is interpolated to a fixed length of 128 samples using linear interpolation function and we use a one-step overlap in order to increase the sample set efficiently. Based on the above, the pre-processing overhead requires only a small amount of basic operations and it is negligible. An example of the gait input signal is shown in Figure 8.2.

8.2.2.1 Early-exit Neural Networks

The gait signals that are captured by the inertial sensors are used as an input for the user authentication. We introduce the use of EENets to process the incoming gait signals. Compared to traditional CNNs, use of EENets reduces the total amount of total computations and Multiply-and-Accumulate (MAC) operations. This is achieved by configuring a number of early-exits inserted within the CNN

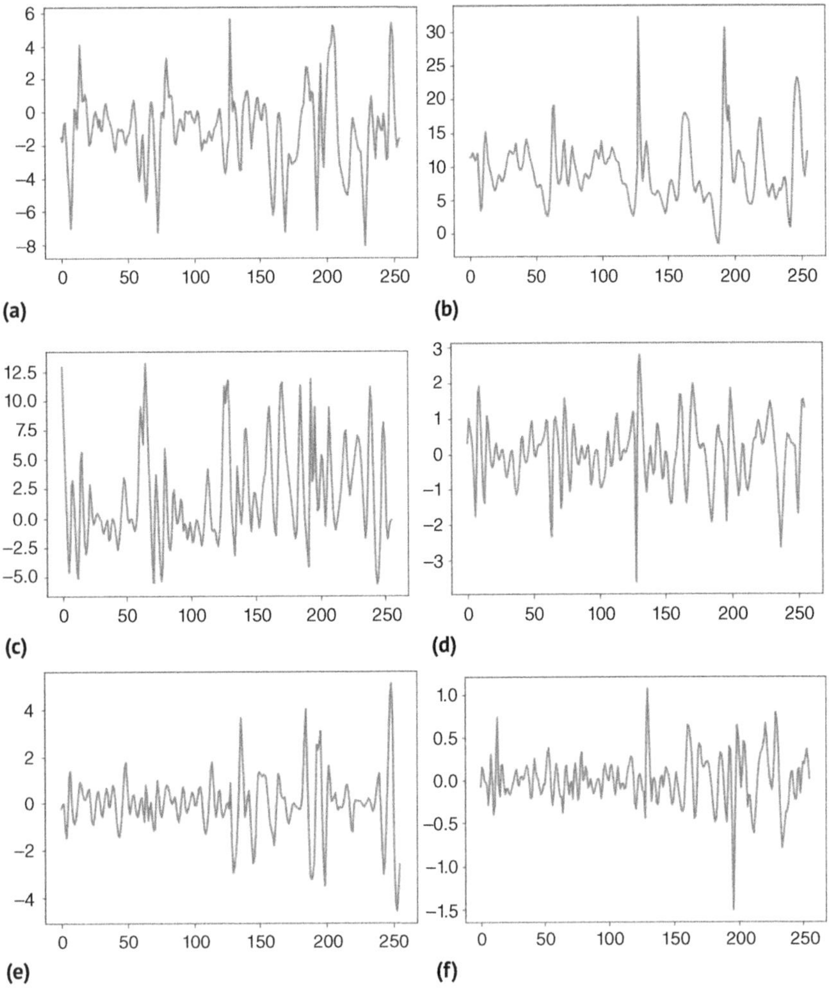

Figure 8.2 Single user gait input example displaying: (a) acc_x, (b) acc_y, (c) acc_z, (d) gyr_x, (e) gyr_y, (f) gyr_z.

of an EENet. The CNN takes the earliest exit for which it reaches a certain level of confidence that its classification decision is correct. As a result, the CNN may not need to run in its entirety in order to produce a prediction. Hence, there is a trade-off between performance and computational cost, making the deployment of EENets suitable on small- and medium-scale device with restricted resources.

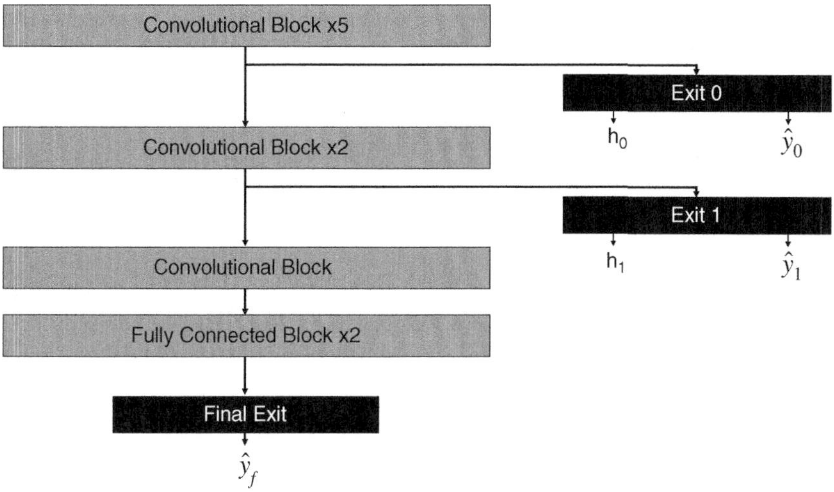

Figure 8.3 Architecture of EENet-lite, our proposed lightweight gait-based authentication model.

A variable number of early exits can be configured and their location in the CNN can be determined heuristically or it can be determined an algorithm. The number and location of the early exits largely depends on the size and architecture of the CNN. In general, a CNN with n convolutional layers can have a maximum of m early exits, where $m \leq n$ at all times.

Every early exit has two distinct branches; *a classification prediction branch* that uses a softmax activation function and *a confidence branch* that uses a sigmoid activation function, as shown in Figure 8.3. The output of the classification prediction branch is the prediction to the classification problem at that particular exit, whereas the output of the confidence branch is the confidence score h_i for exiting at that exit. If the confidence score is higher than the confidence threshold T, then this early exit is taken and the algorithm stops running. Otherwise, the algorithm continues until it reaches an early exit, where $h_i > T$, or it reaches the final exit, whichever comes first.

The training of the EENet is achieved by traditional back-propagation, with ADAM optimization. The model trains both the prediction and confidence branches independently for each early exit. The loss function that is used for the back-propagation during the training takes both prediction and cost into account and is defined as:

$$\mathcal{L} = \mathcal{L}_{\mathcal{P}} + \lambda \mathcal{L}_{\mathcal{C}} \tag{8.1}$$

where $\mathcal{L}_{\mathcal{P}}$ represents the loss from classification prediction and $\mathcal{L}_{\mathcal{C}}$ represents the loss from computational cost. To calculate the classification loss, we consider the prediction vector that is defined as follows:

$$\hat{Y}_0 = h_0 \times \hat{y}_0 + (1 - h_0)[h_1 \times \hat{y}_1 + (1 - h_1)[\cdots \\ h_N \times \hat{y}_N + (1 - h_N) \times \hat{y}_{final}] \cdots] \tag{8.2}$$

$$\hat{Y}_1 = h_1 \times \hat{y}_1 + (1 - h_1)[\ldots \\ h_N \times \hat{y}_N + (1 - h_N) \times \hat{y}_{final}]\ldots] \tag{8.3}$$

$$\hat{Y}_N = h_N \times \hat{y}_N + (1 - h_N) \times \hat{y}_{final} \tag{8.4}$$

$$\hat{Y}_{final} = \hat{y}_{final} \tag{8.5}$$

The $\hat{Y}_i$ represents the output prediction vector produced by all early exits between i and N. The $\hat{y}_i$ is the individual prediction for each exit and h_i is the confidence for each respective prediction. Once we obtain the values from (2)-(5), we use them to calculate the partial loss for classification according to (6):

$$\mathcal{L}_{\mathcal{P}} = -\sum_{k=1}^{K} y_k \times log(\hat{Y}) \tag{8.6}$$

where K is the number of classes that the model classifies for. We get the respective cost values (C_0, C_1,..., C_N, C_{final}) the same way as described in (2)-(5) solving for cost instead of predicted value. Once we have the cost values, they are equal to the loss from computational cost:

$$\mathcal{L}_{\mathcal{C}} = C \tag{8.7}$$

Thus, using (1), (6), (7), one can obtain the value for the loss function for each early exit that is used for back-propagation.

Once training is complete, the EENet is ready to perform inference and generate predictions for each input signal it receives. Each early exit i produces a confidence score h_i and if $h_i > T$, where T is the confidence threshold, then that exit is taken and the local prediction $\hat{y}_i$ becomes the final classification decision $\hat{y}$. The confidence threshold T is set statically and it is the same across all early exits.

8.2.2.2 Design of EENet-lite for IoT Devices

In Section 8.2.2.1, we presented the basic principles of EENets and why they are an appropriate and efficient architecture that achieves high accuracy, while significantly reducing the computational cost of complex classification problems.

The EENet framework can be applied to several DNNs, such as ResNets [26], which are very effective classification algorithms. Most classification algorithms, however, have been designed to process images or video. The shape of such inputs is different from the input that we use for gait-based authentication. We want to be able to extract the maximum amount of features from the input as described in Section 8.2.2, with the minimum number of computations and in the smallest number of layers. For that reason, we developed EENet-lite, a custom gait-based authorization algorithm for lightweight IoT devices.

Our lightweight EENet algorithm needs to be small enough to run on hardware without parallelization support. Therefore, it should not have a large total number of convolutional layers, so that it can efficiently run even when no early exit is taken and the algorithm runs to the end. It also needs to be able to produce sufficiently good accuracy even when the first early exit is taken. As a result, it doesn't add value to configure a large amount of early exits, for instance to have an early exit after each layer. To satisfy these conditions for EENet-lite, we have chosen to implement a total of eight convolutional layers, followed by two fully connected layers. The structure of the convolutional layers is shown in Table 8.2.

Each convolutional layer is combined with a batch normalization layer and a ReLU activation function to form a convolutional block. Given the number of layers on EENet-lite, we implemented two early exits. The first exit can be taken after the fifth convolutional block and the second can be taken after the seventh convolutional block, according to Figure 8.3.

8.2.2.3 Reinforcement Learning with DQN for EENet

As discussed in Section 8.2.2.1, the confidence threshold is set statically for the EENet. Classification performance can vary however, depending on the application, the dataset used or the type of input. That can potentially make the static threshold inefficient and affect the performance-efficiency trade-off. We address this challenge, by introducing a method for dynamic exit selection. The dynamic

Table 8.2 Details of the CNN structure.

Layer	In channels	Out channels	Kernel size	Stride	Padding
Conv 1	1	2	3	(1,2)	1
Conv 2	2	4	3	(1,2)	1
Conv 3	4	8	3	1	1
Conv 4	8	16	3	1	2
Conv 5	16	16	3	1	2
Conv 6	16	32	3	1	2
Conv 7	32	64	3	1	1
Conv 8	64	64	3	1	1

exit selection is not a classification or a clustering problem. Rather, depending on the EENet state (input, accuracy, time, computations executed by EENet, energy required for authentication), an action (exit selection) will be triggered. Considering the ability to perform such computations, we deploy reinforcement learning, to determine the best possible exit selection during runtime.

More specifically, we will use Quality reinforcement learning (Q-learning) in this work. Q-Learning is a model-free stochastic algorithm that can identify optimal action selection. However, due to the convergence issues of traditional Q-learning, we employ deep Q-learning network (DQN) in this work. The use of DQN removes correlations in the observation sequence and smooths the data distribution, addressing the converge issues that traditional Q-Learning may face. Given a set of states S and a set of actions A per state, when an action $a \in A$ is taken, the reinforcement learning agent transitions from state to state. The algorithm has a function that calculates the quality of each state-action combination, assigning a real number to it by using a reward function:

$$Q : A \times S \rightarrow \mathbb{R} \tag{8.8}$$

The DQN keeps getting updated every time that inference is run. By calculating and learning the reward value, DQN learns how to make decisions about the best possible next action and it gets trained to make the best decision possible in every subsequent situation. We ensure that the DQN we use is lightweight, so that it can work well in conjunction with the EENet, avoiding overheads.

Algorithm 1: DQN Overview

Require: Input: (accuracy, time, MAC ops, energy)
Ensure: Output: exit number (n)

```
1: state ← (0, 0, 0, 0)                                ▷ initialization
2: b ← 0
3: while b × b_size ≤ dataset_size do
4:     exit_no ← n                                     ▷ exit number is sent to eenet
5:     accuracy, time, MAC, energy ← input
6:     state ← (accuracy, time, MAC, energy)
7:     R ← f(accuracy, time, energy)
8:     b ← b + 1
9: end while
```

Algorithm 1 outlines the employed DQN and it's functionality based on the accuracy, inference time, # MAC operations and energy consumption of the EENet. We choose these three parameters qualitatively, based on the nature of the application that we examine. Accuracy and inference time is important for authentication, whereas energy as well as inference time is important for the resource restricted IoT device environment. Therefore, we focus on these three parameters to evaluate the performance of the DQN and re-train it, to improve its predictions. As result, in our implementation of the DQN, the state is represented by a tuple that includes the values for accuracy, execution time for inference, the number of MAC operations required for inference and the energy that it takes to authenticate one sample. The action that moves the DQN agent from one state to the next is the exit that the EENet is directed to take. The DQN is a simple sequential neural network. We choose that, because it is smaller and quicker than other neural networks, such as recurrent, residual, etc. We utilize two hidden layers of 24 neurons each. The activation functions are ReLU and we use Adam optimizer. Table 8.3 includes an ablation study of a few different configurations for our DQN. We notice that for a relatively simple sequential neural network the differences in accuracy and execution time are small. Based on that, we choose to use the DQN that produced the smallest execution time and second highest accuracy.

The reward function is expressed by the following equations:

$$R = \alpha \cdot R_a + \beta \cdot R_t + \gamma \cdot R_e \tag{8.9}$$

$$R_x = \begin{cases} r_{n+1}, & \text{if } x > threshold_n \\ r_n, & \text{if } threshold_{n-1} < x \leq threshold_n \\ \vdots & \vdots \\ r_0, & \text{otherwise} \end{cases} \tag{8.10}$$

where x stands for accuracy, inference time or energy and R_x for R_a, R_t or R_e stand for the reward values for accuracy, inference time and energy consumption respectively. In Eq. (8.9), α, β and γ are scaling factors. The scaling factors allow us to determine the balance of the trade-off between accuracy, execution time and energy consumption when we design the method and prioritize according to what is most important to our implementation.

In Algorithm 1, we present the function of the proposed DQN. After initialization (Line 1), the algorithm works in batches (Lines 2-3). For each batch, the DQN

Table 8.3 DQN configuration comparison.

Hidden layers	No. of neurons	Accuracy (%)	Inf. time (ms)
2	24 × 24	85.66	2,699
2	24 × 12	85.33	3.401
3	24 × 24 × 12	85	3.384
2	24 × 6	86.38	3.211
2	12 × 12	85.13	3.664

tells the EENet what action to take (what exit to take) (Line 4). The EENet runs inference for that batch (Line 5) and then sends the state in form of the tuple of accuracy, time, MAC operations and energy consumption back to the DQN (Line 6). Then the reward is calculated according to Eqs. (8.9) and (8.10) (Line 7), and then the DQN moves to the next batch. Based on its previous states and actions, the DQN is trained to dynamically make the optimal decision according the reward function and learns to predict the next best exit from the EENet. At the end of the process, the DQN has worked with the EENet to run inference on the entire testing dataset and we can get the statistics for each element of the state tuple, along with the exit distribution from the EENet for all samples.

8.2.2.4 Intermittent Computing Through Checkpointing

The use of DQN for EENet addresses the challenge of dynamic exit selection and reduces the number of MAC operations performed without trading off the performance. However, one needs to note that medium- and small-scale IoT devices are meant to be versatile and easy to deploy in a variety of different environments. It also indicates that they are often deployed off the electrical grid, powered only by a battery or an energy harvester. This makes such IoT devices vulnerable to non-termination scenarios and power depletion attacks. For that reason, it is important to provide a framework that will allow all power-restricted devices to successfully complete the execution of the code that they run. In cases, when the energy available to them is not sufficient to run their code, provisions need to be made to assure that the progress that they have made is saved. That way, when more energy becomes available, they can continue execution from the point where it was interrupted before. In order to achieve that we implement an intermittent computation technique.

Intermittent computation implementations are mostly applied to microcontrollers. They mostly utilize non-volatile memory (NVM), but there are also techniques that utilize volatile memory [23]. In either case, they are specific to the type of microcontroller and memory that they are developed for. That makes them very efficient, as they directly target specific memory locations to save and load checkpoints. That makes such implementations very efficient, but it also makes it difficult to adapt them to all other hardware.

In contrast, we introduce an intermittent computation implementation that is hardware agnostic, that can run on any device where EENet-lite is deployed to run gait-based authentication. We also want the overhead of our technique to be as small as possible. For that reason, we implement static checkpoints, where we can store the output of certain layers on the system's DRAM, before it is passed on to the next convolutional layer. The reason that we use the DRAM to save the checkpoint is that it is much faster than saving information to the SD card that the Raspberry Pi uses as a main drive. We are taking advantage of the fact that DRAM is able to retain saved information for short amounts of time even in the case of loss of power [23]. There are implementations that use a stochastic approach for the placement of the checkpoints [27], but it is a less reliable approach. Additionally, given the small number of layers in EENet-lite, it is unnecessary to introduce the added overhead of the stochastic algorithm to determine the checkpoint locations.

We build checkpoints into the EENet-lite model so that we can avoid non-termination. When the inference code that produces the authentication prediction starts, it first checks to see if the amount of energy that is available to the system E_{av} is equal or greater to the minimum amount of energy required to get to the next checkpoint or exit during authentication E_{min}. If that is the case, then it checks to see if there is a checkpoint saved or not. If not, it starts executing the EENet from the beginning. If there is a checkpoint stored, it gets loaded and the marker that points to the last saved location is checked. Then execution resumes from the last location where the checkpoint was taken. If at any point the EENet terminates, the process starts over. When the execution reaches a checkpoint location, a new checkpoint is taken. If there was another checkpoint already in place, the new one replaces the old one. Execution then continues until it reaches an exit that the EENet can take with confidence or the final exit. Then, the checkpoint gets deleted and an authentication prediction is produced.

Considering that our model CNN can take an early exit, it is desirable to make the device use one of the early exits and produce an authentication decision, as that ensures that a smaller number of computations is performed compared to running the entire neural network. To inform our decision about the placement of the checkpoints, we consider the number of operations that need to be performed by layer, as shown in Table 8.4.

There is an early exit after layer `Conv 5` and another early exit after layer `Conv 7`. In the interest of creating as balanced a distribution of MAC operations

Table 8.4 EENet-lite layer analysis.

Layer	Input shape	Output shape	MAC	FLOPS
Conv 1	1 × 6 × 256	2 × 6 × 128	27,648	15,360
Conv 2	2 × 6 × 128	4 × 6 × 64	55,296	29,184
Conv 3	4 × 6 × 64	8 × 6 × 64	221,184	113,664
Conv 4	8 × 6 × 64	16 × 8 × 66	1,216,512	616,704
Conv 5	16 × 8 × 66	16 × 10 × 68	3,133,440	1,577,600
Conv 6	16 × 10 × 68	32 × 12 × 70	7,741,440	3,897,600
Conv 7	32 × 12 × 70	64 × 14 × 72	37,158,912	18,643,968
Conv 8	64 × 14 × 72	64 × 14 × 72	74,317,824	37,223,424
FC 1	64	20	2,540	1,280
FC 2	20	2	78	40

as possible going from one entry point to the code to its nearest checkpoint or exit, we consider:

$$MAC(i) \simeq MAC(i+1) \tag{8.11}$$

where $MAC(i)$ is the total number of MAC operations in i and

$$\begin{aligned} i \in [&start \rightarrow checkpoint1, checkpoint1 \rightarrow earlyexit1,\\ &earlyexit \rightarrow checkpoint2, checkpoint2\\ &\rightarrow earlyexit2] \end{aligned} \tag{8.12}$$

Having balanced $MAC(i)$ sections increases the likelihood that each checkpoint or exit is reached. As a result, it is less likely that there will be repeated instances when the device runs out of energy and shuts down. Every time a device runs out of energy, the computations it has performed—between the latest checkpoint and the point of sudden shutdown—are wasted and lost. Therefore, by limiting the times that there is a sudden shutdown, we reduce the total amount of wasted computations. By extension, we reduce the amount of wasted energy and minimize overhead.

The process of generating the checkpoints is described in Algorithm 2. EENet-lite starts running and once the checkpoint location is reached, the first checkpoint is saved. We use a marker (Line 1) to keep track of the location where the state was saved and the checkpoint was created. After that, execution continues. If there are no interruptions, the authentication process (EENet) will keep running until the EENet reaches the next checkpoint or decides to take an exit i.e., authentication process is complete. In the case where the next checkpoint is reached, the marker is increased and the new checkpoint overwrites the previous one (Line 2-5). If an exit is taken, our EENet produces its output and the checkpoint is deleted (Line 6-8).

Algorithm 2: Creating checkpoints

Require: State: output tensor from previous layer
Ensure: Current progress is saved

1: $marker \leftarrow 0$
2: **while** next checkpoint is reached **do**
3:     $marker \leftarrow marker + 1$
4:     $checkpoint \leftarrow (marker, state)$
5: **end while**
6: **if** exit is taken **then**
7:     delete checkpoint
8: **end if**

The process of loading checkpoints is described in Algorithm 3. Once the EENet starts executing, it checks to see if a checkpoint has been created. If so, it gets loaded (Line 2) and the marker gets checked to determine where the checkpoint was made (Line 3). Then, the tensor containing the last saved EENet state is loaded as an input to the next convolutional layer after the checkpoint was made (Line 4-6). If a checkpoint has not been previously created, execution starts from the beginning (Line 7-9).

Algorithm 3: Loading checkpoints

Require: Authentication Request
Ensure: Execution resumes from last saved location

1: **if** Checkpoint exists **then**
2:     $(marker, state) \leftarrow checkpoint$
3:     Check marker
4:     Go to checkpoint location indicated by marker
5:     $layer_{i}nput \leftarrow state$
6:     Execution resumes
7: **else if** No checkpoint exists **then**
8:     Start execution from the beginning
9: **end if**

8.2.3 Experiments

8.2.3.1 Experimental Setup

We have run simulations to evaluate our methodology on two different environments: a PC with parallelization support and a medium-scale IoT device that runs on an ARM processor with no parallelization support. Our PC embeds an Intel i7 processor with 16GB of RAM and an NVIDIA GeForce GTX1050 graphics card. Our IoT device is a Raspberry Pi 4B, with an ARM 64-bit processor and 8GB of memory. For the setup, we used Ubuntu 22.04 LTS and set up a virtual environment with Python 3.9.12, Pip 22.1, Numpy 1.22.3, PyTorch and Torchvision 1.11.0, Scipy 1.8.0, and Tensorflow 2.8.0, built specifically to run on aarch64 architecture.

We compare the performance of the EENet framework to ResNets [26] as they have achieved high classification performance in a variety of applications and they serve as a good classification benchmark. We apply the EENet framework to ResNet18, ResNet34, and ResNet50, producing EENet18, EENet34, and EENet50 respectively. We will also compare all these models to our EENet-lite model.

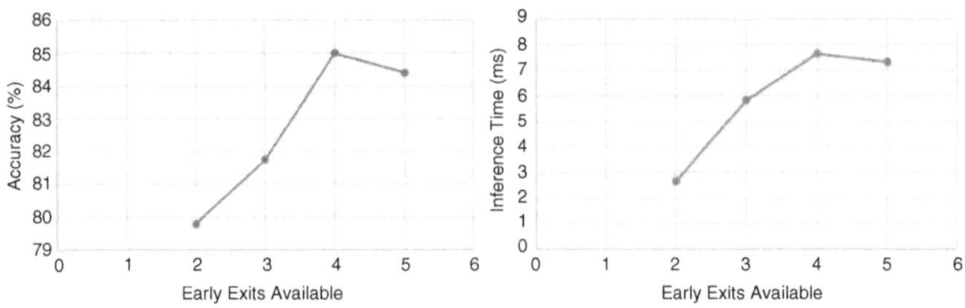

Figure 8.4 Effect of number of available early exits in EENet34 to accuracy and inference time.

To determine how many early exits we should use for the ResNet based EENets, we ran tests to see how the number of exits on EENet34 affected accuracy and performance. The results can be seen in Figure 8.4. For the upcoming tests, we will consider five early exits. For the user authentication comparison, we will use the data from the whuGAIT dataset [2]. The training set consists of more than 65,000 samples and the test set consists of 7,600 samples.

8.2.3.2 Experimental Results

To evaluate the performance of our authentication, we consider accuracy, inference time, and the number of multiply-and-accumulate (MAC) operations. Please note that the computations required for the softmax layer at each exit of the neural network are included in the total amount of MAC operations. The exit distribution also tells us how many of the test samples were directed to take each exit. ResNets do not have any early exits built into them, while EENets has been configured with five possible early exits. Our lightweight EENet-lite model has been configured with two possible early exits, due to its small number of layers. We also show the effect of our DQN model on the performance of different models.

The results of the performance are shown on Table 8.5 and inference times are marked based on our experiments using the PC environment. As expected, there is a trade-off between performance and efficiency. The ResNets achieve higher accuracy scores by 9.6% on average from the EENet counterparts, while the EENets execute 2× faster and need to perform 6× fewer MAC operations than the ResNets. Some of the difference in accuracy on EENets can be made up for by the use of the DQN. The reward function of the DQN can evaluate previous predictions and improve performance dynamically. More specifically, applying the DQN improves accuracy by 8.5%, but it comes with 2.2× more MAC operations and 0.87ms slower inference time.

Table 8.6 shows how the EENet framework compares to Branchynet [15], another methodology that utilizes early exits on a deep neural network. We

Table 8.5 Comparison of EENet vs ResNet.

Model	Accuracy	Exit distribution	Inf. time	Avg. #ops
eenet-lite	74.37%	[7600,0,0]	2.815ms	4.78M
eenet-lite+dqn	85.66%	[2620,2800,2180]	2.699ms	56M
resnet18	91.49%	NA	7.354ms	137.89M
eenet18	79.70%	[0,1215,6332,53,0,0]	4.531ms	26.54M
eenet18+dqn	84.47%	[1210,1150,1230,1290,1400,1320]	5.763ms	49.06M
resnet34	91.71%	NA	12.741ms	279.67M
eenet34	84.36%	[0,0,3308,4292,0,0]	7.319ms	51.94M
eenet34+dqn	85.99%	[1190,940,1300,1470,1370,1330]	6.978ms	86.42M
resnet50	86.55%	NA	18.742ms	331.23M
eenet50	76.83%	[2,6181,1417,0,0,0]	7.023ms	46.9M
eenet50+dqn	85.99%	[990,1140,1230,1520,1410,1310]	9.737ms	97.28M

Table 8.6 Comparison of EENet vs Branchynet.

Model	Accuracy	Inf. time
ResNet	80.70%	70.9ms
Branchy-ResNet [15]	79.17%	37.2ms
EENet	80.51%	12.1ms

Table 8.7 Performance of Eenet-Lite on PC vs ARM.

Model	Accuracy	Inf. time (ARM)	Inf. time (PC)	Avg. #ops
eenet-lite exit1	74.37%	110.41 ms	2.815 ms	4.78 M
eenet-lite exit2	75.24%	228.03 ms	3.277 ms	50.14 M
eenet-lite+dqn	85.66%	254.12 ms	2.699 ms	56 M
eenet34	84.36%	1,244.48 ms	7.319 ms	51.94 M

evaluate both methodologies on the CIFAR10 dataset, and apply them on ResNet110 with two early exits, using a PC with GPU support.

As we can see, EENet can achieve 3× faster inference time than Branchynet, by taking into account computational cost for the calculation of the loss function. At the same time, it can achieve a higher accuracy than Branchynet, comparable to the accuracy of ResNet without early-exit implementation.

Since it is important to create a methodology that works well on IoT devices with more limited resources, Table 8.7 shows how EENet-lite performs on the

Raspberry Pi compared to the PC in terms of inference time. Especially, because the Raspberry Pi uses its ARM processor with an aarch64 architecture without GPU support, its execution cannot be parallelized, and that is most evident when running EENet34. Running EENet-lite with DQN on the Raspberry Pi is 5× faster than EENet34 while also achieving higher accuracy. We can also see the effect of using dynamically chosen exits compared to static exits. Utilizing the DQN increases the authentication performance by more than 10%, while only requiring 5.5% more operations than EENet-lite at the second exit. The difference between running on the ARM processor is that it takes on average 93× longer to run a model on the Raspberry Pi, as expected. When we look at the difference of runtimes for EENet-lite with and without DQN, we notice that the presence of DQN has a different effect on the two platforms. That is caused by the different processor architectures and especially the presence of GPU and parallelization support on the PC.

Energy consumption being important, we tested the energy required to run EENet on Raspberry Pi without the implementation of intermittent computing to get a baseline of the power cost. We tested EENet-lite for early exits 1 and 2, and with DQN support, as well as EENet34 with and without DQN support and ResNet34. Table 8.8 shows how they compare in terms of energy consumption and execution time. As expected, the number of operations required for each of

Table 8.8 Energy Consumption on Raspberry Pi.

Model	Energy (mJ)	Time (ms)	Accuracy
eenet-lite exit1	505.489	110.41	74.37%
eenet-lite exit2	1,043.833	228.03	75.24%
eenet-lite+dqn	1,172.143	254.12	85.66%
eenet34	5,929.222	1,244.48	84.36%
eenet34+dqn	9,486.752	1,991.168	85.99%
resnet34	21,130.711	4,435.11	91.71%

Table 8.9 Memory Usage of EENet.

Model	Avg. Load-store	Parameters stored
eenet-lite exit1	4.006 k	65.832 k
eenet-lite exit2	27.430 k	65.832 k
eenet-lite+dqn	30.37 k	65.832 k
eenet50	652.1 k	23.51 M
eenet50+dqn	4.803 M	23.51 M

these scenarios is reflected on the amount of energy required to complete said operations. It is also more pronounced on execution time, because of the lack of parallelization on smaller devices. The energy savings of taking an early exits are substantial. In general, the energy consumption of EENet-lite is 3.5× to 41.8× lower than ResNets. Table 8.9 shows the memory usage of EENet.

8.3 Sub-threshold Computing for Cryptography

In the era of smart devices and connected technology, the IoT has become an integral part of our daily lives. Ensuring the security of data transmitted between these devices is of utmost importance. Cryptography plays a critical role in ensuring data integrity and protection for IoT devices. However, a significant challenge arises due to the computationally intensive nature of cryptographic algorithms, which leads to high energy consumption on these power-constrained devices.

The primary objective of this work is to implement energy-efficient cryptographic algorithms on IoT devices. A promising approach to achieve this objective is by exploiting undervolting [28]. Undervolting involves operating a device at a voltage below its nominal supply voltage (i.e., sub-threshold computing), which can lead to reduced energy consumption [28–33].

This study has several key contributions. First, we developed an experimental framework to evaluate the effect of undervolting on IoT devices. This involved exploring the impact of undervolting on the functionality of the devices and determining the correlation between the operating temperature of the device and the amount of undervolting it can sustain. Second, we studied the maximum energy savings possible via undervolting for different important cryptographic algorithms running on different IoT devices. Our results show that undervolting led to up to 7% energy savings without changing the cryptography algorithms, which is a significant reduction considering the scale and ubiquity of IoT devices.

8.3.1 Proposed Approach

The primary objective of this task is to implement energy-efficient cryptographic algorithms on IoT devices. A promising strategy to achieve this objective is by exploiting undervolting, which involves operating a device at a voltage below its nominal supply voltage, thereby leading to reduced energy consumption.

To achieve the primary objective, the project will be divided into two main steps: (1) Verify cryptographic operations while undervolting, and (2) Energy saving measurements.

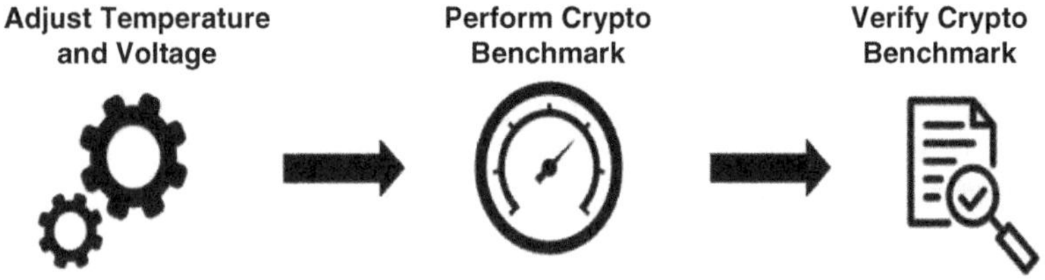

Figure 8.5 Overview of verify cryptographic operations while undervolting.

Figure 8.6 Overview of energy saving measurements.

8.3.1.1 Step 1: Verify Cryptographic Operations While Undervolting

This step involves verifying the cryptography by developing an experimental framework to evaluate the effect of undervolting on IoT devices. This step is shown in Figure 8.5 and include the following sub-tasks:

1. *Undervolt and Adjust Temperature:* The device's voltage will be reduced, and its temperature adjusted to establish the most efficient operational conditions. This will involve exploring the impact of undervolting on the functionality of the devices and determining the correlation between the device's operating temperature and the amount of undervolting it can sustain.
2. *Run Benchmark:* A benchmark will be executed that logs the cipher text of various cryptography algorithms, aiding in identifying the most energy-efficient algorithm.
3. *Verify Cipher Text:* A script will be run to ensure that the cipher texts obtained from the benchmark are correct outputs, ascertaining the integrity and validity of the tested cryptographic algorithms.

8.3.1.2 Step 2: Energy Saving Measurements

This step involves studying the maximum energy savings possible via undervolting for different important cryptographic algorithms running on different IoT devices. This step is shown in Figure 8.6 and include the following sub-tasks:

1. *Undervolt Device:* The device will be undervolted to optimize its operational efficiency.

2. *Collect Power Measurements:* Power measurements will be collected while running the cryptography algorithm to quantify the energy consumed during the operation.
3. *Process Power Measurement:* The collected power measurements will then be processed to determine the power savings achieved.

8.3.2 Experiments

8.3.2.1 Experimental Setup

We ran our experiment using various Raspberry Pi models as shown in Table 8.10. To ensure all devices were tested fairly, each device ran with the same Debian image provided on the Raspberry Pi website. Power measurements on each of the devices were obtained using the UM25C, which is a USB power measurement device that measures real-time current and voltage of the power load coming into the device. Figure 8.7 shows the setup we used to collect our power measurements. Lastly, we used popular cryptography algorithms from PyCryptodome (PyCryptodome: Python package of low-level cryptographic primitives) library in our evaluation, including AES, SHA256, SHA512, and RSA.

8.3.2.2 Results – Verifying Cryptographic Operations While Undervolting

The objective of our experiments was to assess the impact of undervolting on cryptographic operations. To ensure consistency in the experimental procedure, we executed the cryptographic algorithm under test multiple times using the same set of operands (both plaintext and/or key), and we tested 100k unique operand sets.

Our methodology involved the incremental reduction of voltage in small steps (increments of 25 mV). As the voltage was lowered, the specific cryptographic algorithm was continually executed with the same operand set. The primary aim was to identify at which undervolted threshold faults would begin to manifest or if a system freeze would occur prior to the appearance of any faults.

After rigorous testing across all the cryptographic algorithms within our scope (SHA256, SHA512, AES, and RSA), and replicating the experiments on multiple

Table 8.10 Raspberry Pi's used in experiment.

RPi model	Microprocessor	Operating current (mA)	Frequency (GHz)
3B+	Cortex-A53 (ARMv8)	500–800	1.4
3B	Cortex-A53 (ARMv8)	500–800	1.2
2B	Cortex-A7 (ARMv7)	350–400	0.9

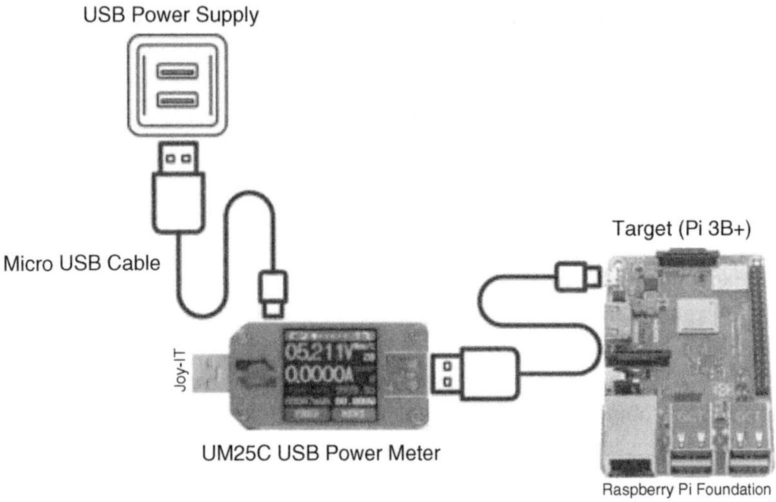

Figure 8.7 Power measurement setup.

Table 8.11 Nominal voltage with respective system freeze undervolting thresholds for each Raspberry Pi device; each undervolting level represented as a decrement of −25 mV.

RPi model	Nominal voltage (V)	Undervolting level for system freeze
3B+	1.3625	4
3B	1.2875	3
2B	1.35	3

devices (Table 8.10), the results are shown in Table 8.11. Through this experiment, we observed the following patterns:

- *Faults in Cryptographic Operations:* Contrary to initial expectations, none of the undervolting levels applied led to faults within the cryptographic operations. Every algorithm executed produced results that were consistent with operations conducted at regular voltage levels (nominal voltage).
- *System Freeze:* Rather than faults, the first significant disruption observed as a consequence of undervolting was a system freeze. This freeze acted as a uniform barrier across all devices and algorithms tested, signaling the limit of operation under undervolting conditions.

In conclusion, our experiments illustrated that undervolting did not compromise the accuracy or reliability of cryptographic operations. Instead, the system's

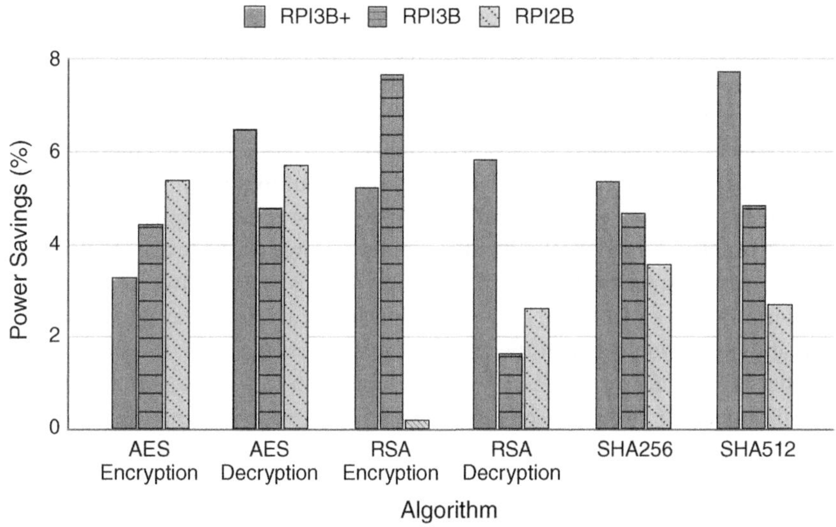

Figure 8.8 Energy savings of running crypto algorithms on Raspberry Pi devices.

threshold was met by a complete cessation of function, manifested as a system freeze, before any operational errors in the cryptographic algorithms could be observed.

8.3.2.3 Results – Energy Saving

In this experiment, our goal is to determine the potential energy savings achievable through undervolting. Each Raspberry Pi device involved in the experiment was undervolted to its maximum tolerable level; to the voltage level just before the device would crash.

To quantify the energy savings, we executed the same cryptographic algorithm on the Raspberry Pi devices, first at their nominal voltage and then at the maximum undervolting level without observing faults; before the crash voltage. The comparative energy consumption under these two scenarios is depicted in Figure 8.8. The results show that by simply changing the supply voltage – without any software or hardware modifications – we achieved energy savings exceeding 7%.

8.4 Edge-based Re-configurable Cryptography

Reconfigurable cryptography is a well-known technique that offloads the high-overhead functions from a resource-constrained device to an edge device to provide gains in performance and energy efficiency. However, they mostly assume that the edge device is trusted. If you do not trust the edge device, one can plausibly apply homomorphic encryption (HE) [34, 35] to enable computation over ciphertext. However, [36] shows that HE has almost four orders of magnitude slow-down even after software optimization.

Another experimental work has been shown in [37] where fully homomorphic encryption (FHE) has been combined with secure multiparty computation (MPC) and trusted execution environment (TEE) to leverage computations inside multiple untrusted servers when the client is a resource-constrained device. As per their experimental results on Intel's SGX platform which is their hardware-based TEE, the performance suffers from three main drawbacks: latency delay due to encryption and decryption between the enclave and the host machine, the small memory size of the enclave (the area where computations are to be performed) and lack of support for operating system (OS) based libraries or system calls inside the enclave. Our observation was that the performance here was hindered due to two layers of crypto operations. The first delay occurred from the encryption/decryption of chunks of data between one or more clients and multiple untrusted servers and the subsequent combining of multiple decrypted results to get the final result (MPC). The second delay came from the bottlenecks of computing inside the SGX enclave. Thus, we decided to look at different options to overcome these appreciable latency delays which can prove to be very costly in terms of energy consumption on a small-scale IoT device.

8.4.1 Background

We start our analysis of edge-based reconfigurable computing by investigating the approach in [38]. In this work, the authors have devised protocols that can help two untrusted small-scale IoT devices mutually authenticate each other followed by securely exchanging data between them. To mutually authenticate each other, they use a group signature scheme based on [39]. For data exchange, initially, they use Diffie Hellman Key Exchange (DHKE) to obtain a shared secret key that can be used for encrypting or decrypting the exchanged data. Next, they add an extra layer of fine-grained security by using attribute-based encryption (ABE) and decryption (ABD) based on [40]. The reconfigurable aspect is introduced by incorporating a security agent (SA) near each end device. These SAs are computationally more powerful than the end IoT devices and act as platforms for executing complex crypto operations. Each SA maintains a public key and a unique secret key obtained from the Global Key Management System (GKMS). This allows only those SAs to successfully decrypt the received data, whose secret key satisfies the attribute policy of encrypted data. The GKMS allows for the allocation and revocation of keys to a SA as per security requirements. During the data exchange process, ABE or ABD computations are offloaded to corresponding SAs, and their results are transmitted back to end devices. With this offloading approach, they reported a reduction of the processing time on the end IoT device by approximately 80% for ABE and ABD.

8.4.2 Proposed Approach

Initially, we evaluate the savings of energy on the end IoT device while offloading ABE or ABD to a powerful edge device. In the first step, we perform ABE and ABD on the end device only. We measure the mean execution time and energy consumption and use that as our baseline. The second step consists of establishing the authenticity of the edge device where ABE and ABD computations are to be offloaded and provision a shared symmetric key to it for future exchange of secure data. The authenticity of the edge device is performed by doing Remote Attestation (RA) using Gramine. The overhead due to RA is a one-time overhead only, where the end device verifies the enclave attributes of the edge device with the Intel Attestation Server (IAS) for correctness before offloading computations to the edge device. For the purpose of simulation only, we assume the shared symmetric key to be stored in advance on both end and edge devices. The final step is to perform offloading of ABE and ABD from end device to edge device with any original or generated plaintext being encrypted using AES-256-CBC cipher before transmitting from one device to another.

8.4.3 Experiments

8.4.3.1 Experimental Setup

The experimental setup shown in Figure 8.9 can be described in three broad categories as shown below.

Hardware: The details of the two end devices used in our experiment are shown in Table 8.12. The edge devices are two independent processes running inside a single desktop machine whose configuration is shown in Table 8.13. The Wi-Fi Router is a Verizon/Fios Home Router G3100 running Verizon's Open Source Software.

Software: We have used two ABE libraries to demonstrate our work. CPABE [41] is the collection of executables of the software implementation of ciphertext policy attribute-based encryption (CPABE) specification. While OpenABE [42] is an open-source package of C++ language-based ABE implementation which can be integrated with a parent software package to perform ABE or ABD. To encrypt plaintext using the AES-256-CBC cipher, we have used the PyCryptodome crypto library [43]. We used Gramine [44] to interface our client code (end device) with the SGX instruction set of the server (edge device). Our implementation is written in Python and can be executed on any Linux-based operating system. We have

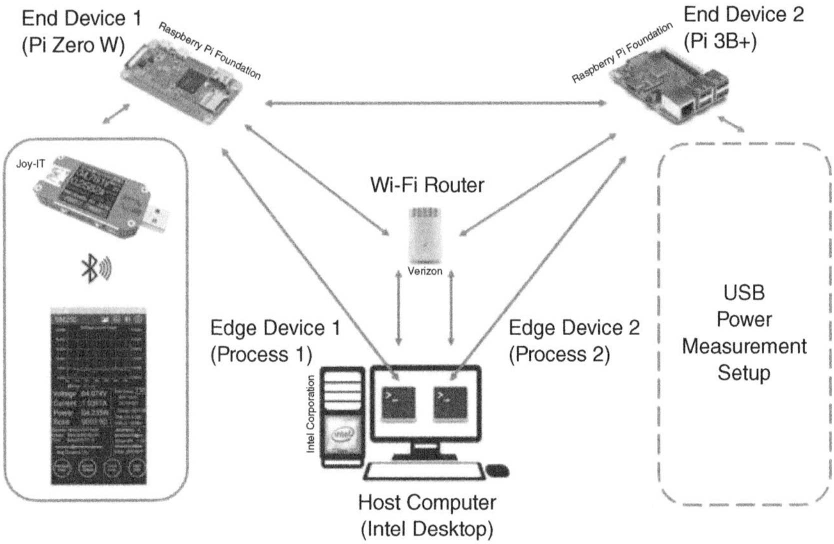

Figure 8.9 Edge-based Crypto offloading.

Table 8.12 End device configuration.

Device (Pi)	RAM (GB)	Freq-uency (GHz)	Bus width (Bits)	Hard-ware	Number of cores	Architecture	OS version (raspberrypi)
Zero W	0.44	1	32	BCM 2835	1	armv6l	6.1.19+
3B+	0.93	1.4	64	BCM 2835	4	aarch64	6.1.21-v8+

Table 8.13 Edge device configuration.

Processor (Intel)	RAM (GB)	Frequency (GHz)	Bus width (Bits)	Number of cores	Architecture	OS version (Ubuntu)
i7-6700	16.12	3.4	64	8	x86_64	20.04.6 LTS

used TCP protocol to transfer data packets from one device to another and our implementation is single-threaded.

Measurement Equipment: We used a UM25C USB Power Meter to measure the mean energy consumed during a crypto operation. The relevant parameters of this device can be found in Table 8.14. We installed its Android application to

Table 8.14 USB power measurement configuration.

Model	Range		Resolution		Accuracy	
	Voltage (V)	Current (A)	Voltage (V)	Current (A)	Voltage (V)	Current (A)
UM25C	4 - 24	0 - 5	10^{-3}	10^{-4}	$\pm$ (0.5% + 2 digits)	$\pm$ (1% + 4 digits)

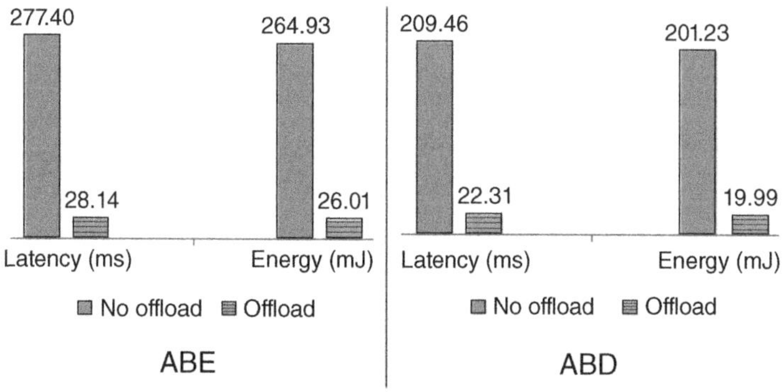

Figure 8.10 Latency and energy consumption on end device for no offloading and offloading of ABE and ABD for plaintext of size 32 B using CPABE library.

capture the instantaneous power readings and exported them to an Excel sheet for offline calculation of energy consumed by a crypto operation. In each of these calculations, the crypto operation was executed a large number of times to capture power readings sufficient enough to calculate the total energy consumed during all crypto operations. This value was then divided by the number of crypto operations to obtain the mean energy consumption during a single crypto operation.

8.4.3.2 Experimental Results

The mean execution time of our implementation for doing RA was 0.53 seconds on the end device. The results of mean execution time and energy consumption for ABE and ABD offloading using the CPABE library for a plaintext of size 32 B are shown in Figure 8.10. From this graph, we can see the savings in both execution time and energy consumption of about 90% for both ABE and ABD. When we executed the offloaded ABE and ABD computations inside the SGX enclave, the end device incurred a huge latency of about 3.5 seconds before receiving the final results from the edge device. Since the end device was idle during this time frame, we don't consider this idle energy consumption while evaluating our results for offloading to an SGX enclave. From Figure 8.11, we can see the savings in both execution time and energy consumption of about 90% for ABE and 94% for ABD

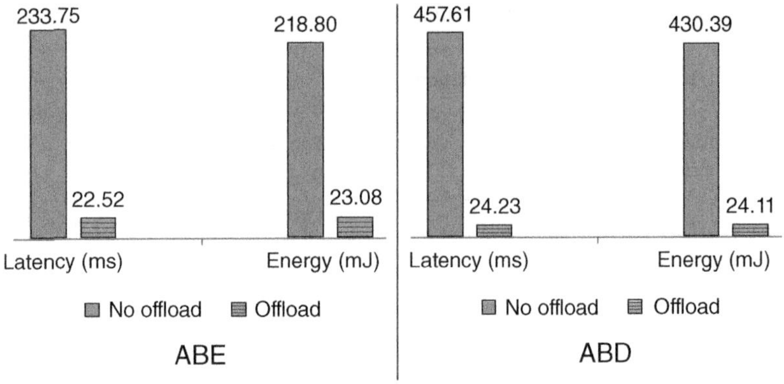

Figure 8.11 Latency and energy consumption on end device for no offloading and offloading of ABE and ABD for plaintext of size 32 B using OpenABE library.

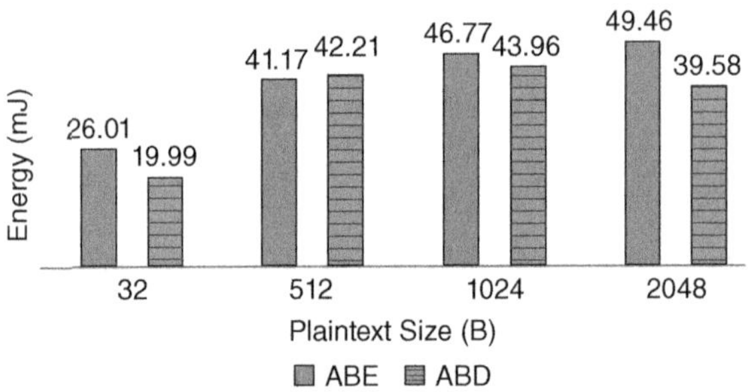

Figure 8.12 Energy consumption of end device for different plaintext sizes for offloading using CPABE library.

using the OpenABE library for a plaintext of size 32 B. We don't find any major difference in both the libraries' energy consumption on the end device. Our results show OpenABE consumes lower energy than CPABE for ABE. While the reverse holds true for ABD. When we varied the plaintext sizes while offloading ABE or ABD using both the libraries, we found a steady increase in the energy consumption of end device as shown in Figures 8.12 and 8.13.

8.5 Key Takeaways

Navigating the complex landscape of security in constrained environments is paramount, especially in the context of the ever-expanding IoT applications. While current cryptographic solutions are designed for robust systems like desktops and servers, they are not feasible for low-power devices, particularly those operating on

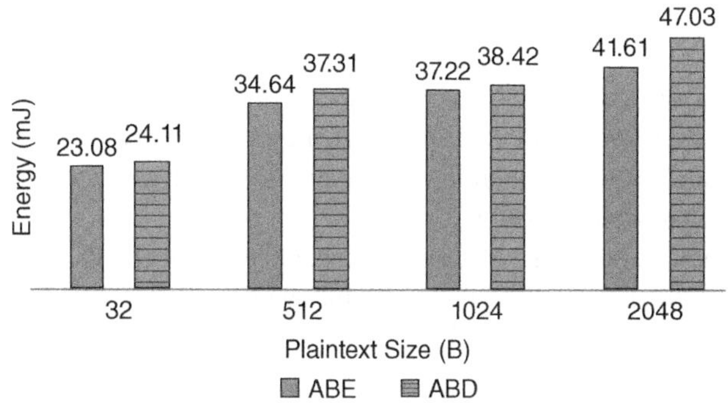

Figure 8.13 Energy consumption of end device for different plaintext sizes for offloading using OpenABE library.

limited battery resources, such as IoT devices. In this chapter, we have made several contributions to address this challenge. Specifically, we first integrated energy harvesting with IoT devices and showcased the feasibility of node authentication with low computational and resource overheads. Additionally, we leveraged undervolting to reduce the power consumption of the on-device computations. Our experiments illustrated that undervolting can be used to lower the power consumption of on-device computations without compromising the reliability of cryptographic operations or the IoT device. Furthermore, we utilize re-configurable cryptography to reduce the computational overhead of traditional cryptography. Our results show that our proposed reconfigurable crypto protocol was able to achieve significant gains in both latency and energy consumption.

This work lays the foundation for future research and development in lightweight cryptography and energy-efficient security solutions, ensuring the safe and reliable operation of the growing ecosystem of low-power devices.

References

1. Kumar, S., Tiwari, P., and Zymbler, M. (2019). Internet of Things is a revolutionary approach for future technology enhancement: A review. *Journal of Big Data*.
2. Zou, Q., Wang, Y., Wang, Q., Zhao, Y., and Li, Q. (2020). Deep learning-based gait recognition using smartphones in the wild. *IEEE Transactions on Information Forensics and Security*, 15: 3197–3212.
3. Cola, G., Avvenuti, M., Vecchio, A., Yang, G.-Z., and Lo, B. (2015). An unsupervised approach for gait-based authentication. In *2015 IEEE 12th International Conference on Wearable and Implantable Body Sensor Networks (BSN)*, pages 1–6.

4 Zouridakis, P. and Pudukotai Dinakarrao, S. M. (2022). Performance-aware lightweight dynamic early-exit-based gait authentication. In *IEEE International Symposium on Circuits and Systems (ISCAS)*, pages 466–470.

5 Kale, A., Sundaresan, A., Rajagopalan, A., Cuntoor, N., Roy-Chowdhury, A., Kruger, V., and Chellappa, R. (2004). Identification of humans using gait. *IEEE Transactions on Image Processing*, 13(9): 1163–1173.

6 Baumfeld, D., Baumfeld, T., Rocha, R., Macedo, B., Raduan, F., Zambelli, R., Silva, T., and Nery, C. (2017). Reliability of baropodometry on the evaluation of plantar load distribution: A transversal study. *BioMed Research International*, 2017: 1–4.

7 Cola, G., Avvenuti, M., Musso, F., and Vecchio, A. (2016). Gait-based authentication using a wrist-worn device. In *Proceedings of the 13th International Conference on Mobile and Ubiquitous Systems: Computing, Networking and Services*, MOBIQUITOUS 2016, page 208–217, New York, NY, USA. Association for Computing Machinery.

8 Li, H., Kadav, A., Durdanovic, I., Samet, H., and Graf, H. P. (2017b). Pruning filters for efficient convnets.

9 Li, H., De, S., Xu, Z., Studer, C., Samet, H., and Goldstein, T. (2017a). Training quantized nets: A deeper understanding.

10 Wess, M., Dinakarrao, S. M. P., and Jantsch, A. (2018). Weighted quantization-regularization in DNNS for weight memory minimization toward HW implementation. *IEEE Transactions on Computer-Aided Design of Integrated Circuits and Systems*, 37(11): 2929–2939.

11 Sutradhar, P. R., Bavikadi, S., Connolly, M., Prajapati, S., Indovina, M. A., Dinakarrao, S. M. P., and Ganguly, A. (2022). Look-up-table based processing-in-memory architecture with programmable precision-scaling for deep learning applications. *IEEE Transactions on Parallel and Distributed Systems*, 33(2): 263–275.

12 P.D., S. M., Lin, J., Zhu, S., Yin, Y., Liu, X., Huang, X., Song, C., Zhang, W., Yan, M., Yu, Z., and Yu, H. (2017). A scalable network-on-chip microprocessor with 2.5d integrated memory and accelerator. *IEEE Transactions on Circuits and Systems I: Regular Papers*, 64(6): 1432–1443.

13 Berestizshevsky, K. and Even, G. (2019). Sacrificing accuracy for reduced computation: Cascaded inference based on softmax confidence. *ArXiv*, abs/1805.10982.

14 Panda, P., Sengupta, A., and Roy, K. (2016). Conditional deep learning for energy-efficient and enhanced pattern recognition. In *2016 Design, Automation Test in Europe Conference Exhibition (DATE)*, pages 475–480.

15 Teerapittayanon, S., McDanel, B., and Kung, H. (2016). Branchynet: Fast inference via early exiting from deep neural networks. In *2016 23rd International Conference on Pattern Recognition (ICPR)*, pages 2464–2469.

16 Figurnov, M., Collins, M. D., Zhu, Y., Zhang, L., Huang, J., Vetrov, D., and Salakhutdinov, R. (2017). Spatially adaptive computation time for residual networks. In *2017 IEEE Conference on Computer Vision and Pattern Recognition (CVPR)*, pages 1790–1799.

17 Wu, Z., Nagarajan, T., Kumar, A., Rennie, S., Davis, L. S., Grauman, K., and Feris, R. (2018). Blockdrop: Dynamic inference paths in residual networks. In *Proceedings of the IEEE Conference on Computer Vision and Pattern Recognition*, pages 8817–8826.

18 Veit, A. and Belongie, S. J. (2019). Convolutional networks with adaptive inference graphs. *International Journal of Computer Vision*, 128: 730–741.

19 Wang, X., Yu, F., Dou, Z.-Y., and Gonzalez, J. (2018). Skipnet: Learning dynamic routing in convolutional networks. *ArXiv*, abs/1711.09485.

20 Maeng, K., Colin, A., and Lucia, B. (2017). Alpaca: Intermittent execution without checkpoints. *Proceedings of the ACM on Programming Languages*, 1(OOPSLA).

21 Van Der Woude, J. and Hicks, M. (2016). Intermittent computation without hardware support or programmer intervention. In *Proceedings of the 12th USENIX Conference on Operating Systems Design and Implementation*, OSDI'16, page 17–32, USA. USENIX Association.

22 Maeng, K. and Lucia, B. (2018). Adaptive dynamic checkpointing for safe efficient intermittent computing. In *13th {USENIX} Symposium on Operating Systems Design and Implementation ({OSDI} 18)*, pages 129–144.

23 Williams, H., Jian, X., and Hicks, M. (2020). Forget failure: Exploiting SRAM data remanence for low-overhead intermittent computation. In *Proceedings of the Twenty-Fifth International Conference on Architectural Support for Programming Languages and Operating Systems*, ASPLOS '20, page 69–84, New York, NY, USA. Association for Computing Machinery.

24 Ahmed, S., Bhatti, N. A., Alizai, M. H., Siddiqui, J. H., and Mottola, L. (2019). Efficient intermittent computing with differential checkpointing. In *Proceedings of the 20th ACM SIGPLAN/SIGBED International Conference on Languages, Compilers, and Tools for Embedded Systems*, LCTES 2019, page 70–81, New York, NY, USA. Association for Computing Machinery.

25 Demir, E. (2019). Early-exit convolutional neural networks. Master's thesis, Middle East Technical University.

26 He, K., Zhang, X., Ren, S., and Sun, J. (2016). Deep residual learning for image recognition. In *2016 IEEE Conference on Computer Vision and Pattern Recognition (CVPR)*, pages 770–778.

27 Colin, A. and Lucia, B. (2018). Termination checking and task decomposition for task-based intermittent programs. In *Proceedings of the 27th International Conference on Compiler Construction*, CC 2018, page 116–127, New York, NY, USA. Association for Computing Machinery.

28 Chippa, V. K., Mohapatra, D., Roy, K., Chakradhar, S. T., and Raghunathan, A. (2014). Scalable effort hardware design. *IEEE Transactions on Very Large Scale Integration (VLSI) Systems*, 22(9): 2004–2016.

29 Islam, M. S., Alouani, I., and Khasawneh, K. N. (2023a). Securevolt: Enhancing deep neural networks security via undervolting. *IEEE Transactions on Computer-Aided Design of Integrated Circuits and Systems.*

30 Islam, M. S., Omidi, B., Alouani, I., and Khasawneh, K. N. (2023b). VPP: Privacy preserving machine learning via undervolting. In *2023 IEEE International Symposium on Hardware Oriented Security and Trust (HOST)*, pages 315–325. IEEE.

31 Islam, M. S., Alouani, I., and Khasawneh, K. N. (2021b). Stochastic-HMDS: Adversarial resilient hardware malware detectors through voltage over-scaling. *arXiv preprint arXiv:2103.06936.*

32 Islam, S., Alouani, I., and Khasawneh, K. N. (2021c). Lower voltage for higher security: Using voltage overscaling to secure deep neural networks. In *2021 IEEE/ACM International Conference On Computer Aided Design (ICCAD)*, pages 1–9. IEEE.

33 Islam, M. S., Alouani, I., and Khasawneh, K. N. (2021a). Enhancing hardware malware detectors' security through voltage over-scaling. In *2021 5th ACM SIGARCH Workshop on Cognitive Architectures.*

34 Fan, J. and Vercauteren, F. (2012). Somewhat practical fully homomorphic encryption. *Cryptology ePrint Archive.*

35 Paillier, P. (1999). Public-key cryptosystems based on composite degree residuosity classes. In *International conference on the theory and applications of cryptographic techniques*, pages 223–238. Springer.

36 Reagen, B., Choi, W.-S., Ko, Y., Lee, V. T., Lee, H.-H. S., Wei, G.-Y., and Brooks, D. (2021). Cheetah: Optimizing and accelerating homomorphic encryption for private inference. In *2021 IEEE International Symposium on High-Performance Computer Architecture (HPCA)*, pages 26–39. IEEE.

37 Yakupoglu, C. and Rohloff, K. (2022). PREFHE, PREFHE-AES and PREFHE-SGX: Secure multiparty computation protocols from fully homomorphic encryption and proxy reencryption with AES and intel SGX. In *Security and Privacy in Communication Networks.*

38 Hsu, R.-H., Lee, J., Quek, T. Q. S., and Chen, J.-C. (2017). Reconfigurable security: Edge-computing-based framework for IOT. *IEEE Network*, 32: 92–99.

39 Boneh, D., Boyen, X., and Shacham, H. (2004). Short group signatures. In *Annual International Cryptology Conference.*

40 Goyal, V., Pandey, O., Sahai, A., and Waters, B. (2006). Attribute-based encryption for fine-grained access control of encrypted data. *IACR Cryptology ePrint Archive*, 2006: 309.

41 Bethencourt, J., Sahai, A., and Waters, B. (2007). Ciphertext-policy attribute-based encryption. *2007 IEEE Symposium on Security and Privacy (SP '07)*, pages 321–334.

42 OpenABE. https://github.com/zeutro/openabe.
43 PyCryptodome: Python package of low-level cryptographic primitives. https://github.com/Legrandin/pycryptodome.
44 Gramine. https://gramine.readthedocs.io/en/stable/index.html.
45 Invensense MPU-9250 Product Specification Revision 1.1. https://invensense.tdk.com/wp-content/uploads/2015/02/PS-MPU-9250A-01-v1.1.pdf.

9

Securing Smart Microgrid with Distributed Energy Resources

Saikat Das, Salih Sarp, Zhifang Wang and Yanxiao Zhao

Department of Electrical and Computer Engineering, Virginia Commonwealth University, USA

9.1 Introduction

The modern power system has evolved rapidly with the growing demand for electricity since reliable and scalable energy plays a crucial role for today's interconnected world [1]. We've also observed the rapid adaptation of renewable energy sources such as solar and wind and their widespread applications across various sectors in our daily lives [2]. Without a doubt, integration of renewable sources and their smart utilization will continue to play a significant role in our critical societal infrastructures. An exemplar of this utilization is the smart microgrid, which leverages NextG networks to enhance power grid resilience and reliability, optimize energy generation sources, and facilitate the integration of renewable and clean energy [3].

A typical microgrid comprises interconnected loads, Distributed Energy Resources (DERs), and intermediate energy storage units, operating as a cohesive controllable unit within the grid. Microgrids have gained significant traction as an increasingly favored choice for diverse applications. These applications span from energizing remote communities to offering crucial backup power support for vital facilities like hospitals and data centers [4]. A microgrid, in essence, represents a compact power network capable of functioning autonomously or in conjunction with the primary power grid. It typically comprises a mixture of distributed energy resources (DERs), including solar panels, wind turbines, and energy storage systems. Additionally, intelligent controls and monitoring setups are integrated to

6G Security: New Frontiers in Mobile Network Technologies and Verticals, First Edition.
Edited by Luiz A. DaSilva.

facilitate effective energy management. The decentralized character inherent in microgrids not only enhances their resilience and reliability but also contributes to curbing greenhouse gas emissions and operational expenses [5].

Although numerous benefits exist for microgrid infrastructure, unpredictable characteristics of renewable energy sources introduce complexities in the effective functioning of microgrids. To tackle these complexities, Energy Management Systems (EMS) assume a pivotal role in enhancing the operational efficiency of microgrids. They achieve this by harmonizing diverse energy resources and maintaining equilibrium between energy supply and demand, thereby optimizing microgrid performance [6].

Furthermore, the security of microgrids is still in its early stages. There is a lack of analytical approaches to quantitatively assess the effect of cyber-attacks on the operation of microgrids [7]. In contrast to legacy power grids, micro grids involve bidirectional power and information flow [8–10]. That's why micro grids require additional mechanisms to be operated. Some of the challenges that microgrids encounter include ensuring seamless integration of diverse energy sources, managing fluctuating energy supply from renewable sources, coordinating effective energy storage strategies, maintaining grid stability in the presence of variable loads, and addressing cybersecurity concerns related to prevention, detection, and countering malicious manipulation of grid functionality as a result of the interconnected nature of modern microgrid communication systems [11].

Moreover, there are various threats against microgrids. These threats can disrupt the continuous operation of the microgrid or exert control by tampering with data. The manipulation can occur at various levels of the data processing, with energy forecasts being particularly vulnerable. The manipulation of energy forecasts has the potential to cause significant disruptions in power systems and market operations. The extent of these manipulations become severe depending on the criticality of the application. Consequently, it becomes imperative to develop strategies for the detection and mitigation of these cyberattacks, in order to eliminate or minimize the undesirable effects on social, economic, and technical consequences that may arise. This book chapter uses solar energy forecasting as a use case to evaluate the impact of adversarial attacks and proposes an Artificial Intelligent (AI)-driven forecasting model to decrease the effect of deceptive data attacks. Specifically, an Artificial Neural Networks (ANN)-based forecasting network is developed. The model's effectiveness is examined under Fast Gradient Method (FGSM), which is a popular adversarial attack method. The efficacy of the ANN-based forecasting model is measured using metrics Root Mean Squared Error (RMSE), Mean Squared Error (MSE), and Mean Absolute Error (MAE). The outcomes show the pressing need to develop effective approaches for detection and mitigation, with the aim of proactively averting potential damage arising from manipulated energy forecasts.

This book chapter provides a comprehensive review of microgrids, energy management systems and cybersecurity of microgrids. It also provides a brief description of widely used cyber-attacks. The attacks include Denial of Service (DoS), Jamming, Man in the Middle and Adversarial attacks. This book chapter also provides a description of proposed microgrid structure and forecasting mechanism that can effectively increase the security of the microgrid. Section 9.2 offers a concise introduction to the microgrid concept. Elaboration on the energy management system is furnished in Section 9.3. The security aspect of a microgrid is presented in Section 9.4. The proposed model is thoroughly examined in these sections and the final conclusions are encapsulated in Section 9.5.

9.2 Microgrid: Overview

A microgrid serves as a contemporary, compact electrical power grid infrastructure designed to enhance efficiency, reliability, and the integration of renewable energy sources. It can be defined as a network comprising energy storage systems, distributed energy resources, and loads, all within specific boundaries that function as a unified controllable entity [12]. Essentially, a microgrid embodies a system facilitating a bidirectional flow of both power within the electrical network and information within the communication network. This dual communication, encompassing power and information, brings forth numerous benefits, including effective network monitoring, predictive failure analysis, and decreased maintenance expenses [13]. By harnessing advanced control methodologies and real-time monitoring, microgrids can operate autonomously, detached from the primary grid, while also seamlessly transitioning to grid-connected mode when needed. The integration of renewable energy sources and energy storage systems within a microgrid further contributes to emission reduction, ensuring a dependable and sustainable energy source.

9.2.1 Microgrid Components

A microgrid comprises various components that work together to provide a reliable and sustainable power supply. These components include [14, 15]:

Distributed Energy Resources (DERs): DERs are decentralized power generation sources located within the microgrid. They can include renewable energy technologies like solar panels, wind turbines, biomass generators, and small-scale hydroelectric systems. DERs generate electricity locally and contribute to the power supply of the microgrid.

Energy Storage Systems (ESS): ESSs store excess energy generated by DERs during periods of low demand and supply it during periods of high demand. Common types of ESSs include batteries, flywheels, and pumped hydro storage.

Energy storage enhances the stability and reliability of the microgrid by balancing supply and demand fluctuations and improving the integration of intermittent renewable energy sources.

Loads and Demand Side Management (DSM): Loads in a microgrid refer to the electrical consumption from various sources, including residential buildings, commercial entities, and industrial parks. These loads can vary in terms of power demand, duration, and characteristics, and they form the basis for energy consumption within the microgrid. DSM involves strategies to manage and optimize electricity consumption in response to price signals or grid conditions. This helps in reducing peak demand and enhancing grid stability.

Control and Monitoring Systems: These systems ensure the efficient and reliable operation of the microgrid by monitoring and controlling the power flow, voltage, frequency, and other parameters. They use advanced algorithms and communication technologies to coordinate the operation of DERs, ESSs, and other components, optimizing their performance and responding to changes in demand or grid conditions.

Smart Meters and Sensors: Smart meters are installed at consumer sites to measure electricity consumption accurately. They enable real-time data collection and support demand-response programs. Various sensors monitor parameters such as voltage, current, temperature, and humidity across the microgrid. They provide critical data for control algorithms, fault detection, and maintenance planning.

Switches and Protection Devices: Switchgears are used for isolating, controlling, and protecting the electrical components of the microgrid. It ensures safe operation and helps prevent cascading failures during faults. Circuit breakers, fuses, and relays safeguard the microgrid from overloads, short circuits, and other electrical faults, minimizing disruptions.

Grid Interconnection Equipment and Point of Common Coupling (PCC): Grid tie inverters convert the DC output from renewable energy sources into AC suitable for the microgrid and the main utility grid. Interconnection protection mechanisms ensure seamless transitioning between grid-connected and islanded modes while adhering to safety standards. The PCC is the electrical connection point between the microgrid and the utility system. It is typically located at the low-voltage bus of the substation transformer. The PCC serves as the interface between the microgrid and the main utility grid, allowing for bi-directional power flow and the exchange of electricity when needed.

Communication Infrastructure: Reliable communication networks are essential for real-time data exchange among microgrid components and controllers. These networks support monitoring, control, and coordination functions.

Cybersecurity Infrastructure: Robust cybersecurity measures protect the microgrid from potential cyber threats, ensuring data integrity, confidentiality, and operational safety. Intrusion Detection Systems continuously monitor the network for any unauthorized access or malicious activities and raise alarms when detected.

Energy Market Integration: Microgrids can participate in energy markets by buying and selling electricity. Integration with energy markets allows them to optimize their operation economically.

By integrating and coordinating these components, microgrids offer greater control, resilience, and efficiency in meeting electricity needs, supporting renewable energy integration, and enhancing the overall sustainability of the power system.

9.3 Energy Management System

An Energy Management System (EMS) is a system of computer-aided tools used by operators of electric utility grids to monitor, control, and optimize the performance of the generation or transmission system. Also, it can be used in small-scale systems like microgrids. In microgrid systems, an EMS plays a crucial role in ensuring reliable functionality, maximizing renewable energy penetration, and optimizing cost and economic efficiency in the electricity market. To achieve these goals, EMSs must address dispatch optimization problems by considering available production and storage capacity, market data, real-time stats, and operational constraints. Additionally, they rely on production and consumption forecasting information to achieve their techno-economic and environmental objectives [16]. The EMS uses communication networks to coordinate the operation of various energy resources within microgrids, including renewable energy sources, generators, and energy storage systems. In our study, we have used an optimization-based energy management system for our microgrid operation.

9.3.1 Objective Function

The primary objective of a microgrid EMS optimization is the reduction of operational costs linked with microgrid functioning. This optimization is achieved while simultaneously upholding a balanced equilibrium between the microgrid's electric power generation and consumption, and adhering to the system's operational limitations. Here, the cost entails two fundamental elements: the cost of procuring grid electricity at variable prices, a measure adopted to meet the microgrid's demand, and the expenditure incurred due to the charging and discharging of the Battery Energy Storage System (BESS). Through the process of optimization, our objective is to reduce these expenses and thereby increase the economic efficiency of the microgrid's operation. Our study specifically entails the optimization of the microgrid's operating expenses over a 24-hour period, encompassing considerations of variable grid electricity costs as well as the cost incurred in maintaining the battery system.

$$minimize\left(\sum_{t=0}^{N}(C_{\text{grid}}(t)\times P_{\text{grid}}(t))\times \Delta t+\sum_{t=0}^{N}(P_{\text{batt}}(t))\times C_{\text{batt}}\times \Delta t\right) \tag{9.1}$$

Here, $C_{\text{grid}}(t)$represents the cost of grid electricity during time t, $P_{\text{grid}}(t)$ represents the power flow from the grid to the microgrid at time t,$P_{\text{batt}}(t)$ represents the power flowing into or out of the battery energy storage system (BESS) within the microgrid at time t, C_{batt} represents the operational and maintenance cost of the BESS per unit of electric power, and Δt represents the specific time slot. In our study, we have data for every time slot (Δt) and we are optimizing the operational cost of the microgrid for the next 24-hour horizon in every time slot.

9.3.2 Constraints

The EMS optimization in this study involves a constrained optimization problem with two distinct types of constraints. These constraints encompass power balance and input/output power to the battery, serving as equality constraints within the optimization process.

$$P_{\text{pv}}(t)+P_{\text{grid}}(t)+P_{\text{batt}}(t)=P_{\text{load}}(t) \tag{9.2}$$

$$E_{\text{batt}}(t)=E_{\text{batt}}(t-1)+P_{\text{batt}}(t)\times \Delta t \tag{9.3}$$

Here $P_{\text{pv}}(t)$ represents the PV output power at time t, $P_{\text{load}}(t)$ represents the total load that needs to be served at time t, $E_{\text{batt}}(t)$ represents the stored energy of the BESS at time t. In Eq. (9.2), the power balance equation ensures the load is served by the available generation. In Eq. (9.3), the equation relates the battery energy with the charging/discharging rate at every time slot.

The operating limits of the components of the microgrid are the inequality constraints of the optimization: $X_{\min}\leq X\leq X_{\max}$. Here X represents P_{grid}, P_{batt}, and E_{batt}, $X_{\min}$ is the lower boundary, and $X_{\max}$ is the upper boundary, of the parameters.

9.3.3 Simulation Model

In this simulation model, we have assumed that we have historical data regarding the weather information, the customer load information, and the electricity market price of the specific microgrid location. These data have served as the basis for forecasting the 24-hour horizon in our EMS optimization. In our study, we utilized an open source dataset [17] that provided 24-hour ahead forecasted data for every fifteen minutes using a classic deterministic technique, Recurrent Neural Network

(RNN) with the keras python library. This dataset served as the direct input for our EMS. In our model, we have used this 24-hour horizon forecasted data in every 30 minutes step directly from the dataset. Additionally, the EMS incorporated other inputs such as the system parameters and constraints of the microgrid, battery energy storage system parameters, and the battery state of charge, among others. Figure 9.1 shows the overall working diagram of the proposed simulation model, which consists of three main parts. The first part involves collecting the dataset of weather information specific to the region, customer load information, and electricity market data. These datasets are processed using appropriate forecasting techniques to generate the 24-hour horizon data for Renewable Energy (RE) generation, load profile, and variable electricity prices. These forecasted data serve as inputs for the second part of the model, the EMS. The EMS, based on an

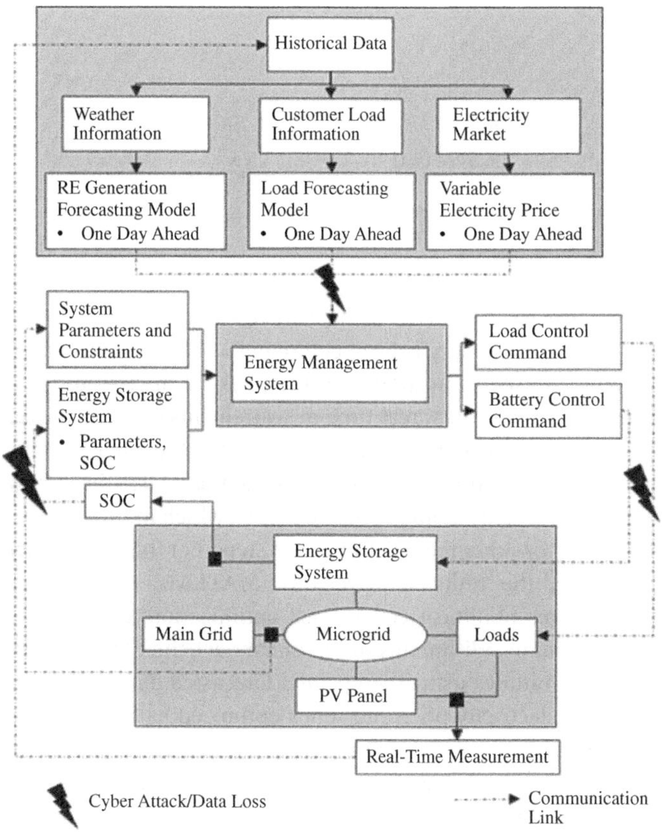

Figure 9.1 Working flow of the simulation model.

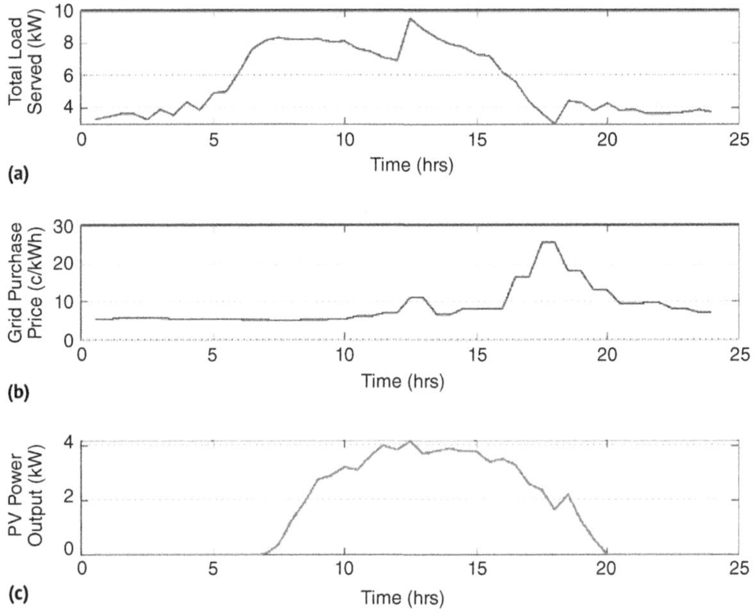

Figure 9.2 (a) Actual load profile in the microgrid, (b) variable grid electricity price, (c) PV generation.

optimization approach, utilizes the forecasted data, system parameters, and constraints to evaluate the optimal battery control command and load control command. These commands are then transmitted to the physical microgrid's local controllers for execution. Concurrently, real-time measurements of loads and RE generations obtained from sensors in the microgrid are sent to the database. This allows for the updating of the next time steps' forecasted data, as well as the State of Charge (SOC) of the Battery Energy Storage System (BESS), which in turn updates the microgrid's parameters for subsequent optimization steps. For this constrained optimization, we have used the 'fmincon' solver from MATLAB's Optimization Toolbox. 'Fmincon' performs nonlinear constrained optimization and supports linear and nonlinear constraints. Within this simulation model, the EMS performs the optimization every 30 minutes using the updated forecasted data of RE generation, load and variable electricity price and sends optimized battery charging discharging command data every thirty minutes. To illustrate, let's consider the scenario at 12:30 AM, where we engage in optimizing operational costs based on forthcoming 24-hour data and subsequently dispatch optimized control instructions to the microgrid controllers. Following this, at the next time step of 1:00 AM, we once again optimize operational expenses by integrating the updated dataset for the forthcoming 24 hours. Subsequently, we transmit the updated optimized control directives to be executed by the microgrid component controllers. This ensures that the microgrid operates optimally based on the latest available information and allows for effective control and coordination of the battery and load in response to changing conditions.

Figure 9.2 illustrates across 24 hours an actual load profile in the microgrid, the variable grid electricity price, and the PV generation.

Data loss in communication systems can have vital impacts on the operation of any system that relies on data transfer for control or decision-making. In a microgrid, where energy generation and consumption need to be carefully managed and coordinated, data loss can result in suboptimal control decisions or even system instability. For example, if battery control commands are lost, the microgrid may not be able to properly utilize its energy storage resources, leading to reduced efficiency and increased operating costs. In addition, if data loss occurs during critical events such as sudden load changes or system failures, the microgrid may not be able to respond appropriately, potentially resulting in blackouts or other disruptions. Therefore, it is important to consider the impact of data loss in communication channels when designing and operating microgrids. In our simulation model, we have simulated probabilistic data loss in battery command data to assess the economic impact of microgrid operation. Probabilistic data loss refers to the random loss of data that can occur in communication channels or storage devices. One way to model this type of loss is through probability distributions, such as the uniform distribution. The uniform distribution assumes that the probability of losing data is equal across all possible values. In other words, each possible data value has the same likelihood of being lost. This distribution can be used to simulate the effects of data loss in a system and to quantify the potential impact on the system's performance. By incorporating probabilistic data loss models into simulation tools, researchers and engineers can better understand the behavior of complex systems under uncertain conditions, and design more robust and reliable systems.

9.3.4 Findings and Interpretation

In this section, we conducted simulations to assess the effects of increasing data loss in battery command across various levels of renewable energy penetration within a small microgrid context. The microgrid considered in our simulations features a peak load of 10.2 kW and consists of a PV system as a distributed energy resource along with a battery energy storage system. To analyze the economic implications on microgrid operation, we incorporated different PV penetration rates (10%, 50%, 100%). For the 10% PV penetration scenario, a 15-kWh battery energy storage system with a maximum charging/discharging rate of 1 kW was employed. In the case of 50% and 100% PV penetration, a 75-kWh battery energy storage system with a maximum charging/discharging rate of 5 kW was utilized. To explore the influence of data loss, we simulated varying amounts of data loss

(0%, 10%, 30%, 50%, 100%) in battery command for each penetration scenario. Multiple simulations were carried out for each case, and the outcomes were meticulously compared and analyzed.

Figure 9.3 provides insight into a specific scenario involving a 10% data loss rate within a microgrid with 50% PV penetration. In Figure 9.3a, the battery system's state of charge over the course of the day is depicted. Figure 9.3b illustrates the distribution of the total load, showcasing how it is met through a combination of PV

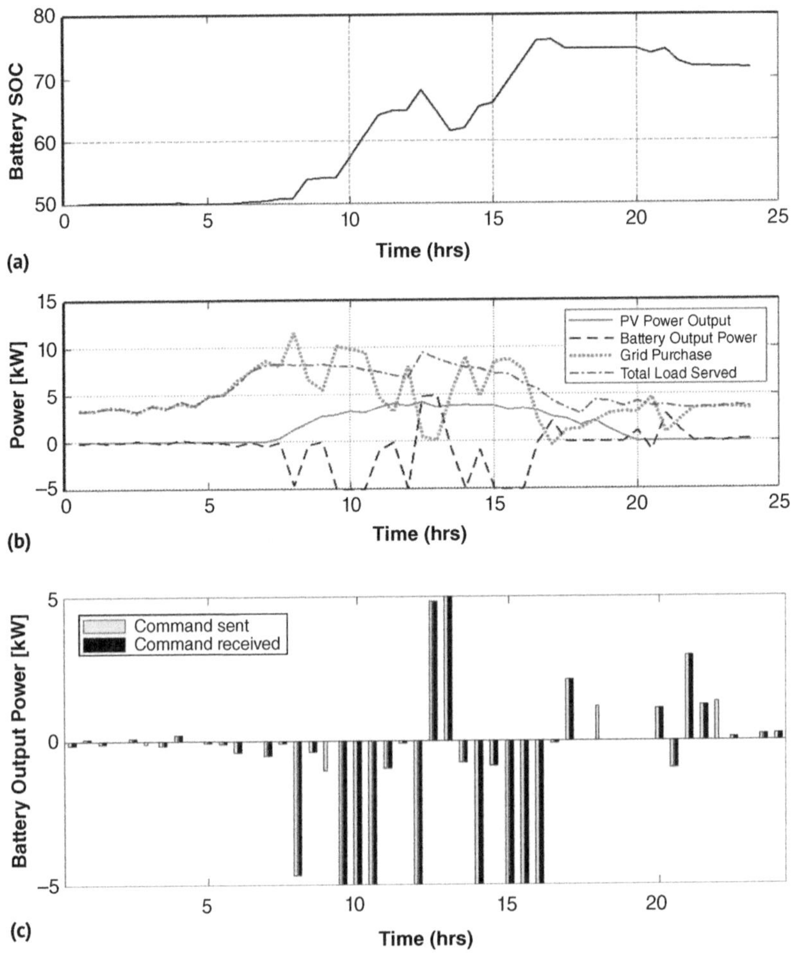

Figure 9.3 A specific case of 50% PV penetration scenario with 10% data loss in the microgrid (a) State of charge of the BESS. (b) Optimized power profiles of PV generation, battery output, grid electricity, and total load over time. (c) Battery charging/discharging command.

generation, grid electricity, and battery output power throughout the day. Additionally, Figure 9.3c illustrates the effect of varying data loss levels on the charging-discharging schedule command of the battery system. In instances of data loss, as shown in Figure 9.3c, the battery remains idle, neither charging nor discharging during the affected time slot. The system awaits the next command to proceed. This observation underlines the impact of data loss on the microgrid's ability to effectively manage its energy resources and respond to load requirements.

Figure 9.4 illustrates the varying grid electricity costs of the microgrid under the same level of data loss. In each setup, with a 10% PhotoVoltaic (PV) penetration and data loss amounts of 10%, 30%, and 50%, fifteen different simulations were conducted. There are instances where a 10% data loss leads to higher operational costs compared to a 50% data loss. This outcome is primarily influenced by the timing of data loss. If the battery command is lost during a period of high-priced grid electricity, when the battery should be discharged to meet the load demand, it significantly raises the operational cost of the microgrid. Consequently, even with a lower data loss percentage, the loss occurring during critical time slots can result in elevated operational costs. This highlights the importance of considering the timing and context of data loss in assessing its impact on microgrid operation.

In Figure 9.5 and Table 9.1, we can see the average grid electricity cost for different PV penetration with varied amounts of data loss. We can notice that high PV penetration results in less grid electricity cost in microgrid operation as it should. Also, with increasing data loss amount, the average grid electricity cost increased in most cases.

For 10% PV penetration in the microgrid, 10%, 30%, 50%, and 100% data loss result in on average 0.8%, 1.3%, 3.9%, and 5.3% increase in grid electricity cost,

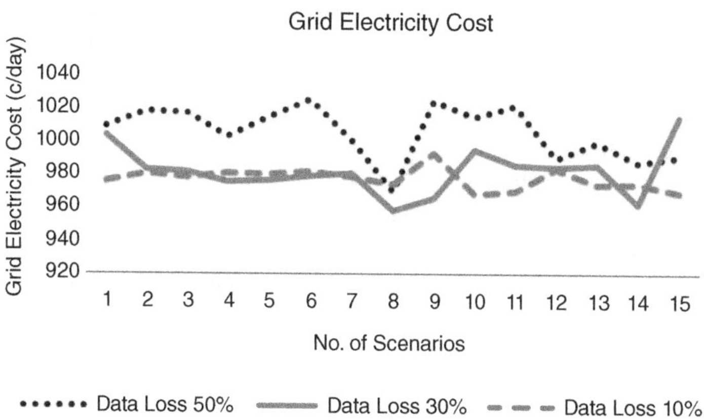

Figure 9.4 Comparison of grid electricity cost for the microgrid with 10% PV penetration with varied amounts of data loss.

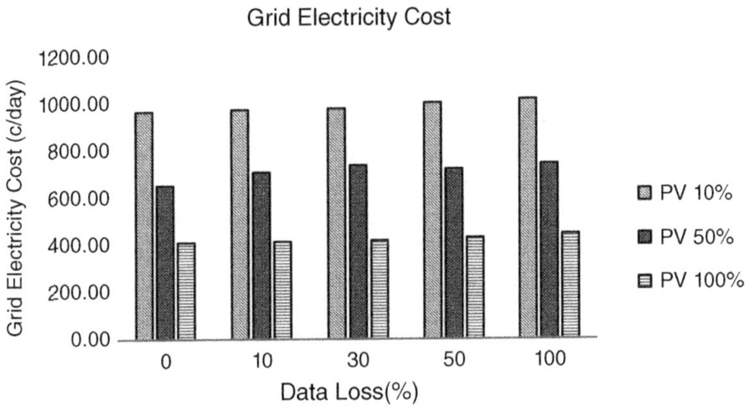

Figure 9.5 Comparison of average grid electricity cost for varying PV penetration and data loss rate in microgrid.

Table 9.1 Comparison of average grid electricity cost for varying PV penetration and data loss rate in microgrid.

Data loss % **PV%**	**Grid electricity cost (c/day)**				
	0	**10**	**30**	**50**	**100**
10	969.64	977.63	982.33	1007.86	1021.10
50	656.12	710.55	739.41	723.31	746.65
100	414.53	417.04	421.50	432.63	448.88

respectively. For 50% PV penetration in the microgrid, 10%, 30%, 50%, and 100% data loss result in on average 8.3%, 12.7%, 10.2%, and 13.8% increase in grid electricity cost respectively. For 100% PV penetration in the microgrid, 10%, 30%, 50%, and 100% data loss result in on average 0.6%, 1.7%, 4.4%, and 8.3% increase in grid electricity cost, respectively.

Despite the intuitive expectation that less data loss would always result in lower operational costs in a microgrid, this is not always the case. The timing of the data loss plays a crucial role in cost minimization. Additionally, the microgrid operates on a 24-hour ahead optimization basis, where control commands are generated using forecasted data for the next 24 hours. For instance, at noon, the optimization considers forecasted data 24 hours ahead to determine the optimal control commands. Consequently, when calculating the operational cost for a single day, it may not reflect the entire picture, as higher costs on a particular day may be part of an optimized set of actions within the 24-hour horizon.

9.4 Cybersecurity of Microgrids

In a more interconnected energy system, the security of microgrids bears critical importance with the increasing integration of renewable sources such as solar and wind [18]. Microgrids offer numerous benefits to the localized energy landscape including resilience, energy efficiency and sustainable energy without depending on the main grid. However, ensuring the security of microgrids and its operation against the cyber threats have become paramount to maintain these small-scale power networks within communities and industries [19]. As a second goal of this chapter, we have investigated microgrids from a cybersecurity perspective. The increase in numbers and types of microgrid devices inevitably leads to a wider surface of attack whereas the impact of successful attacks is becoming increasingly severe as more critical loads are connected. Therefore, the security of microgrids is critical and new strategies should be developed.

9.4.1 Types of Cyber Threats

Potential security threats include Denial of Service (DoS), jamming, eavesdropping, spoofing attacks, and others. These malicious behaviors should be identified to avoid their harm in the first place.

9.4.1.1 DoS Attacks

The goal of DoS attack is to make a computer system or a network unavailable to the users by exhausting it with a flood of traffic including resource requests. The concept of DoS attacks goes back to the 1990s with sending oversized packets to a target system for its crash. Also, the TCP handshake process was targeted to explode resources which is known as "SYN Flood". Then more sophisticated Distributed Denial of Service (DDoS) attacks emerged by using a vast number of computers. Coordinated attack launches affected many highly used government, industry and organization websites. Till today, different techniques were developed to magnify the impact of DoS attacks. Microgrids rely on continuous communication and digital control of the energy generation and its distribution. DoS attacks can have serious consequences on the functionality and reliability of the system [20]. A DoS attack may overwhelm the energy management system of a microgrid which results in unresponsive or slow down response time. This can lead to loss of control and remote monitoring on the operation of the microgrid which causes energy instability, imbalance and disruptions. The connected critical infrastructure may experience severe energy problems that may impact the essential services and result in serious revenue losses.

9.4.1.2 Jamming Attacks

Jamming attacks target wireless communication and disrupt its functioning over a specific area. A microgrid utilizes wireless communication for transmitting data between sensors, controllers and central energy management systems. A jamming attack can interfere with these communications and cause data losses, delays, and corruption [21]. The control and monitoring of a microgrid can be compromised where accurate and timely data exchange are critical. This can lead to operational instability, inefficiencies, and potential energy imbalances. Moreover, delayed communication due to jamming can impede rapid responses during emergencies, while attackers exploit vulnerabilities, potentially gaining unauthorized access. This also increases operational costs due to diagnostics, restoration, and equipment replacement.

9.4.1.3 Man in the Middle (MITM) Attacks

MITM attacks such as spoofing and eavesdropping are coined to refer to unauthorized interception and monitoring of a communication. Attackers can capture sensitive and critical information of microgrid components. Energy production, consumption, distribution and operation of the microgrid may lead to privacy breach which may result in command manipulations. As a result of these manipulations, control commands can be altered to affect energy production, and distribution. The exchanged data between sensors and centralized energy management systems can be falsified that leads to inaccurate decisions about energy generation, distribution and consumption [22].

9.4.1.4 Adversarial Attacks

Adversarial attacks are used to manipulate the models used within a microgrid such as power and demand forecasting AI models. Recent research has demonstrated that existing Deep Neural Network (DNN) models are susceptible to carefully crafted attacks [23]. These attacks involve perturbing input instances in a manner that maximizes the likelihood of incorrect decision-making, thereby leading to erroneous predictions. There are two types of scenario for the adversarial attacks namely, whitebox and blackbox attacks. In a whitebox attack, the attacker possesses full access to the target model, including both its architecture and parameters. On the other hand, a blackbox attack occurs when the attacker lacks access to the model itself, limited only to observing the outputs produced by the targeted model. There are different types of techniques to realize these kinds of attacks like, the Fast-Gradient Sign Method (FGSM), Projected Gradient Descent (PGD), Basic Iterative Method (BIM), and Carlini&Wagner (C&W) attacks [24].

9.4.2 Detection of Security Threats

There are multiple ways to detect attacks on a microgrid. One of the models proposed in here benefits from Machine Learning (ML) techniques. Artificial Intelligence (AI) methodologies like anomaly detection and categorization can prove invaluable, particularly as traditional techniques may struggle to identify new types of attacks. The general overview for identification of security threats is shown in Figure 9.6.

To harness the potential of AI techniques, an extensive dataset comprising both malicious and legitimate data packets can be annotated for dynamic model training purposes. Following this, the development of a classifier capable of distinguishing between benign and malicious network flows will be undertaken. Initially, established machine learning methods such as logistic regression, gradient boosting, k-nearest neighbors, random forest, and SVM are applied and evaluated to gauge the viability of conventional ML approaches [25]. Subsequently, cutting-edge Deep Learning (DL) models like Convolutional Neural Networks (CNN), Artificial Neural Networks (ANNs) and autoencoders can be harnessed to create a competing classifier, given the recent recognition of DL's promise in the area of anomaly detection. The AI-powered system learns typical network behavior through the analysis of archived benign data packets, enabling the identification of novel attack types by scrutinizing network traffic for irregular data patterns [25].

In light of the inherent challenge in storing significant amounts of electric energy, maintaining real-time power supply and load equilibrium emerges as critical factors for upholding the stability of the power system. Another application of AI-based security measures involves dynamic energy demand forecasting to mitigate spoofing attacks. These attacks involve the malicious manipulation of one or more microgrid attributes. For example, in the event of a spoofing attempt, disparities between dynamically forecasted values and falsified data can be detected, prompting the system to raise an alert for reevaluation of the manipulated information.

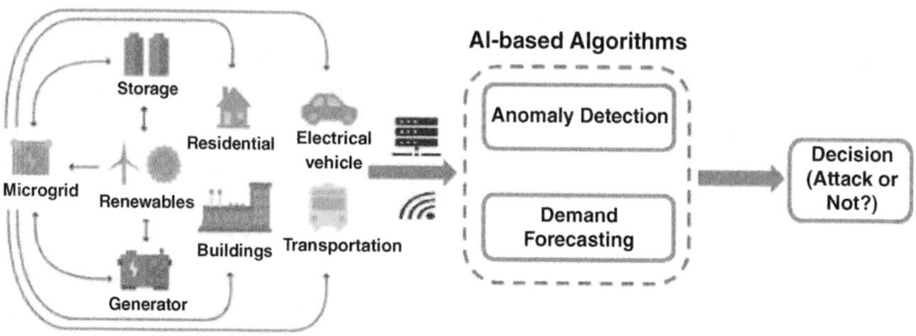

Figure 9.6 Identification of potential security threats of a microgrid.

9.4.3 Adversarial Machine Learning Attack Against Solar PV Power Generation Forecasting

Assessing the resilience of forecasting models is crucial through adversarial ML attacks or as known as deceptive data attacks targeting PV power generation predictions. In this subsection, we use solar PV power generation forecasting as an example to evaluate adversarial machine learning attacks against it. The AI-based solar PV power generation forecasting model is first introduced, followed by applications of adversarial attack against it.

9.4.3.1 AI-based Forecasting Model

ML techniques have garnered substantial popularity within the domain of solar power forecasting. Ongoing research concentrates on harnessing AI-driven ML models to achieve resilient solar power generation prediction. This study specifically delves into the experimentation and assessment of ANN-based model's performance, both with and without the incorporation of adversarial attacks or deceptive data attacks. The inception of ANN algorithms was influenced by the biological neurons, aiming to emulate human cognitive functions including learning and decision-making. This emulation process entails mathematical modeling within the software domain.

Typically composed of perceptrons, activation functions, input, output, and hidden layers, ANNs employ synapses to convey signals represented as real numbers from one layer to the next. The output of each layer is determined through a weighted sum of inputs. Most ANN algorithms undertake data processing via division into primary categories: training and testing datasets. Some models also integrate a validation dataset to refine the classifier's architecture. It is noteworthy that the applicability of ANN models extends beyond classification tasks.

Throughout the training phase, weights and the associated threshold within a given ANN are initialized as random values. These parameters are iteratively adjusted during training to converge to similar outcomes as the labeled data. Activation functions govern neuron behavior and output, utilizing normalization functions within limits of 0 to 1 or −1 to 1. Activation functions can take the form of linear, unit step, sigmoid, or hyperbolic tangent, based on specific requirements, functioning as gatekeepers. If the same number exceeds the threshold boundaries set by the specific activation function, the corresponding neuron's pathway is activated to relay the number originating from the preceding neurons. Then, this signal broadcasting and propagation, spanning from the input layer to the output layer via hidden layers, is repeated.

Figure 9.7 illustrates a broad overview of the forecasting model. The envisioned framework centers on forecasting models for solar power generation using ANNs and encompasses a solitary adversarial attack, specifically FGSM. The framework

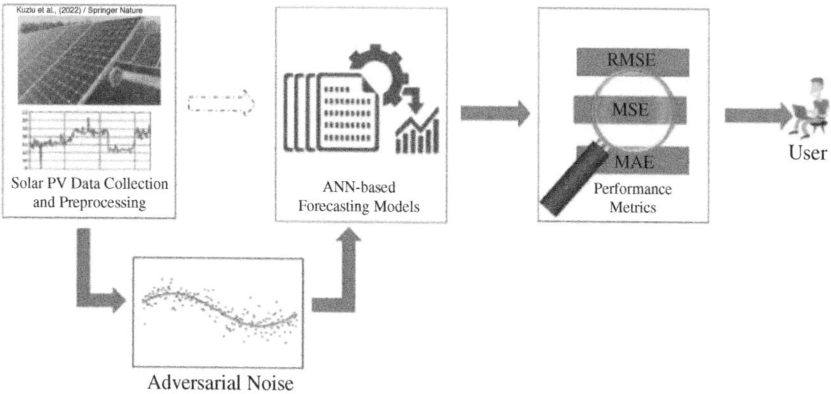

Figure 9.7 Framework for solar forecasting model. *Source*: Kuzlu et al. (2022)/ Springer Nature.

operates through the subsequent sequence of four steps: (1) Gathering and Processing Solar PV Data, (2) Introduction of Adversarial Noise, (3) Implementation of ANN-based Forecasting Models, and (4) Assessment of Performance Metrics.

- Gathering and Processing of Solar PV Data: The initial stage involves the accumulation of data from openly accessible sources or user-driven uploads. The framework securely retains time-series data, specifically pertaining to solar power generation. Subsequently, an enhancement in data quality is pursued through pre-processing, promoting robust analysis and informed decisions. This process includes rectification of absent or incomplete values, identification of anomalies, and data normalization. For this part of the study, data are sourced from a publicly available dataset.
- Introduction of Adversarial Noise: Within this framework, the popular FGSM constitutes one of the primary and most established adversarial attack models. It orchestrates the creation of adversarial noise aimed at misleading solar power generation models through adept manipulation of inputs with slight discrepancies. This step is performed in instances involving adversarial attacks.
- Implementation of ANN-based Forecasting Model(s): ANNs find application in an array of domains encompassing classification, prediction, and regression analysis. The framework encompasses an ANN-based model designed for forecasting solar power generation. This phase encompasses both training and subsequent testing stages.
- Evaluation of Performance Metrics: This critical step revolves around the assessment and quantification of model performance, including scenarios both with and without adversarial attacks. The focus is on the model's precision, with metrics such as Mean Squared Error (MSE), Root Mean Squared Error (RMSE), and Mean Absolute Error (MAE) utilized for measurement.

9.4.3.2 Application of Adversarial Attacks

In this part of the study, it is presumed that solar PV generation, both observed and forecasted, remains at "0" during the evening and nighttime periods. FGSM method is utilized to test the reliability of the forecasting models. (FGSM) [26] stands out as an early and widely adopted adversarial attack in ML. FGSM leverages the derivative of the model's loss function concerning the input image to ascertain the direction in which the pixel values of the input image should be adjusted to minimize the model's loss function. Once this direction is determined, it concurrently alters all pixels in that direction to maximize the prediction's loss value. For a model employing the classification loss function denoted as $L(\theta, x, y)$, where θ signifies the model's parameters, x represents the benign input fed to the model (in our context, a sample input image), and y_{true} is the true label of our input, we can generate adversarial samples using the following formula:

$$x^* = x + \epsilon \cdot \text{sign}(\nabla_x L(\theta, x, y_{\text{true}})) \tag{9.4}$$

Here, x represents the original input, x^* signifies the adversarial input, L corresponds to the MSE loss function, θ denotes the weights of the AI model, and ϵ governs the extent of perturbation applied.

An additional notable aspect of FGSM is its prioritization of speed over optimality. This implies that its design doesn't aim to generate the minimum essential adversarial perturbation. Moreover, the method's effectiveness is relatively limited at lower ϵ values when compared to other attack methodologies.

9.4.4 Findings and Interpretation

Solar PV power generation forecasting holds immense significance due to its important role in ensuring the seamless integration of solar panels into the power grid, depending on the reliability and predictability of power generation. In this part of the study, we present the efficacy of the ANN-based model's performance, assessed through metrics encompassing RMSE, MSE, and Mean Absolute Error (MAE). The performance of the ANN model has been computed and tabulated in Table 9.2, revealing RMSE (0.0874), MSE (0.0076), and MAE (0.0425) scores. A lower score denotes superior model performance in the evaluation process.

The depicted graphs in Figure 9.8a–c portray the observed and forecasted data alongside FGSM-generated deceptive inputs for ϵ values of 0.01, 0.03, and 0.20,

Table 9.2 Evaluation metric.

Model	RMSE	MSE	MAE
ANN	0.0874	0.0076	0.0425

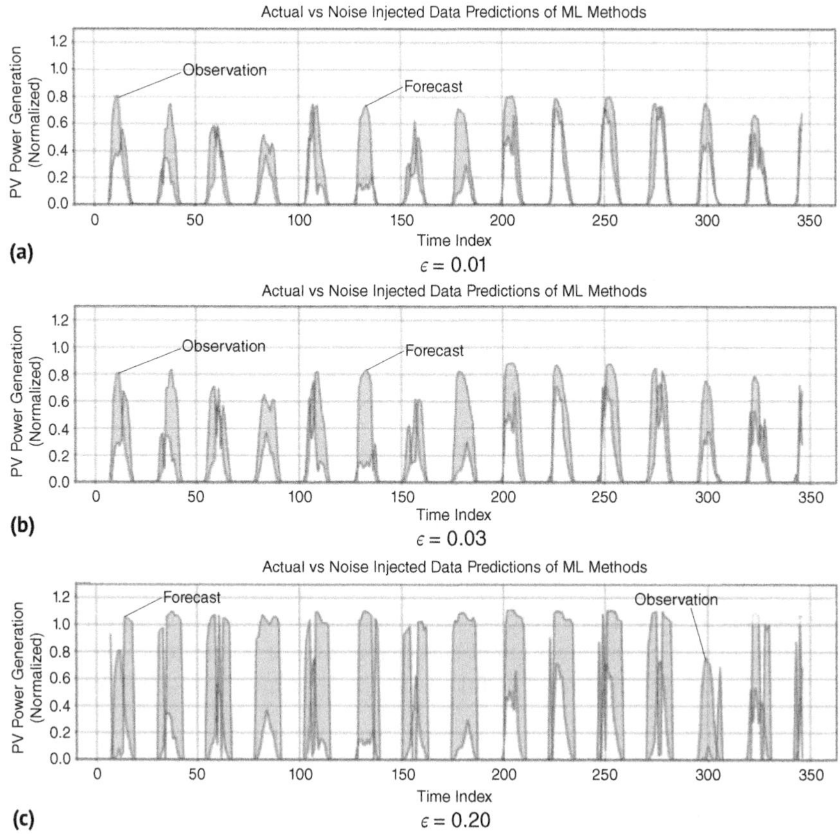

Figure 9.8 AI model predictions for the test and malicious inputs.

respectively. The observed data is represented by the red line, while the forecasted data is indicated by the green line. Evidently, Figure 9.8a–c illustrates that a substantial disruption in the AI model outcomes occurs when the attack power is significant, particularly in instances with a high ϵ and low solar PV generation.

An RMSE of up to 0.1 is considered acceptable within this context. As depicted in Figure 9.9, the solar PV forecasting model demonstrates resilience against FGSM attacks with $\epsilon = 0.01$. Conversely, the outcomes highlight the susceptibility of the solar PV forecasting model to FGSM attacks. For instance, the RMSE performance outcome with $\epsilon = 0.5$ reflects vulnerability approximately 8.85 times higher to FGSM attacks.

The outcomes indicate that in the absence of any deceptive data attacks, particularly FGSM, the ANN-based model yields commendable performance when forecasting solar PV power generation. The metrics stand at 0.0874, 0.0076, and 0.0425

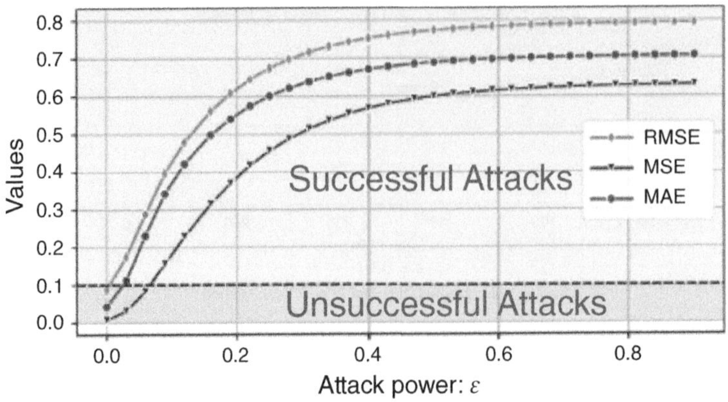

Figure 9.9 AI Model Prediction performance changes with different ϵ values.

for RMSE, MSE, and MAE, respectively. Nevertheless, the model exhibits remarkable susceptibility to FGSM attacks. For instance, the RMSE value can surge significantly to 0.793 under the influence of a substantial deceptive data attack, characterized by $\epsilon = 0.5$. Figure 9.9 encapsulates the impact of executed deceptive data attacks on the precision of energy forecasts. It delineates the boundary between successful and unsuccessful attacks, as determined by RMSE, MSE, and MAE metrics.

9.5 Key Takeaways

The microgrid projects have been initiated to decentralize the power generation and distribution which enhances the reliable energy supply. This book chapter uncovered the complicated challenges that surround the security of smart microgrids, emphasizing the significance of securing data integrity and communication networks. The discussion covered potential vulnerabilities arising from data-related cyber-attacks, specifically focusing on data loss. Through a simplified simulation model, the economic implications of such attacks were investigated. These findings emphasize the importance of considering data loss mitigation strategies and robust optimization techniques in microgrid energy management systems. By addressing data loss and/or data attacks challenges and incorporating reliable forecasting techniques, microgrid operators can enhance the efficiency and resilience of their systems. In conclusion, the exploration of securing smart microgrids with distributed energy resources has shed light on the critical importance of safeguarding these dynamic and interconnected energy systems. As the global energy landscape undergoes transformation towards sustainability and efficiency, microgrids have emerged as an indispensable component. This chapter examined the interconnected components that constitute a microgrid, highlighting their collective role in enhancing reliability, resilience, and the integration of renewable energy sources. Furthermore, the chapter highlighted the essential role of the microgrid's Energy Management System (EMS) in optimizing its operation. By minimizing costs and balancing supply and demand, the EMS becomes an integral tool in achieving economic efficiency.

Incorporating case studies involving varying renewable energy penetration levels, the chapter explored the effects of data loss and deceptive data attack impacts on microgrid operations. The insights obtained from these scenarios provided valuable understanding for enhancing microgrid resilience against data-related threats. The insight gained on the vulnerability of energy forecasting models, particularly in the context of solar PV power generation, shows the criticality of deceptive data attacks. The research underscores the significance of accurate energy predictions for stable power systems and market operations. Employing an AI-based model, the study assesses model performance against adversarial attacks, showcasing both robustness and susceptibility. The findings emphasize the need for effective detection and mitigation strategies to prevent potential damages caused by manipulated energy forecasts.

As the field of securing smart microgrids with distributed energy resources continues to evolve, it is evident that innovation in defense mechanisms, robust communication protocols, and advanced detection techniques will be imperative. The future will see the integration of more sophisticated threat models and the pursuit of strategies to mitigate the impact of attacks on the entire energy ecosystem.

References

1 Astier, N. and Ovaere, M. (2022). Reliability standards and generation adequacy assessments for interconnected electricity systems. *Energy Policy*, 168: 113131.

2 Talaat, M., Farahat, M., and Elkholy, M. (2019). Renewable power integration: Experimental and simulation study to investigate the ability of integrating wave, solar and wind energies. *Energy*, 170: 668–682.

3 Yoldaş, Y., Önen, A., Muyeen, S., Vasilakos, A. V., and Alan, I. (2017). Enhancing smart grid with microgrids: Challenges and opportunities. *Renewable and Sustainable Energy Reviews*, 72: 205–214.

4 Lagrange, A., de Simón-Martín, M., González-Martínez, A., Bracco, S., and Rosales-Asensio, E. (2020). Sustainable microgrids with energy storage as a means to increase power resilience in critical facilities: An application to a hospital. *International Journal of Electrical Power & Energy Systems*, 119: 105865.

5 Hamidieh, M. and Ghassemi, M. (2022). Microgrids and resilience: A review. *IEEE Access*, 10: 106059–106080.

6 Thirunavukkarasu, G. S., Seyedmahmoudian, M., Jamei, E., Horan, B., Mekhilef, S., and Stojcevski, A. (2022). Role of optimization techniques in microgrid energy management systems – a review. *Energy Strategy Reviews*, 43: 100899.

7 Khalil, S. M., Bahsi, H., Ochieng'Dola, H., Korõtko, T., McLaughlin, K., and Kotkas, V. (2023). Threat modeling of cyber-physical systems-a case study of a microgrid system. *Computers & Security*, 124: 102950.

8 Uzair, M., Li, L., Eskandari, M., Hossain, J., and Zhu, J. G. (2023). Challenges, advances and future trends in ac microgrid protection: With a focus on intelligent learning methods. *Renewable and Sustainable Energy Reviews*, 178: 113228.

9 Pérez-Guzmán, R. E., Salgueiro-Sicilia, Y., and Rivera, M. (2017). Communication systems and security issues in smart microgrids. In *2017 IEEE Southern Power Electronics Conference (SPEC)*, pages 1–6. IEEE.

10 De Dutta, S. and Prasad, R. (2019). Security for smart grid in 5g and beyond networks. *Wireless Personal Communications*, 106: 261–273.

11 Altaf, M. W., Arif, M. T., Islam, S. N., and Haque, M. E. (2022). Microgrid protection challenges and mitigation approaches – a comprehensive review. *IEEE Access*, 10: 38895–38922.

12 DoE, U. (2014). Office of electricity delivery and energy reliability. *National Electric Transmission Congestion Study—Draft for Public Comment.*

13 de Souza, A. C. Z. and Castilla, M. (2019). *Microgrids design and implementation.* Springer.

14 Farhangi, H. (2016). *Smart microgrids: Lessons from campus microgrid design and implementation.* CRC Press.

15 Katiraei, F., Iravani, R., Hatziargyriou, N., and Dimeas, A. (2008). Microgrids management. *IEEE Power and Energy Magazine*, 6(3): 54–65.

16 Faia, R., Lezama, F., Soares, J., Pinto, T., and Vale, Z. (2024). Local electricity markets: A review on benefits, barriers, current trends and future perspectives. *Renewable and Sustainable Energy Reviews*, 190: 114006.

17 Dumas, J. (2019). Liege microgrid open data. https://www.kaggle.com/datasets/jonathandumas/liege-microgrid-open-data

18 Villanueva-Rosario, J. A., Santos-García, F., Aybar-Mejía, M. E., Mendoza-Araya, P., and Molina-García, A. (2022). Coordinated ancillary services, market participation and communication of multi-microgrids: A review. *Applied Energy*, 308: 118332.

19 Yusupov, Z., Yaghoubi, E., and Soyibjonov, V. (2023). Reducing the vulnerability in microgrid power systems. *Science and Innovation*, 2(A5):166–175.

20 Chen, X., Zhou, J., Shi, M., Chen, Y., and Wen, J. (2022). Distributed resilient control against denial of service attacks in dc microgrids with constant power load. *Renewable and Sustainable Energy Reviews*, 153: 111792.

21 Arbab-Zavar, B., Golestan, S., Vasquez, J. C., and Guerrero, J. M. (2023). Exploring communication architectures in microgrids: Applications and scenarios. *IEEE Industrial Electronics Magazine*.

22 Kuzlu, M., Sarp, S., Catak, F. O., Cali, U., Zhao, Y., Elma, O., and Guler, O. (2022). Analysis of deceptive data attacks with adversarial machine learning for solar photovoltaic power generation forecasting. *Electrical Engineering*, pages 1–9.

23 Catak, F. O., Kuzlu, M., Sarp, S., Catak, E., and Cali, U. (2022). Mitigating attacks on artificial intelligence-based spectrum sensing for cellular network signals. In *2022 IEEE Globecom Workshops (GC Wkshps)*, pages 1371–1376. IEEE.

24 Kuzlu, M., Catak, F. O., Zhao, Y., Sarp, S., and Catak, E. (2023). Security and privacy concerns in next-generation networks using artificial intelligence-based solutions: A potential use case. In *Wireless Networks: Cyber Security Threats and Countermeasures*, pages 205–226. Springer.

25 Sarp, S., Kuzlu, M., Cali, U., Elma, O., and Guler, O. (2021). Analysis of false data injection impact on ai based solar photovoltaic power generation forecasting. *arXiv preprint arXiv:2110.09948*.

26 Goodfellow, I. J., Shlens, J., and Szegedy, C. (2014). Explaining and harnessing adversarial examples. *arXiv preprint arXiv:1412.6572*.

10

Securing Autonomous Vehicles with Machine Learning

Shiyi Liu, Tanmoy Sen and Haiying Shen

Department of Computer Science, University of Virginia, USA

10.1 Introduction

In recent years, the field of AVs has undergone remarkable progress, leading to a transformative impact on the transportation industry. These vehicles, equipped with state-of-the-art sensors and intelligent systems, heavily rely on vast amounts of data to make well-informed decisions and ensure safe navigation. However, the sheer volume of data and the need for real-time processing and decision-making present significant challenges. Traditional data communication approaches can be inefficient and may not meet the latency requirements for safe vehicle operation. The advent of NextG networks, the next generation of communication infrastructure, has expanded the potential of leveraging ML for AVs. In this context, AVs generate vast amounts of sensor data, which are transmitted via NextG networks to the cloud, where machine learning (ML) algorithms process and interpret this data, enabling critical decision-making.

ML plays a crucial role in the advancement of AVs. It is the backbone of numerous applications in the AV domain, including perception, prediction, planning, and control. In perception, ML algorithms are used to interpret sensor data and understand the surrounding environment, identifying objects such as other vehicles, pedestrians, and road signs. In prediction, ML helps anticipate the future actions of other road users, such as whether a pedestrian will cross the road or a vehicle will change lanes. In planning and control, ML is used to decide the vehicle's path and control its movements smoothly and safely. Furthermore, ML can

6G Security: New Frontiers in Mobile Network Technologies and Verticals, First Edition.
Edited by Luiz A. DaSilva.

be used to optimize traffic flow, reduce congestion, and improve fuel efficiency. In the context of NextG networks, ML for AVs introduces both promising opportunities and critical considerations. The integration of ML with this advanced network infrastructure offers the potential for improved connectivity, low-latency communication, and efficient data exchange among AVs. However, it also gives rise to significant challenges pertaining to the safeguarding of privacy, integrity, and security throughout the learning process and data sharing mechanisms.

The primary objective of this research study is to investigate the complexities associated with securing ML service through the NextG network for AVs. This study aims to develop and analyze potential adversarial threats that may compromise the efficacy of ML in the AV domain. In pursuit of this goal, the study delves into the development and examination of cutting-edge adversarial attack techniques. Understanding these adversarial attacks is crucial for securing ML in NextG networks. By studying these attacks, we can gain insights into the potential weaknesses of our systems and develop effective countermeasures. This knowledge can help us design more robust ML algorithms and implement stronger security protocols, ensuring that the AVs can operate safely and efficiently in the context of the NextG networks.

In response to the developed adversarial attacks and identified risks, this study also provides discussion about potential defense mechanisms and strategies aimed at mitigating these potential attacks. The results and conclusions from this study will significantly contribute to the advancement of AV technologies, ensuring the protection of user security and instilling confidence among stakeholders. Consequently, this research can potentially create a safer and more efficient transportation ecosystem.

In summary, this study underscores the importance of understanding and mitigating adversarial threats in the context of ML for AVs within the NextG networks environment. The development of adversarial attacks and the discussion of potential defense mechanisms are vital for ensuring the robustness and security of NextG-enabled ML systems for AVs. This study can serve as a crucial step towards the practical realization of secure and efficient AVs operating within the NextG network infrastructure. The insights derived from this study aim to significantly contribute to the advancement of AV technologies, fostering a safer and more efficient transportation ecosystem. As such, this study serves as a valuable resource for researchers, practitioners, and policymakers in the field, emphasizing the need for continuous vigilance, research, and innovation in the face of evolving adversarial threats.

10.2 Machine Learning for AVs

ML has emerged as a transformative technology with wide-ranging applications in transportation. At its core, ML involves the development of algorithms that allow computers to learn from and make decisions or predictions based on data.

This ability to learn from data makes ML a powerful tool for a variety of tasks, from image recognition and natural language processing to predictive analytic and decision-making. In the context of AVs, ML plays a pivotal role in enabling these vehicles to navigate complex environments safely and efficiently. AVs are equipped with a multitude of sensors, including cameras, LiDAR, and radar, which generate vast amounts of data. ML algorithms are used to process and interpret this data, allowing the vehicle to perceive its environment, recognize objects, predict the behavior of other road users, and make informed decisions. The advent of NextG networks has further expanded the potential of leveraging ML for AVs. NextG networks offer improved connectivity, low-latency communication, and efficient data exchange, which are crucial for the real-time operation of AVs. The integration of ML with NextG networks can enable more advanced functionalities in AVs, such as cooperative driving and vehicle platooning.

The integration of ML models, AVs, and NextG networks is a multifaceted process involving several key components. Firstly, the sensors on the AVs generate vast amounts of data, which are transmitted to the cloud via NextG networks. In the cloud, ML algorithms process and interpret this data, enabling the understanding of the vehicle's environment and facilitating decision-making. Secondly, AVs communicate with each other and with the cloud server through NextG networks. This communication allows the vehicles to share information and coordinate their actions, thereby enhancing their ability to navigate complex environments. The ML models in the cloud can leverage this shared information to make more informed decisions. Thirdly, the low-latency communication offered by NextG networks enables real-time operation of AVs. The ML models, deployed in the cloud, utilize this low-latency communication to make real-time decisions based on the latest sensor data and shared information from other vehicles. For instance, an ML model in the cloud can predict the behavior of a pedestrian based on the latest sensor data received from an AV, and send back instructions to adjust the vehicle's path accordingly in real-time. In this setup, the ML models are not housed within the AVs themselves, but rather in the cloud, and are accessed by NextG. This allows for more powerful processing capabilities and greater storage capacity than would be possible within the vehicle, while the use of NextG networks ensures fast and reliable communication between the AVs and the cloud.

In summary, ML, particularly in the context of the NextG networks, is a cornerstone of AVs, enabling the vehicles to perceive their environment, predict the behavior of other road users, make informed decisions, and cooperate with other vehicles. In this integrated system, ML models are trained and deployed in the cloud. AVs continuously generate and send sensor data to the cloud via NextG networks to make queries for specific decisions. However, this data transmission process can be manipulated, introducing potential vulnerabilities. These manipulations can lead to adversarial attacks, where malicious inputs are designed to deceive the ML models, causing them to produce incorrect outputs. Therefore, realizing these ML applications for AVs in NextG network requires addressing various challenges, such as ensuring the robustness and reliability of ML algorithms in the face of uncertainties and dynamic changes in the environment.

10.3 Adversarial Attacks Against AVs

10.3.1 Overall Introduction

An adversarial attack in the context of ML refers to a strategic and calculated attempt to deceive ML models deployed in the cloud by providing malicious input. This pernicious input is not arbitrary. This malicious input is far from random; it is carefully crafted to manipulate and exploit the ML models. These manipulations cause the model to produce incorrect outputs, known as 'adversarial examples', which are transferred to the cloud. Adversarial examples, in the realm of machine learning, can be analogously viewed as the computational counterpart to optical illusions in human perception, misleading the model into erroneous interpretations, inducing the model to perceive something that isn't there or misinterpret what is. The consequence of adversarial examples can be severe, especially when considering their potential impact on real-world applications of machine learning.

AVs heavily rely on ML algorithms for numerous tasks, including object detection, predicting the movement of other road users, and making crucial driving decisions. Consequently, an adversarial attack that causes an AV to misinterpret a traffic sign, misjudge the distance to another vehicle, or fail to detect a pedestrian, can potentially result in catastrophic outcomes. In addition, in the NextG scenario, AVs often rely on vehicle-to-vehicle (V2V) and vehicle-to-infrastructure (V2I) communications for decision making. Adversarial attacks could target these communication channels, manipulating the shared information to disrupt traffic flow or cause accidents. For example, an adversarial attack could manipulate the data shared between vehicle and cloud server to create the illusion of a non-existent traffic jam, causing unnecessary traffic congestion.

Furthermore, the low-latency communication offered by NextG networks, while crucial for the real-time operation of AVs, also provides a larger window of opportunity for adversarial attacks. Given the speed at which data is processed and decisions are made in this environment, there is less time to detect and mitigate adversarial attacks before they can cause harm.

Therefore, comprehending adversarial attacks, their implications, and devising strategies to detect and mitigate them is of paramount importance, especially as we increasingly entrust ML algorithms with critical tasks. Safeguarding the integrity and reliability of ML systems for AVs becomes imperative in ensuring the safety and viability of autonomous driving technologies.

One prevalent form of adversarial attack involves the introduction of "perturbations" or precise alterations to the input data provided to a model. These alterations are skillfully designed to be imperceptible to humans while significantly impacting the model's predictions or classifications. In the context of a AV scenario, an attacker might subtly modify the pixel values in an image fed into the vehicle's object recognition system. These alterations could be crafted in a manner that causes the system to no longer recognize a pedestrian or misclassify a stop sign as a yield sign. Although seemingly inconspicuous to humans, such alterations can result in critical mistakes for AVs, such as failure to stop for a pedestrian or disregard of traffic rules, leading to severe consequences.

Another form of adversarial attack is physical-world attacks, which transcend the digital realm and manifest in the real world. These attacks involve physically modifying real-world objects to deceive ML models. For instance, an attacker may slightly alter the color or shape of a stop sign. While still easily recognizable to a human driver as a stop sign, an AV's object detection system could be deceived by this subtle modification, potentially leading to hazardous driving situations. This type of attack emphasizes that adversarial threats are not confined to the digital domain, necessitating the implementation of robust defensive mechanisms to counter such risks effectively.

A more subtle form of adversarial attack involves the injection of carefully crafted and misleading data into a model's training set. The malicious data, also known as "position," is deliberately designed to subtly shift the model's learned patterns in favor of the attacker. For example, an attacker might introduce images into the training set where stop signs are consistently labeled yield signs. Subsequently, when the model is trained on this poisoned dataset, it learns the erroneous association, leading it to misclassify actual stop signs as yield signs during deployment. The insidious nature of poisoning attacks, coupled with their potential to cause significant harm, presents a substantial threat to the safety and reliability of AVs. Hence, robust defenses against such attacks are imperative to ensure the trustworthy operation of AVs in real-world scenarios.

10.3.2 Adversarial Attack Categories

Adversarial attacks in the context of AVs can be broadly categorized into two main types: white-box attacks and black-box attacks. These attacks exploit vulnerabilities in ML models and can have severe implications for AV safety and reliability.

In a white-box setting, the adversary has complete knowledge of the target ML model's architecture, parameters, and training data deployed in a cloud server. With this access, attackers can directly apply optimization-based techniques to generate adversarial examples. A well-known method is the Fast Gradient Sign Method (FGSM) [1], which perturbs input data in the direction that maximizes model error by leveraging gradients of the loss function. Another example is Projected Gradient Descent (PGD) [2], which iteratively applies FGSM with smaller step sizes to maintain imperceptible perturbations while amplifying adversarial impact.

In contrast, black-box attacks assume limited or no access to the internal information of the target model. The attacker interacts with the model only by sending inputs and observing outputs. Even with this restriction, such attacks can remain effective by exploiting the transferability property of adversarial examples – adversarial samples crafted using a substitute model can still mislead the target model. Alternatively, attackers can employ optimization-based query methods, where repeated queries are used to approximate gradients and generate adversarial examples.

In the remainder of this chapter, we focus on black-box adversarial attacks, particularly those leveraging optimization techniques.

10.3.3 Adversarial Attacks

This section sheds light on our work on emerging threat of adversarial attacks against ML models for AVs, which can maliciously manipulate the inference process and compromise the integrity of the decisions made by the Deep Neural Network (DNN) model deployed in the cloud. Four adversarial attacks targeting AVs are discussed in this section: Advanced black-box Adversarial Attack (A^3) [3], Context-aware Black-box Adversarial Attack (CBAA) [4], Suspicion-free Boundary Black-box Adversarial (SBBA) attack [5], and Fast Black-box Adversarial (FBA) [6]. Each attack exhibits distinct characteristics and techniques for perturbing the learning process through data manipulations of AVs. To facilitate a comprehensive understanding, a comparison of these attacks is provided in Table 10.1. As the threat of adversarial attacks continues to evolve, the exploration of effective defense strategies and secure learning methodologies becomes imperative to ensure the safe and reliable operation of AVs.

10.3.3.1 Threat Model

Assuming an external attacker seeks to launch a black-box adversarial attack against the DNN model utilized in the safety-related application for a victim vehicle. In the context of black-box adversarial attacks, the primary objective is to generate adversarial inputs that remain in close proximity to the original inputs, as measured by the $L2$ distance metric, while simultaneously achieving a high attack success rate. Formally, the objective function for this black-box adversarial attack can be defined as follows:

$$\min_{\vec{x}}(||\vec{x} - \vec{x_0}||_2 + c \cdot f(\vec{x}, t_0)), \tag{10.1}$$

where $c > 0$ is a regularization parameter, t_0 is the target class, and $f(\vec{x}, t_0)$ measures the level of unsuccessful adversarial attacks. The function f is further defined as:

$$f(\vec{x}, t_0) = \max([F(\vec{x})]_{t_0} - \max_{i \neq t_0}[F(\vec{x})]_i, -\tau), \tag{10.2}$$

In this context, t_0 represents the ground truth label of the target class for $\vec{x_0}$, $F(\vec{x})$ is the model output represented as a probability vector for each class, $[F(\vec{x})]_i$

Table 10.1 Summary of introduced adversarial attacks.

Method	Year	Features	Performance
A^3	2020	Efficiency in time-sensitive AV scenarios, reducing queries and perturbation	Requires up to 84.12% fewer queries and 57.67% less perturbation, with 94.87% higher success rates than existing black-box adversarial attacks
CBAA	2021	Context-awareness, analyzing specific influential spatio-temporal signals	Requires 43% fewer queries and 53% less time than existing adversarial attacks
SBAA	2021	Undetectability by detection methods, uses DNN model's output to design the adversarial perturbation	Outperforms previous adversarial attacks with a 56% higher success rate under detection methods, 238% less time to launch the attacks, 76% less perturbation, and 257% fewer queries to the DNN model
FBA	2021	Efficiency and unnoticeability based on the driving time-series signals	Requires significantly less time and fewer queries to generate a successful attack, adds a smaller amount of perturbation, making it less noticeable

denotes the i-th coordinate in vector $F(\vec{x})$, $[F(\vec{x})]_{t_0}$ is the probability of the ground truth label t_0, and τ is a predefined threshold for successful attacks.

In black-box attacks, gradient information of the DNN model is typically unknown, making conventional gradient estimation methods, like gradient descent, inapplicable. As a result, non-gradient optimization techniques such as the finite difference method, zeroth optimization, and random gradient estimation are utilized. The ZOO method [7], for example, employs the finite difference method to estimate the gradient for the i-th input coordinate, as follows:

$$\hat{g}_i = \frac{f(\vec{x} + h\vec{e}_i) - f(\vec{x} - h\vec{e}_i)}{2h}, \tag{10.3}$$

adaptive moment estimation (ADAM) based zeroth-order optimizer is then used to iteratively update each coordinate to minimize the number of queries. A successful attack can then be verified when the predicted class label matches the target class label and the $L2$ distance is below a defined threshold. This process continues until $\vec{x}$ converges or a predefined max iteration L has been reached. This approach provides a systematic way to perform adversarial attacks on deep learning models.

10.3.3.2 Assumptions

This section focuses on the adversarial attacks targeting DNN-based driving maneuver classifiers. The driving maneuver classifiers are designed to recognize various driving actions, encompassing left turn (LT), maintaining speed (MS), right turn (RT), slowing (SL), speeding up (SP), stationary (STD), and decelerating to stop (STP). These DNN-based models are trained, deployed, and regularly updated by vehicle vendors to adapt to diverse environmental conditions.

The fundamental premise of an adversarial attack model is based on the assumption that an attacker seeks to manipulate the final output class label of a given input, deliberately causing misclassification. The objective is to deceive the DNN-based driving maneuver classifier into classifying a particular driving maneuver as a different target driving maneuver. A successful attack occurs when the classifier erroneously classifies the targeted driving maneuver.

In this attack scenario, the attacker vehicle denoted as v_i, utilizes driving signals collected during time steps $(l - p)$ to l to predict the driving maneuver of the victim vehicle, represented as v_j, for each subsequent time period from l to $(l + T)$. Once these predicted signals are obtained, the adversary employs them to craft adversarial signals. These adversarial signals are then introduced into the victim DNN-based driving maneuver classifier during the time period from l to $(l + T)$. The objective is to deceive the classifier into misclassifying the driving maneuver during this period, ultimately leading to potential adversarial effects on the AV's decision-making process and safety. Assumptions for the attack generation process are as follows:

1. AVs do not collaborate with others to interpret or classify the received driving maneuvers from other vehicles, providing immunity against Sybil attacks.
2. The scenario assumes a data-driven driving maneuver classifier model. The communication network of the AVs can handle high traffic congestion without information loss between sender and receiver vehicles or cloud server.
3. It is assumed that the DNN model API is accessible by the fraudulent vehicle, or that the model utilized by both the malicious and victim's vehicles provides the same prediction results for identical inputs.
4. The attacker does not have information regarding the internal architecture of the victim's DNN model and cannot generate a substitute DNN model with high prediction accuracy.

10.3.3.3 A^3

The A^3 approach is specifically designed to enhance the efficiency of adversarial attacks by generating unidentifiable poison data in the context of AVs. A^3 combines binary search and optimization algorithms to reduce the number of queries to the DNN model and the amount of perturbation, thereby improving time efficiency and success rates. Experimental evaluations show that A^3 requires significantly fewer queries and less perturbation, with much higher success rates than existing black-box adversarial attacks. It also maintains a high level of human imperceptibility and prediction confidence. The detailed design of A^3 is shown as follows.

As mentioned above, in black-box adversarial attacks, the optimization process cannot be directly solved because the attacker lacks access to the internal architecture of the targeted DNN. Existing methods, such as Zeroth Order Optimization (ZOO) [7], utilize zero-gradient stochastic coordinate descent techniques for optimization. However, these techniques may demand a large number of queries and might not be time-efficient. To address these challenges and improve the efficiency of adversarial attacks, the A^3 method incorporates two major steps:

1. **Binary space partition.** The method first identifies the boundaries of perturbation necessary for a successful attack by partitioning the space of the problem (i.e., the temporal and signal spaces) into binary sections. This helps reduce the search space and consequently increases the speed of the attack.
2. **Zeroth-order conditional gradient descent.** After obtaining the reduced adversarial perturbation space from the binary space partition, zeroth-order conditional gradient descent approach is used for fast solution search in multi-dimension datasets. The idea is to estimate the gradient of the output function of the DNN model, which is not directly accessible in black-box scenarios.

For binary space partition, let's denote a coordinate by j and a signal by i. In each iteration, a gradient is generated and added to the value of a randomly chosen signal in the coordinate to generate $z_{i,j}$, while ensuring that $z_{i,j} \in [x_{\min}, x_{\max}]$. The perturbation boundaries are first identified horizontally along the time coordinate; then, the perturbation boundaries are identified for different signals separately. The process continues until the length of the reduced perturbation space T reaches a certain threshold θ_t, and the perturbation space of each signal reaches a certain threshold θ_i. Subsequently, the search space is reduced from $Y_{N,T}$ to $Y_{N,\tilde{T}}$.

After obtaining the reduced adversarial perturbation space $Y_{N,\tilde{T}}$, the algorithm utilizes the zeroth-order conditional gradient descent approach to find the optimal solution for the perturbation within $Y_{N,\tilde{T}}$. To do this, the algorithm generates a list of standard Gaussian matrices $u = [u_1, u_2, ..., u_m]$, where $u_i = I_{N \times T}$ and m is a constant.

Next, to estimate an average gradient G_a^j, the attacker computes the average gradient of function $F()$ using the generated Gaussian matrices u:

$$G_a^j(z, \epsilon, u) = \frac{1}{m}\sum_{i=1}^{m} \frac{F(z + au_i, \epsilon) - F(z, \epsilon)}{a} u_i \quad (10.4)$$

where ϵ is a small value within the range $[0, 1]$ and a is a constant.

The algorithm then computes the current gradient y using the minimum of M average gradients $[G_a^{\bar{1}}, G_a^{\bar{2}}, ..., G_a^{\bar{M}}]$:

$$y = \arg\min_{M} \bar{G}_a \quad (10.5)$$

After that, the algorithm updates the adversarial input z using the gradient y and a constant weight β:

$$z = (1 - \beta)z + \beta y; \quad \beta \in [0, 1] \tag{10.6}$$

Finally, the algorithm generates a query to the DNN model using the newly generated input z. This process iterates until the desired adversarial attack is achieved. The A^3 method efficiently combines binary search and optimization techniques, resulting in reduced query numbers and less perturbation required to achieve successful adversarial attacks in AV scenarios.

10.3.3.4 CBAA

The CBAA method is designed to enhance the success of generated perturbed data in AV scenarios. CBAA is context-aware and works by analyzing specific driving signals at certain time points that have a higher impact on the DNN output. These influential spatio-temporal factors vary in different traffic contexts. CBAA first generates the perturbation only on the influential spatio-temporal signals for each context offline. Experimental evaluations show that CBAA requires 43% fewer queries and 53% less time than existing adversarial attacks. The design of CBAA is inspired by two primary observations:

- Some driving signals follow certain patterns and influential spatial signals. The magnitude of their patterns in different contexts may be different.
- A significant spatial driving signal at different time windows may have a different impact on the DNN output in a given context, and the influential temporal time windows at different contexts may be different.

Based on the identified observations, it is evident that certain influential context-aware spatio-temporal portions in a driving maneuver input $x_{N \times T}$ significantly contribute to the DNN model's classification process. To identify these influential portions, two main steps are used:

- Knowledge-based Identification. In the first step, the CBAA method identifies knowledge-based significant driving signals for each type of maneuver and their corresponding significant temporal portions through data analysis. This process involves analyzing the driving signals in the dataset to determine which signals and temporal windows have a notable impact on the DNN model's classification.
- Significance Calculation. In the second step, CBAA calculates the significance of each driving signal using Eq. (10.7). Given the confidence score $C(x)$ of input x as $P(y|x)$, where y is the output label of input x. We define the

significance of an input portion as the deviation of the confidence score ($\delta C(\cdot)$) after changing input x to input x' as follows:

$$\delta C(x) = \frac{C(x)}{1 - C(x)} - \frac{C(x')}{1 - C(x')} \tag{10.7}$$

If the significance is larger than a threshold, we consider a driving signal as an influential signal.

Next, the input matrix $x_{N \times T}$, representing the driving maneuver, is partitioned into N row vectors, each corresponding to the time series of a specific driving signal. To identify influential time windows within each driving signal, a sliding window approach with decreasing window size is employed. For every influential driving signal, a systematic exploration of different combinations of modified inputs is conducted in a sequential manner. These modified inputs are carefully constructed to contain relevant portions of one of the knowledge-based significant driving signals. As a result of this process, the algorithm identifies and designates the influential spatio-temporal portions as $x_{1,a}, x_{2,b}, \ldots, x_{n,c}$, where $1, 2, \ldots, n$ correspond to the N types of driving signals, and $a, b, \ldots, c$ represent the specific numbers falling within the range of $[1, T - p]$, where p is the number of sliding windows. We then generate the offline adversarial signal portion $O_{i,j}$ for each maneuver i in a certain context j using a Variational Auto-encoder (VAE) base on the influential portion.

Finally, the attacker applies the zeroth method to reduce the size of offline perturbation. In this step, the goal is to minimize perturbation size z_0 and generate the adversarial input z (i.e., $z = x + z_0$) where the output of z is the target class label. This process is divided into four steps.

- Stochastic Gradient Calculation. The zeroth method calculates the stochastic gradient of the function $G_a(\cdot)$ with respect to the perturbation z_0, given the modified input $x + z_0$. The stochastic gradient is obtained by considering small perturbations z_0 and estimating the change in the output of the DNN model using finite differences. The goal is to find the direction that minimizes the function $G_a(\cdot)$.
- Average Stochastic Gradient. The stochastic gradient of $G_a(x + z_0)$ is derived h times, and the average stochastic gradient $\bar{G}_a$ is calculated. This process helps in obtaining a more accurate estimation of the gradient direction.
- Current Gradient Calculation. The algorithm then calculates the current gradient γ by selecting the minimum value from the average stochastic gradients $\bar{G}_a$. The current gradient γ represents the direction that leads to the most significant reduction in the perturbation.

- Gradient Update. Finally, the algorithm updates the perturbation z_0 based on the current gradient γ. The perturbation is modified using the formula $z_0 = (1 - \alpha)z_0 + \alpha\gamma$, where α is a constant. This updating process ensures that z_0 moves in the direction of the current gradient, thereby reducing the size of the perturbation.

By applying the zeroth method, the CBAA optimizes the adversarial perturbation efficiently, resulting in successful adversarial attacks with reduced perturbation sizes. This perturbation size minimization step plays a critical role in enhancing the success rate and effectiveness of adversarial attacks in AV scenarios while maintaining a high level of human imperceptibility and prediction confidence.

10.3.3.5 SBBA

The SBBA method is designed to enhance adversarial attacks in AV scenarios so that the generated perturbed sample can be undetectable by detection methods. It uses the DNN model's output to design the adversarial perturbation and formulates the attack design problem as a goal-satisfying optimization problem with constraints. The optimization problem is solved using the Bayesian optimization method [8], with a Gaussian process used to model the posterior distribution of the DNN model. A gradient estimation technique is used to reduce the solution searching time. Experimental evaluations show that SBBA outperforms previous adversarial attacks with a 56% higher success rate under detection methods, 238% less time to launch the attacks, 76% less perturbation, and 257% fewer queries to the DNN model to verify the attack success. The detailed design of SBAA is presented as follows.

In this approach, the primary objective is to find an adversarial input that maximizes the probability of the target class label while evading detection by defense mechanisms. The problem is mathematically formulated as follows:

$$\text{maximize}_{z_0}\ [f(x + z_0)]_t \tag{10.8}$$

$$\text{subject to } f(x + z_0) = t \tag{10.9}$$

$$\bar{G}_{y,t}(x + z_0) \le \tau_{y,t} \tag{10.10}$$

$$v(x + z_0) = 0 \tag{10.11}$$

$$\phi(x + z_0) \le \gamma_t \tag{10.12}$$

$$m(x + z_0) = 0 \tag{10.13}$$

$$q(x + z_0, c) = 0 \tag{10.14}$$

Here, $[f(x + z_0)]t$ represents the classification probability of the target class label, and $f(x + z_0) = t$ indicates that the predicted class label should be the target label. Additionally, the condition $\bar{G}y, t(x + z_0) \le \tau_{y,t}$ ensures that the probability of class confusion is kept below a certain threshold. The remaining conditions ensure that the adversarial input does not trigger attack detection methods, encompassing various defense mechanisms.

To address this optimization problem effectively, Bayesian optimization is employed, utilizing Gaussian process as a surrogate model and knowledge gradient as an acquisition function. Bayesian optimization is a suitable approach when the gradient information is not directly accessible, which is often the case in black-box scenarios where the inner workings of the model are not fully known. Leveraging the derivative of the Gaussian process enables the calculation of the gradient of the mean function, which in turn facilitates the determination of the adversarial perturbation efficiently. This optimization process significantly enhances the efficiency and success of generating adversarial inputs that maximize the target class label while evading detection.

10.3.3.6 FBA

The FBA method is designed to enhance adversarial attacks in AV scenarios, so that it can be efficient and unnoticeable based on the driving time-series signals. It consists of two steps for the adversarial signal generation: offline and online. In the offline step, FBA finds the influential input portion for each maneuver as the offline adversarial signal portion. In the online step, FBA replaces the influential input portion with the offline adversarial signal portion and smooths the signals. This input is then used as the initial solution to find the optimal perturbation online using the zeroth-order gradient descent method. This approach significantly reduces the time to find the optimal perturbation. Experimental evaluations show that FBA requires significantly less time and fewer queries to generate a successful attack compared to other methods. It also adds a smaller amount of perturbation, making it less noticeable. The detailed design of FBA is presented as follows.

The design of FBA follows a process involving the identification of influential input portions, the creation of an offline adversarial signal group, and the use of a zeroth-order conditional gradient descent approach [7] to obtain optimal adversarial signals.

To identify influential input portions, the method employs a scoring technique based on confidence scores similar to CBAA. The score, denoted as $\delta C(x_i)$, is calculated using the confidence scores of the original input x and the newly generated input x_i. The influential portion is determined by selecting the segment with the highest $\delta C(x_i)$. The detailed calculation is as follows:

$$\delta C(x_i) = \frac{C(x)}{1 - C(x)} - \frac{C(x_i)}{1 - C(x_i)} \tag{10.15}$$

After identifying the influential input portion, the next question is how we can combine the influential input portions to produce the final offline adversarial signal portion. For this purpose, we use the curve fitting method for polynomial regression. The function to predict signal value of $x_{n,t}$ is given as:

$$f_n^O(t) = b_n^0 + \sum_{q=1}^{Q} b_n^q t; \quad \text{where } t = 1, 2, \dots, l \tag{10.16}$$

Here, f_n^O is the function to predict signal value, Q is the degree of the polynomial curve, and b_n^q is the q-th coefficient of x_n. We then use the obtained per maneuver coefficients b_n^q to create the offline adversarial signal based on different values of time t.

In the online procedure, the attacker leverages the pre-computed offline adversarial signal and the predicted benign signal input to generate an effective adversarial input. The process involves several key steps to ensure the successful generation of adversarial signals.

Initially, the attacker identifies the influential input portion from the predicted benign signal input. Subsequently, the attacker combines this influential portion with the corresponding pre-computed offline adversarial signal.

The optimization of the adversarial signal is performed using the zeroth-order conditional gradient descent approach, commencing with the pre-computed offline adversarial signal as the starting point. The stochastic gradient $Ga(z_0, \delta, u)$ is calculated, wherein u represents a standard Gaussian vector, a is a randomly chosen number, and $F(z, \delta) = f(z)$, with δ being an error term added to input z.

$$Ga(z_0, \delta, u) = \frac{F(x + z_0 + au, \delta) - F(x + z_0, \delta)}{a} u \tag{10.17}$$

The current gradient γ is then determined as the minimum average stochastic gradient $G_{\bar{a}}$ derived from multiple stochastic gradients.

$$\gamma = \min G_{\bar{a}} \tag{10.18}$$

Finally, the gradient z_0 is updated using a weighted average of the previous value and the current gradient, where α is a constant non-negative value controlling the update process.

$$z_0 = (1 - \alpha)z_0 + \alpha\gamma \tag{10.19}$$

The obtained adversarial input z (i.e., $(x + z_0)$) effectively replaces the benign input signal x and is ready to be utilized for adversarial attacks in the AV system. By efficiently combining offline and online procedures, this approach facilitates the generation of adversarial signals with high success rates and minimum detection probability, contributing to the robustness and security of AVs against potential adversarial attacks.

10.4 Performance Evaluation

This chapter aims to present experimental evaluation of our proposed methodologies. These experiments involve the implementation of four distinct attack methods, as discussed in Section 10.3.3, along with two comparison methods. To ensure the validity and relevance of the evaluation, we leverage two distinct real driving datasets that encompass a diverse range of driving scenarios. These datasets capture various real-world driving events and maneuvers, providing a robust foundation for assessing the performance and effectiveness of the proposed approaches. Through these experiments, we seek to gain valuable insights into the vulnerabilities of ML application for AVs and demonstrate the necessity of defense strategies in enhancing the security and privacy of the learning process in the context of NextG.

10.4.1 Experimental Settings

The first dataset [4] is collected from 25 volunteers driving in city traffic, comprising a wide array of signals such as velocity, acceleration, jerk, steering angle, steering velocity, gas pedal pressure, brake pedal pressure, turn light signals, and engine revolutions per minute (RPM). The second dataset is obtained from ten volunteers driving along four different paths, containing a total of 51 different signals [9]. Overall, the second dataset consists of 94401 records, encompassing various traffic events and driving maneuvers.

The datasets are carefully curated to focus on four specific driving signals: speed, brake pedal pressure, gas pedal pressure, and steering wheel angle. These signals are chosen as they effectively differentiate the seven driving maneuvers considered in this research. For each maneuver prediction, a sequence of 1400 time steps is employed, and the aforementioned four signals are utilized to output the driving maneuvers.

By leveraging these comprehensive datasets and specific driving signals, the experimental evaluation aims to thoroughly assess the performance and robustness of the proposed attack methods in the context of AVs, shedding light on potential vulnerabilities and defense strategies for real-world driving scenarios.

In our evaluations, we selected the MLSTM-FCN DNN model [10] as the target model. This particular model has demonstrated high prediction accuracy for multivariate time series classification tasks, making it a suitable choice for our experiments.

To compare the performance of the four attack methods described in Section 10.3.3, we employed two comparison methods as baselines. These methods are as follows:

- ZOO [7]. ZOO is an optimization-based approach that aims to minimize adversarial perturbations by solving an optimization problem using the ADAM optimizer.
- AutoZOOM [11]. AutoZOOM utilizes the averaged random gradient estimation method with the ADAM optimizer. Additionally, it incorporates a time-series auto-encoder consisting of three LSTM layers in both the encoder and decoder modules to generate adversarial perturbations.

These comparison methods serve as baselines to gauge the effectiveness and efficiency of our proposed attack methods in comparison to existing state-of-the-art techniques. By conducting a comprehensive evaluation against these baselines, we can gain valuable insights into the strengths and weaknesses of our proposed methods and their potential practical applications in real-world scenarios involving AVs.

10.4.2 Experimental Evaluations

This section presents the results of the experimental evaluation, focusing on three key metrics: the number of queries, required time, and attack success rate. These metrics are critical in assessing the performance and effectiveness of the adversarial attack methods. By analyzing the number of queries, we gain insights into the efficiency and resource requirements of each approach in generating successful adversarial attacks. The required time metric provides valuable information on the speed of the attack methods. Lastly, the attack success rate offers an evaluation of the effectiveness of adversarial attack methods. Through the examination of these metrics, we can ascertain the strengths and limitations of the attack approaches, enabling us to make informed decisions on optimizing the security and efficiency of ML in the dynamic landscape of AVs.

Number of queries. Figure 10.1 illustrates the average number of queries required to generate a successful attack for each method. The results show that A^3, CBAA, SBBA, FBA, ZOO, and AutoZOOM require an average of 145.45, 24.51, 71.24, 27.75, 920.44, and 820.27 queries, respectively. Notably, all the methods introduced in this chapter outperform the comparison methods in terms of the number of queries needed.

A^3 achieves a significant improvement over the comparison methods due to several key factors. Firstly, the optimization problem in A^3 is designed to minimize the perturbation magnitude, leading to a more continuous search space and faster perturbation generation. Secondly, the binary partition method employed

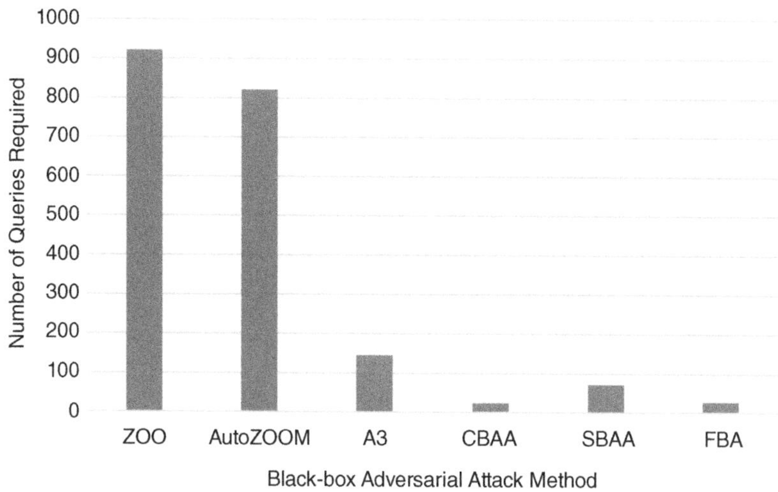

Figure 10.1 Average number of queries per successful attack.

in A^3 efficiently identifies the perturbation boundaries for different driving signals and time windows. This allows for a more targeted and effective search for the optimal perturbation. Lastly, A^3 utilizes the zeroth-order stochastic gradient descent approach, which is adept at searching for solutions in high-dimensional datasets. This contributes to a reduced number of required queries to the DNN model. CBAA also demonstrates a lower number of queries needed due to its context-aware spatio-temporal perturbation generation. By generating perturbations offline based on significant portions of the data, CBAA minimizes the number of queries required during online perturbation generation. Similarly, SBBA requires fewer queries than AutoZOOM and ZOO, primarily because Bayesian optimization, employed in SBBA, is more efficient in finding optimal solutions compared to the non-gradient optimization algorithms used in AutoZOOM and ZOO. Lastly, FBA has a similar number of queries to the CBAA method because of the offline perturbation generation method.

In summary, the proposed methods in this chapter demonstrate their effectiveness in minimizing the number of queries required, making them more time-efficient and practical for real-world adversarial attack scenarios involving AVs.

Required Time. Figure 10.2 illustrates the average time taken to generate a successful attack for each method. The average times for ZOO, AutoZOOM, A^3, CBAA, SBAA, and FBA are recorded as 243.53s, 205.55s, 58.42s, 16.5s, 87.01s, and 9.23s, respectively. It is evident from the results that our proposed methods, namely A^3, CBAA, SBAA, and FBA, significantly outperform the comparative methods ZOO and AutoZOOM in terms of time efficiency. Several factors contribute to these improvements:

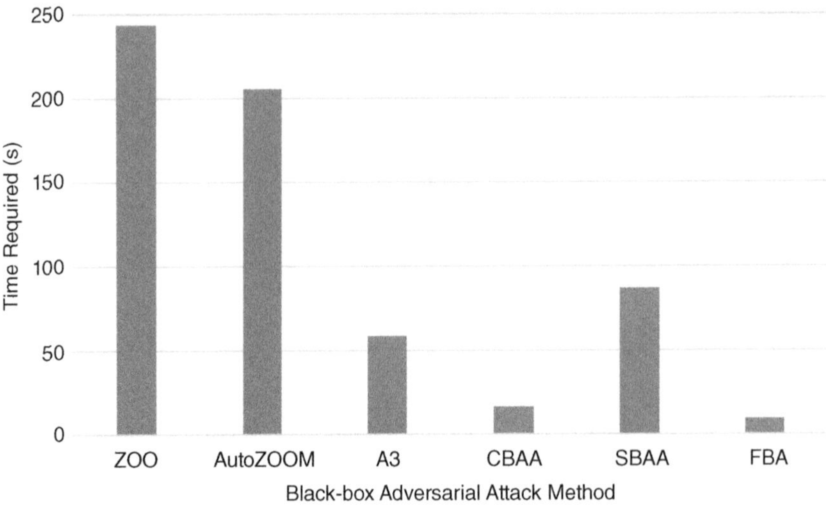

Figure 10.2 Average time required for different black-box attacks.

Firstly, A^3 optimizes its objective function to minimize the perturbation amount, resulting in a continuous search space and faster perturbation generation. The binary partition method employed by A^3 to identify perturbation boundaries for different driving signals and time windows further streamlines the process. Additionally, A^3 leverages the zeroth-order stochastic gradient descent approach to efficiently solve the optimization problem, leading to faster search for solutions in high-dimensional datasets. These optimizations collectively contribute to A^3's speed and significantly reduce the time required to launch a successful attack. Secondly, CBAA's context-aware spatio-temporal perturbation generation strategy substantially reduces the time needed for successful attacks. By generating perturbations offline based on a significant portion of the data, CBAA minimizes the need for online perturbation generation, effectively reducing the overall attack time. Similarly, SBAA's utilization of Bayesian optimization allows for faster finding of optimal solutions compared to the non-gradient optimization algorithms employed by ZOO and AutoZOOM. This speed advantage enables SBAA to accomplish successful attacks in less time. Lastly, similar to the CBAA, FBA also employs an offline generation of perturbations, resulting in shorter time requirements for successful attacks.

In conclusion, the methods introduced in this study demonstrate superior time efficiency compared to the comparative methods. The specialized optimization techniques and context-aware perturbation generation strategies implemented by A^3, CBAA, SBAA, and FBA significantly reduce the time needed to generate successful adversarial attacks against DNN driving maneuver classification model during various driving scenarios.

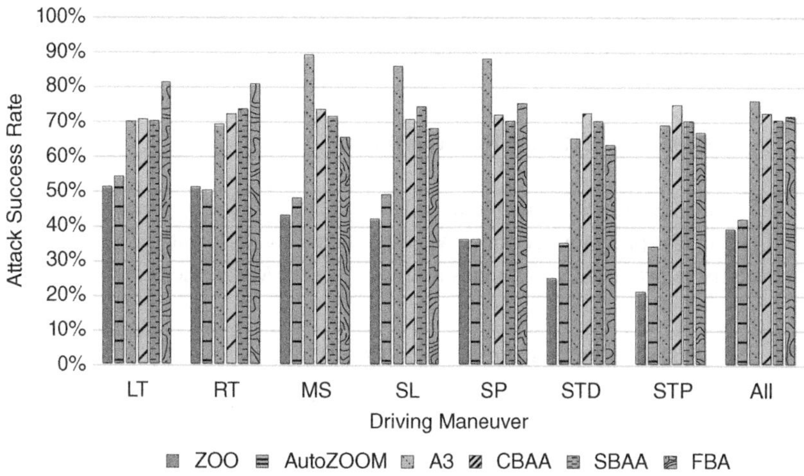

Figure 10.3 Attack success rate for different black-box attacks.

Attack Success Rate. Figure 10.3 presents the success rate of each attack method for different driving maneuvers, including left turn (LT), right turn (RT), maintaining speed (MS), slowing (SL), speeding up (SP), stopped (STD), and stopping (STP). The success rates of ZOO, AutoZOOM, A^3, CBAA, SBAA, and FBA are evaluated and compared for each maneuver. From the results, it is evident that A^3 consistently outperforms ZOO and AutoZOOM across all driving maneuvers. Notably, A^3 exhibits significantly higher success rates in maintaining speed and speeding up scenarios, achieving 89% and 88% success rates, respectively. This superiority can be attributed to the specialized optimization and perturbation generation techniques employed by A^3, which enhance its effectiveness in dynamic driving scenarios.

CBAA and SBAA show comparable success rates, slightly outperforming ZOO and AutoZOOM in most maneuvers but falling short of A^3's proficiency. CBAA demonstrates a slightly higher success rate in right turn and stopping scenarios, indicating its effective spatio-temporal perturbation generation technique for these specific maneuvers. FBA, despite achieving a lower success rate in the maintaining speed scenario (65.5%), displays impressive performance in left turn and right turn scenarios, achieving success rates of 81.2% and 80.5%, respectively. This suggests that FBA's strength lies in tackling complex maneuvers, such as turning, owing to its effective offline perturbation generation step.

Considering the overall success rates, A^3 outperforms the other methods, with FBA closely behind. These results highlight that these methods, each employing distinct techniques, can significantly improve the success rate of adversarial attacks on AVs during various driving maneuvers. The higher success rates achieved by A^3 and FBA particularly underscore the efficacy of their specialized strategies in overcoming the challenges posed by real-world driving scenarios.

10.5 Possible Defense Mechanisms

This section describes the possible approaches to tackle the adversarial attacks introduced in the previous sections. We also discuss the associated challenges to deploying these proposed approaches in the context of ML in the NextG era.

10.5.1 Approaches Against Adversarial Attacks

In response to the growing threats of adversarial attacks on ML systems for AVs, researchers have developed various defensive mechanisms. These defenses aim to counteract the negative impact of the attacks, ensuring the integrity and robustness of the inference process in ML-based AV applications. In this section, defense approaches include defensive distillation [12], input reconstruction [13], adversarial detection [14], and feature squeezing [15] are introduced.

Defensive distillation, proposed by Papernot et al. [12], is a defense mechanism that aims to enhance the robustness of ML models, particularly in the context of adversarial attacks. The process involves training a secondary model, also known as a distilled model, on the output probabilities of a primary model, rather than using the traditional hard labels. By doing so, the distilled model learns to match the output probabilities of the primary model, effectively capturing the uncertainty and subtlety in the relationships between different classes. The key advantage of defensive distillation lies in the fact that it forces the model to learn more nuanced relationships between classes. Traditional hard labels only provide discrete and categorical information about the correct class, making the model susceptible to small perturbations that may change its output drastically. However, when using probabilities as target labels, the model is compelled to develop a more refined understanding of the decision boundaries between classes. This subtle knowledge makes it challenging for an attacker to find specific input directions that lead to misclassification, as small perturbations in the input space are unlikely to cause significant changes in the output probabilities. In the context of AVs, defensive distillation can be leveraged to enhance the resilience of the ML models against adversarial attacks. A secondary model can be trained using the softened outputs of the primary model. The softening of labels refers to the use of probabilities as target labels, allowing the model to learn not only how to distinguish different classes but also to grasp the underlying relationships between these classes. Consequently, when faced with an adversarial attack that aims to manipulate the perception of traffic signs, for example, defensive distillation can aid the secondary model in resisting such manipulation. Even if the attacker subtly alters the appearance of a stop sign to make it resemble a yield sign, the secondary model's softened understanding of the relationship between these signs can help it maintain the correct classification, ensuring the safety and integrity of the AV's decision-making process.

Input reconstruction, proposed by Gu and Rigazio [13], is another defense mechanism aimed at bolstering the robustness of ML models against adversarial attacks. This method employs an autoencoder, a type of neural network architecture, to learn a compressed representation of the input data, often referred to as the latent space. The autoencoder is composed of two main components: an encoder and a decoder. The encoder maps the input data into the compressed latent space, while the decoder reconstructs the original input data from this compressed representation. The crux of this defense lies in the behavior of the autoencoder during the reconstruction process. When adversarial noise or perturbations are added to the input data, these perturbations are often not faithfully captured in the compressed latent space. As a result, when the reconstructed output is generated from the compressed representation, the adversarial perturbations tend to be removed or significantly reduced. This phenomenon is attributed to the autoencoder's ability to capture the most essential and dominant features of the data in the compressed representation while discarding less relevant details, including the adversarial noise. To defend against adversarial attacks, the reconstructed output from the autoencoder is used as the input to the ML model instead of the potentially perturbed original input. By doing so, the model is shielded from the adversarial perturbations since they have been effectively eliminated during the reconstruction process. Consider a scenario in which an adversarial attacker manipulates some data in a vehicle's dataset with the intention of distorting it's decision-making. By leveraging input reconstruction with an autoencoder, these manipulations are identified and removed during the reconstruction process.

Adversarial detection methods involve techniques to identify and flag suspicious model updates [14]. This can be achieved by monitoring the statistical properties of the model updates and flagging those that deviate from the norm. For instance, if a model update results in a drastic decrease in the model's performance or changes the model's decision boundary significantly, it could be an indication of an adversarial attack. In the context of ML-based AV applications, consider a situation where an adversarial attack targets a vehicle and modifies its model update to disrupt the learning process. Anomalies in the update, detected using the adversarial detection system, can be a clear sign of an attack.

Feature squeezing works by reducing the amount of information available in the input data, thereby limiting the space that an adversary can exploit [15]. For example, reducing the color depth in images, a form of feature squeezing, can mitigate the effect of small perturbations in pixel values, a common method used in adversarial attacks. Thus, by simplifying the input data, feature squeezing can make adversarial perturbations ineffective and protect the model from being deceived. In the context of ML-based AVs, feature squeezing can be adapted to sensor data such as LiDAR measurements used in object detection systems. One possible approach is "binning", a form of feature squeezing that reduces the precision of the LiDAR data. Instead of using the exact distance values provided by the LiDAR sensor, the range of possible distances is divided into a predetermined number of bins. Each data point is then represented by the bin it falls into. This squeezing process reduces the granularity of the LiDAR data, making it more difficult for adversarial perturbations to create misleading point clouds. By training the model with this binned data, it becomes more robust against adversarial attacks aiming to manipulate the detailed point cloud data. Consequently, this can enhance the resilience of the ML system in a AV environment.

Each of these defense mechanisms has its strengths and weaknesses, and often a combination of them offers the best protection against adversarial perturbation attacks. It is crucial to regularly reassess and update these defenses to deal with evolving adversarial threats effectively.

10.5.2 Additional Measures for Securing ML for AVs in NextG

Securing ML for AVs in NextG is indeed a complex task, given the sensitivity of the data involved and the need to ensure the integrity, privacy, and robustness of the learning process. The key considerations outlined above are essential steps toward achieving a secure and reliable ML-based AV environment in NextG. However, it is crucial to acknowledge that this field of research is continuously evolving, and additional security measures may be required to address emerging threats and vulnerabilities. Following are some additional measures that can be taken:

- Differential Privacy [16]. By implementing differential privacy techniques, individual vehicle data can be protected during the ML process, preventing the disclosure of sensitive information. Differential privacy adds noise to the data or model before sharing them, preventing the inference of sensitive information from specific data points.
- Secure Communication [17]. Establishing secure communication channels between AVs and the cloud server through NextG is paramount in ensuring data transmission confidentiality and integrity during ML training and inference. Encryption protocols such as Secure Socket Layer (SSL) and Transport Layer Security (TLS) play a critical role in achieving this goal.
- Trusted Execution Environment (TEE) [18]. Leveraging hardware-based security solutions, such as TEEs, constitutes an effective approach to safeguarding the privacy and integrity of both the model and data in ML for

AVs in the NextG scenario. TEEs offer a secure and isolated area within the hardware environment, where critical computations and sensitive data operations take place. By utilizing TEEs, ML systems can ensure that the computation processes occur within a protected enclave, shielded from external interference or unauthorized access. This fortified environment guarantees the confidentiality of the model and data, preventing any potential breaches or data leaks. Furthermore, TEEs ensure the integrity of computations, assuring that the executed algorithms remain free from tampering or malicious modifications.

- Data Anonymization [19]. AVs can adopt data anonymization techniques as a proactive measure to safeguard user privacy and minimize the risk of data leakage. Data anonymization involves the removal or modification of personally identifiable information (PII) from the training data, rendering the data points anonymous and unlinkable to specific individuals. By anonymizing the data, sensitive information such as names, addresses, and other personally identifiable details are obscured or replaced with pseudonyms. As a result, the data becomes dissociated from individual identities, making it challenging for any unintended disclosure of personal information during the inference process, and thus making it harder for the attackers to identify a specific victim.
- Access Control and Authorization [20]. The incorporation of access control mechanisms plays a pivotal role in ensuring the integrity and security of the ML process for AVs. These mechanisms are imperative to restrict participation in the collaborative learning system solely to authorized entities. To achieve this, robust authentication and authorization protocols are employed, as gatekeepers to verify the identities and permissions of the vehicles.
- Model Verification and Validation [21]. Ensuring the integrity of the ML process is crucial. It involves verifying the accuracy and security of the DNN inference model in the cloud, detecting potential adversarial attacks, and conducting robust testing and evaluation. These measures help maintain the reliability and effectiveness of the system.
- Regular Security Updates [22]. Keeping the software and systems up to date with the latest security patches and updates is crucial to address any known vulnerabilities and protect against emerging threats.
- Generative Intelligence. The emergence of generative AI, exemplified by powerful language models like GPT [23] and autonomous AI agents such as AutoGPT [24], presents novel opportunities for enhancing the security of ML systems in AVs. These advanced AI technologies possess the ability to understand, generate, and analyze vast amounts of data, making them potentially valuable tools for identifying and mitigating security vulnerabilities in ML

systems. By leveraging the generative intelligence, researchers and developers can explore innovative approaches to securing AVs. For instance, these generative AI systems could be deployed to simulate realistic traffic scenarios and generate synthetic sensor data for testing and validating the robustness of AV systems against adversarial attacks. By exposing AVs to a wide range of simulated attacks and edge cases, these generative AI can help identify and patch vulnerabilities before they can be exploited in real-world situations. As the complexity and sophistication of AVs in NextG networks continue to grow, the integration of generative AI as additional security measures holds significant promise for ensuring the safety, reliability, and trustworthiness of these critical technologies.
- Adversarial Machine Learning. Adversarial ML [1] focuses on identifying and mitigating threats arising from intentionally crafted inputs designed to mislead ML models, such as small perturbations to sensor data that cause incorrect perception or decision-making. For AVs in NextG environments, adversarial attacks can target perception models (e.g., object detection, traffic sign recognition) or communication-based decision systems, potentially causing unsafe maneuvers. Defensive strategies, including adversarial training, feature squeezing, model ensemble methods, and input reconstruction, can improve robustness against such attacks. Integrating adversarial robustness testing into the ML lifecycle ensures that AV models remain resilient even under worst-case scenarios.

It is imperative to acknowledge that the security of ML system for AVs in the context of NextG remains an active and evolving research area. As new security challenges emerge, additional measures may be required to effectively safeguard the learning process and ensure the robustness of the system.

10.6 Key Takeaways

Securing ML for AVs in the context of NextG is a paramount concern to safeguard the privacy, integrity, and reliability of the learning process. This study extensively explores various facets of securing ML, with a specific focus on the challenges and solutions concerning adversarial attacks in the context of NextG-enabled ML system for AVs. Recent research in this domain is presented to illustrate these attacks' potential impact on ML model's decision-making and performance.

To counter adversarial attacks, defensive distillation, input reconstruction, adversarial detection, and feature-squeezing techniques are recommended. These defenses fortify the ML model against adversarial attacks, enhancing the overall resilience of the ML inference.

However, there remain challenges and future research directions to address in securing ML for AVs in the NextG network. Key areas of focus include the scalability and efficiency of secure communication protocols, emerging privacy-preserving techniques, robust defense mechanisms and the importance of standardization and interoperability.

In conclusion, securing ML for AV in the NextG network is an ongoing endeavor that demands integrating state-of-the-art technologies. By addressing the challenges and implementing cutting-edge security solutions, we can facilitate the adoption of ML in NextG networks, thereby fostering safer, more efficient, and privacy-conscious AV systems. This collaborative approach ensures the continued advancement of AVs and reinforces public trust in the transformative potential of ML in shaping the future of transportation.

References

1 Goodfellow, I. J., Shlens, J., and Szegedy, C. (2014). Explaining and harnessing adversarial examples. *arXiv preprint arXiv:1412.6572.*

2 Deng, Y. and Karam, L. J. (2020). Universal adversarial attack via enhanced projected gradient descent. In *Proceedings of the 2020 IEEE International Conference on Image Processing (ICIP)*, pages 1241–1245. IEEE.

3 Sarker, A., Shen, H., Sen, T., and Uehara, H. (2020). An advanced black-box adversarial attack for deep driving maneuver classification models. In *Proceedings of the 2020 IEEE 17th International Conference on Mobile Ad Hoc and Sensor Systems (MASS)*, pages 184–192. IEEE.

4 Sarker, A., Shen, H., and Sen, T. (2021a). A context-aware black-box adversarial attack for deep driving maneuver classification models. In *Proceedings of the 2021 18th Annual IEEE International Conference on Sensing, Communication, and Networking (SECON)*, pages 1–9. IEEE.

5 Sarker, A., Shen, H., and Sen, T. (2021b). A suspicion-free black-box adversarial attack for deep driving maneuver classification models. In *Proceedings of the 2021 IEEE 41st International Conference on Distributed Computing Systems (ICDCS)*, pages 786–796. IEEE.

6 Sarker, A., Shen, H., Sen, T., and Mendelson, Q. (2021c). Efficient black-box adversarial attacks for deep driving maneuver classification models. In *Proceedings of the 2021 IEEE 18th International Conference on Mobile Ad Hoc and Smart Systems (MASS)*, pages 536–544. IEEE.

7 Chen, P.-Y., Zhang, H., Sharma, Y., Yi, J., and Hsieh, C.-J. (2017). Zoo: Zeroth order optimization based black-box attacks to deep neural networks without training substitute models. In *Proceedings of the 10th ACM Workshop on Artificial Intelligence and Security*, pages 15–26.

8 Snoek, J., Larochelle, H., and Adams, R. P. (2012). Practical Bayesian optimization of machine learning algorithms. *In Proceedings of the Advances in Neural Information Processing Systems*, 25.

9 Kwak, B. I., Woo, J., and Kim, H. K. (2016). Know your master: Driver profiling-based anti-theft method. In *Proceedings of the 2016 14th Annual Conference on Privacy, Security and Trust (PST)*, pages 211–218. IEEE.

10 Karim, F., Majumdar, S., Darabi, H., and Harford, S. (2019). Multivariate LSTM-FCNs for time series classification. *Neural Networks*, 116: 237–245.

11 Tu, C.-C., Ting, P., Chen, P.-Y., Liu, S., Zhang, H., Yi, J., Hsieh, C.-J., and Cheng, S.-M. (2019). Autozoom: Autoencoder-based zeroth order optimization method for attacking black-box neural networks. In *Proceedings of the AAAI Conference on Artificial Intelligence*, volume 33, pages 742–749.

12 Papernot, N., McDaniel, P., Wu, X., Jha, S., and Swami, A. (2016). Distillation as a defense to adversarial perturbations against deep neural networks. In *Proceedings of the 2016 IEEE Symposium on Security and Privacy (SP)*, pages 582–597. IEEE.

13 Gu, S. and Rigazio, L. (2014). Towards deep neural network architectures robust to adversarial examples. *arXiv preprint arXiv:1412.5068*.

14 Feinman, R., Curtin, R. R., Shintre, S., and Gardner, A. B. (2017). Detecting adversarial samples from artifacts. *arXiv preprint arXiv:1703.00410*.

15 Xu, W., Evans, D., and Qi, Y. (2017). Feature squeezing: Detecting adversarial examples in deep neural networks. *arXiv preprint arXiv:1704.01155*.

16 El Ouadrhiri, A. and Abdelhadi, A. (2022). Differential privacy for deep and federated learning: A survey. *IEEE access*, 10: 22359–22380.

17 Dastres, R. and Soori, M. (2021). A review in recent development of network threats and security measures. *International Journal of Information Sciences and Computer Engineering*.

18 Quoc, D. L. and Fetzer, C. (2021). SecFL: Confidential federated learning using tees. *arXiv preprint arXiv:2110.00981*.

19 Murthy, S., Bakar, A. A., Rahim, F. A., and Ramli, R. (2019). A comparative study of data anonymization techniques. In *Proceedings of the 2019 IEEE 5th Intl Conference on Big Data Security on Cloud (BigDataSecurity), IEEE Intl Conference on High Performance and Smart Computing,(HPSC) and IEEE Intl Conference on Intelligent Data and Security (IDS)*, pages 306–309. IEEE.

20 Mundhenk, P., Paverd, A., Mrowca, A., Steinhorst, S., Lukasiewycz, M., Fahmy, S. A., and Chakraborty, S. (2017). Security in automotive networks: Lightweight authentication and authorization. *ACM Transactions on Design Automation of Electronic Systems (TODAES)*, 22(2): 1–27.

21 Carson, J. S. (2002). Model verification and validation. In *Proceedings of the Winter Simulation Conference*, volume 1, pages 52–58. IEEE.

22 Cirne, A., Sousa, P. R., Resende, J. S., and Antunes, L. (2022). IoT security certifications: Challenges and potential approaches. *Computers & Security*, 116: 102669.

23 Brown, T., Mann, B., Ryder, N., Subbiah, M., Kaplan, J. D., Dhariwal, P., Neelakantan, A., Shyam, P., Sastry, G., Askell, A., et al. (2020). Language models are few-shot learners. *Advances in Neural Information Processing Systems*, 33: 1877–1901.

24 Gravitas, S. and contributors (2023). Auto-GPT: An autonomous GPT-4 experiment. https://github.com/Significant-Gravitas/Auto-GPT. Accessed: 3/22/2024.

Index

Note: Page numbers in *italics* and **bold** refers to figures and tables respectively.

6G Security: New Frontiers in Mobile Network Technologies and Verticals, First Edition.
Edited by Luiz A. DaSilva.

o

p

Printed and bound by CPI Group (UK) Ltd, Croydon, CR0 4YY

23/01/2026

14811948-0001